Organic Spectroscopy
Workbook

Organic Spectroscopy Workbook

TOM FORREST

JEAN-PIERRE RABINE

MICHEL ROUILLARD

A John Wiley & Sons, Ltd., Publication

Print ISBN: 978-1-119-99379-7
ePDF ISBN: 978-1-119-97594-6

Typeset in 9.5/11.5pt Times Roman by Laserwords Private Limited, Chennai, India
Printed and bound by CPI Group (UK) Ltd, Croydon, CR0 4YY

C9781119993797_150623

Contents

Preface

This workbook arose out of a long standing project for the development of computer based programs to assist students in learning how to deduce the structures of organic compounds from spectroscopic data. The ability to elucidate chemical structures is built largely on implicit heuristics and informal experience accumulated by solving many examples. The accumulation of this experience is greatly facilitated by these computer based programs, http://spectros.unice.fr/. However, the knowledge that is so arduously obtained at the computer may evaporate rapidly unless some reinforcement mechanism is in place.

This workbook was originally conceived as a way to provide the computer users with a vehicle for making a lasting record of their learning experience. However, we recognized that a similar situation occurs for students who prefer to undertake their study without using computer-based exercises. They could also benefit from a workbook where they could get assistance in solving spectral problems and have a record that could be used for review and reinforcement. The resulting workbook was designed so that it can be used with or without the computer based program. The computer exercises, which are freely available online to any user, provide more detailed explanations and supplementary graphical material for the structural determinations.

For each problem in the workbook we provide four spectra together with a detailed analysis of the spectra and an explanation of the derivation of the structure of the compound.

The level of detail provided in the solution decreases gradually, until near the end of the set of exercises, very little or no analysis is given. The solutions to all of the exercises are provided, but no structural formulae are shown on the work area of the exercise, so that the student may follow along the analysis and draw the structural components as they are revealed.

There are many advanced techniques available in modern spectrometers that may be employed to assist in the solution of more difficult structural problems. In these exercises we have chosen to give low resolution electron impact mass spectra because these spectra are readily available, and are rich in structural information. Modern mass spectrometry employs a multitude of other ionization techniques that can be used in particular circumstances to provide specific information, but these are beyond the scope of an introductory spectral exercise book. In the field of NMR spectroscopy there are numerous sophisticated pulse sequences available that can generate various multidimensional spectra which are used for more difficult structural problems. These techniques are particularly useful for large complex structures, but they are aids that still require the user to have an understanding of the fundamentals of spectroscopic interpretation, which is the goal of these exercises.

We hope that students will find that using this workbook will prove to be a pleasant and rewarding experience.

Acknowledgements

We have used the resources of our university libraries extensively in preparing this workbook, but have found that we made even more extensive use of the resources available on the internet. We are indebted to the many people who have taken the time to place their data at the disposal of the scientific community in this manner. This easy access to spectral data has been of immense assistance to us in the preparation of this workbook. The availability of such a large amount of primary data allowed us to find very good model compounds which gave more specific support for the assignments than possible from the general correlation tables of secondary publications.

It is impossible to list all of the internet sites that we visited during the preparation of the manuscript; however, we would like to specifically acknowledge the providers of those sites that we have found we were consistently using. These are for the most part sites provided by institutions, universities and chemical suppliers:

NIST, National Institute of Standards and Technology, Chemistry Webbook;
National Institute of Advanced Industrial Science and Technology (AIST, Japan), Spectral Database for Organic Compounds;

Sigma-Aldrich Corporation, NMR and IR spectra;
Acros Organics, IR spectra;
University of Wisconsin, Chemistry Department, H. J. Reich, compilation of NMR chemical shifts and coupling constants.

We are grateful for the support of:

Université de Nice Sophia-Antipolis;
Université Numérique Thématique:UNISCIEL (Université des Sciences en Ligne)
Dalhousie University

How to Use this Workbook

The exercises give spectral charts and verbal analyses of the spectra, but the structural diagrams that would complete the picture are missing. Although annotated structures would be very helpful in the spectral assignments, they are deliberately omitted to avoid a premature revelation of the answer. You don't want to see the solution before you have had the opportunity to think about the problem. You may want to use a small sheet of paper to cover the conclusion/summary section as you try to solve each structure. There is adequate white space available on the pages to allow you to draw the appropriate structures, and to make notes on the spectral assignments. Complete structural drawings are available in the Answers section of the workbook. Each exercise has a second number in brackets which identifies the number by which the compound can be located in the Answers section, and in the computer-based exercises. The structures are not presented in the same order as the exercises, thus lessening the possibility of an unwanted exposure of the answer to the next exercise.

There is no single method that provides the best way to solve these problems. There is usually an abundance of information that allows many paths to the solution. However, it is unlikely that the complete structure will be revealed intact, so the general approach is to derive the partial structures of the molecule and to connect them to form the complete structure. The order of the spectral analysis provided in each exercise is not presented as the best way to approach the problem. Many people do a preliminary scan of the spectra to look for significant features, identify the responsible structural component, then verify its presence using other spectra. For example, a carbonyl group that stands out in the IR spectrum can be verified by the peak at the appropriate chemical shift in the ^{13}C NMR spectrum. As one moves from spectrum to spectrum like this, it is imperative that an organized approach be maintained and that good notes are preserved.

It is recommended that you write down the partial structures as you derive them. You can always erase them if you make a mistake or, better still, simply draw a mark through them, and make notes to ensure that you have them for review and can learn from your mistakes. Take the opportunity to make detailed notes before you leave each exercise. As clear as it will be in your mind at that moment, it is amazing how the details evaporate with time.

All the ^{13}C NMR DEPT spectra are DEPT-135 spectra in which methylene carbons show a negative signal and quaternary carbons show no signal. Methyl and methine carbons show up as normal positive signals.

All the mass spectral charts are low-resolution electron impact mass spectra (EIMS). These spectra frequently provide the molecular weight of the compound, and the high degree of fragmentation also offers valuable information to those with knowledge of reactivity of organic compounds. In some examples, the results of high-resolution mass spectra (HRMS) are provided which can be used to obtain the molecular formula directly.

The exercises are organized somewhat in the order of complexity of the molecule and level of explanation provided. You will find very simple spectra and detailed explanations at the beginning, and eventually more complex exercises with less in the way of explanation. For the last 20 exercises, only general hints and suggestions are provided. In these exercises the ^{1}H NMR spectral data is provided in standard text format to show the coupling constants where appropriate. These data should be compared to the spectral pattern, so that you could write such a verbal description of the spectrum.

The resource material you require to analyse the spectra is provided in tables and algorithms in separate data sections for each of the spectroscopy types.

The exercises presented in this work book are also available in an on-line program, Multispectroscopy, which you may access at spectros.unice.fr. This program provides more detailed explanations, as well as elaborate diagrams and graphics, to help to explain the spectral analysis and structure determination. You may wish to take advantage of this additional resource, while you are using this workbook.

Organic Spectroscopy Workbook, First Edition. Tom Forrest, Jean-Pierre Rabine, and Michel Rouillard.
© 2011 John Wiley & Sons, Ltd. Published 2011 by John Wiley & Sons, Ltd.

Preliminary Observations

It is easy to get some information from a quick glance at some spectra. There are some typical shapes that give a good indication of particular structural features. It is essential that these deductions are verified by detailed inspection of the spectrum, and with data from other spectra.

Infrared

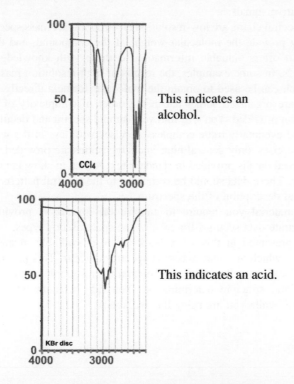

This indicates an alcohol.

This indicates an acid.

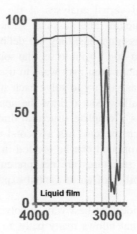

This indicates an alkene.

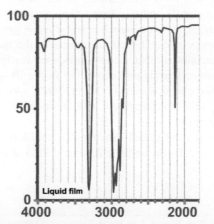

This indicates a terminal alkyne.

Organic Spectroscopy Workbook, First Edition. Tom Forrest, Jean-Pierre Rabine, and Michel Rouillard.
© 2011 John Wiley & Sons, Ltd. Published 2011 by John Wiley & Sons, Ltd.

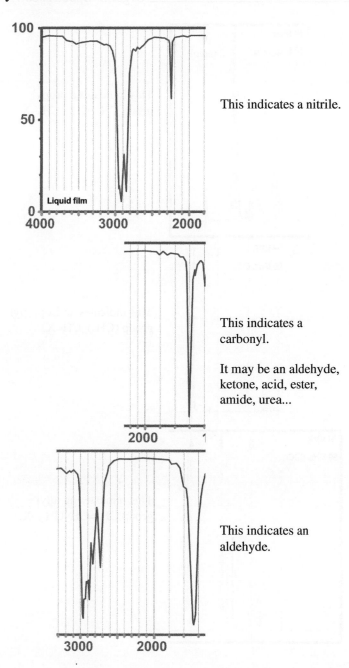

This indicates a nitrile.

This indicates a carbonyl.

It may be an aldehyde, ketone, acid, ester, amide, urea...

This indicates an aldehyde.

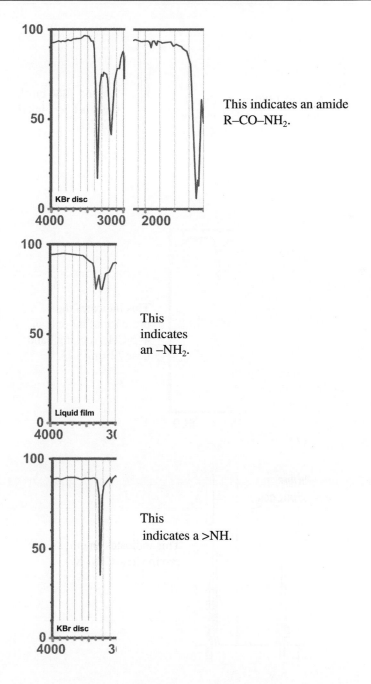

This indicates an amide R–CO–NH$_2$.

This indicates an –NH$_2$.

This indicates a >NH.

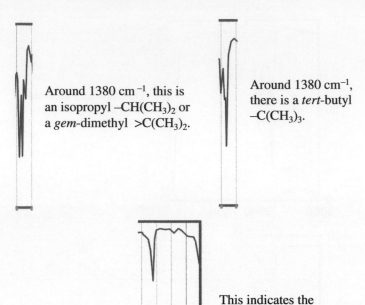

Around 1380 cm^{-1}, this is an isopropyl –CH(CH$_3$)$_2$ or a *gem*-dimethyl >C(CH$_3$)$_2$.

Around 1380 cm^{-1}, there is a *tert*-butyl –C(CH$_3$)$_3$.

This indicates the presence of more than 4–5 CH$_2$, rocking band near 720cm^{-1}.

500

^{1}**H NMR**

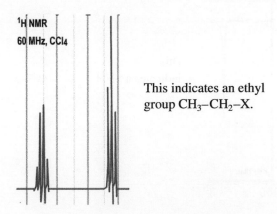

This indicates an ethyl group CH$_3$–CH$_2$–X.

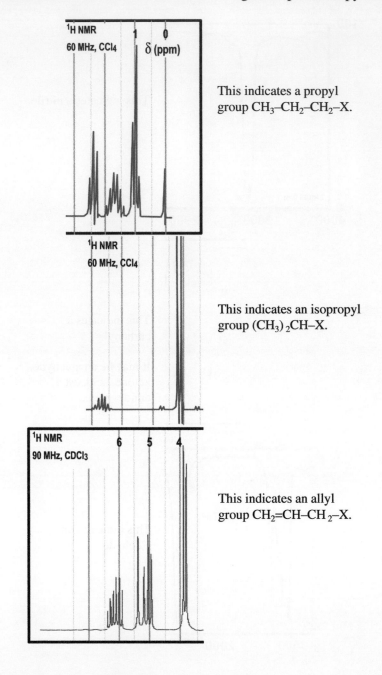

This indicates a propyl group CH$_3$–CH$_2$–CH$_2$–X.

This indicates an isopropyl group (CH$_3$)$_2$CH–X.

This indicates an allyl group CH$_2$=CH–CH$_2$–X.

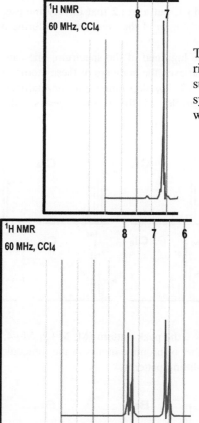

This indicates a benzene ring with one alkyl substituent, or a benzene symmetrically substituted with alkyl groups.

This indicates a benzene para-disubstituted by two different groups (not alkyl groups) which gives an AA'XX' system.

^{13}C NMR

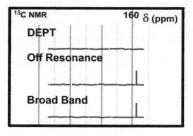

This indicates a quaternary carbon >C<.

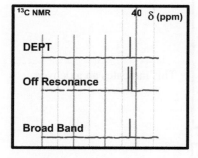

This indicates a methine >CH–.

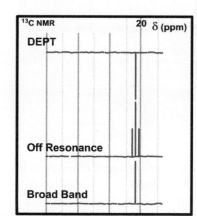

This indicates a methylene –CH$_2$–.

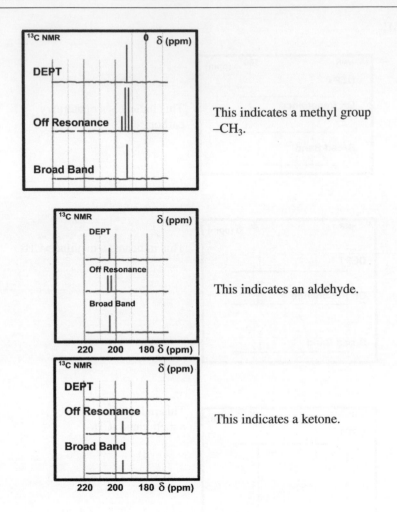

This indicates a methyl group –CH$_3$.

This indicates an aldehyde.

This indicates a ketone.

Mass Spectra

The molecular ion peak, M, is not generally the highest mass peak observed in the spectrum. Normally one finds higher mass ions due to isotope peaks:

– M+1 is normally due to the presence of the ^{13}C isotope of carbon, 1.1% of M per C, or in the case where silicon is present, it makes a strong contribution to M+1 (5.1% of M per Si). Sulfur also contributes to M+1 but it is easily detected by M+2.

– M+2 is generally less intense than M+1, but a M+2 that is very intense, perhaps even more intense than M, gives a good indication of the presence of chlorine or bromine.

By looking at the isotope patterns in the high end of the spectrum, one can usually detect the presence of chlorine and bromine, and even the numbers of these atoms in the molecule.
   For chlorine, ^{35}Cl (75%) and ^{37}Cl (25%), depending on the number of Cl atoms present, one has distinctly different profiles of the isotopic peaks (M, M+2, M+4, M+6, etc.).

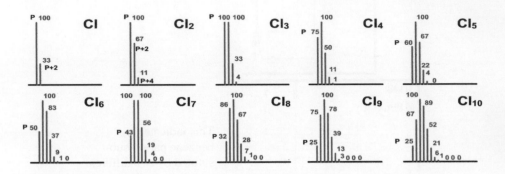

For bromine, ^{79}Br (50%) and ^{81}Br (50%), the peak pattern (M, M+2, M+4, M+6, etc.) is quite distinctive, depending on the number of bromine atoms in the molecule. Visually, the pattern looks similar to simple NMR splitting patterns.

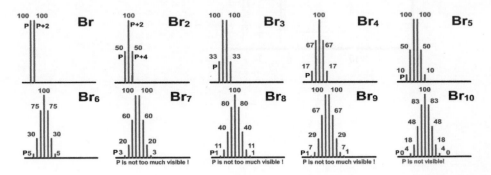

For chlorine and bromine in the same molecule, the isotope peak profile is more complex but one can still see distinctive patterns of isotopic peaks (M, M+2, M+4, M+6, etc.).:

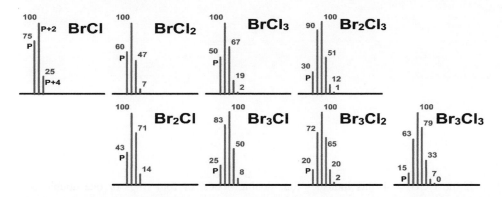

For sulfur, ^{32}S 100% M, ^{33}S 0,8% M+1, ^{34}S 4,4% M+2.

In general, the contribution of sulfur to M+1 is largely masked by the contribution from carbon ^{13}C (1.1% per carbon). Only the peak M+2 gives a good indication of the number of sulfur atoms, but only in the absence of chlorine or of bromine and only if the peak M is sufficiently intense, so that the intensity of peak M+2 can be read with some precision.

Centres of Unsaturation

If you have a molecular formula for the compound, don't forget to take the formula into account in all your deductions. The first thing to do is to calculate the number of centres of unsaturation of the compound.

Number of centres of unsaturation

Knowing the formula **CxHyOzNn**, it is possible to determine the number of centres of unsaturation of the compound NCU:

$$NCU = (2x + 2 + n - y)/2$$

If the formula contains halogens, count them like hydrogens.
  If the NCU is less than 4, the compound cannot have a benzene ring.
  If the NCU is 0, the compound cannot have a double bond, or a carbonyl or a ring.
  You need an NCU of at least 2 to have a compound containing a triple bond.

CxHy: no oxygen, no nitrogen

The functional group of the compound cannot be alcohol, ketone, aldehyde, phenol, acid, ester, anhydride, amide, amine, nitrile...
  If the NCU = 0, it is an alkane.
  If the NCU = 1, this indicates one ring or one double bond (alkene).
  If the NCU = 2, this indicates two double bond (alkene), or two rings or one double bond and one ring or one triple bond (alkyne).

If the NCU = 4, this indicates four double bonds, or four rings or one double bond and three rings, or two double bonds and two rings, or three double bonds and one ring (benzene ring).

CxHyO: one oxygen, no nitrogen

The functional group of the compound cannot be acid, ester, anhydride, amide, amine, nitrile...
  If the NCU = 0, it is an alcohol or an ether.
  If the NCU = 1, this indicates one ring or one double bond (>C=C< alkene or >C=O, ketone, aldehyde).
  If the NCU = 4, this indicates four double bond, or four rings, or one double bond (>C=C< or >C=O) and three rings, or two double bonds (>C=C< or >C=O) and two rings, or three double bonds (>C=C< or >C=O) and one ring (it may be a benzene ring).
  If the NCU $\geq$ 5, you may have a benzene ring with a double bond (>C=C< or >C=O) or a ring...

CxHyO2: two oxygens, no nitrogen

The functional group of the compound cannot be anhydride, amide, amine, nitrile...
  If the NCU = 0, it is an alcohol or an ether, or both.
  If the NCU = 1, this indicates one ring or one double bond (>C=C< or >C=O) or an acid RCOOH or an ester RCOOR'.

Organic Spectroscopy Workbook, First Edition. Tom Forrest, Jean-Pierre Rabine, and Michel Rouillard.
© 2011 John Wiley & Sons, Ltd. Published 2011 by John Wiley & Sons, Ltd.

CxHyO₃: three oxygens or more, no nitrogen

The functional group of the compound cannot be amide, amine, nitrile...
   If the NCU = 2, it may be an ahydride RCOOCOR'.

CxHyON: one oxygen, one nitrogen

The functional group of the compound cannot be acid, ester, anhydride...
   If the NCU = 1, it may be an amide RCON<

CxHyN: no oxygen, one nitrogen

The functional group of the compound cannot be acid, ester, anhydride, amide...
   If the NCU = 0, it is an amine.
   If the NCU = 2, it may be a nitrile.

Exercises 1–100

Exercise 001 (001)

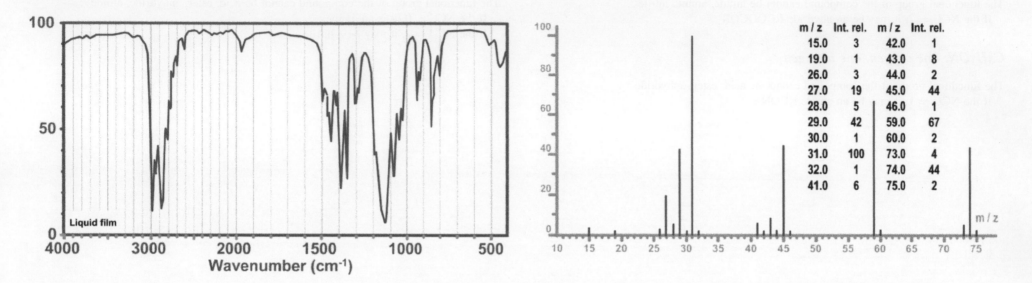

m/z	Int. rel.	m/z	Int. rel.
15.0	3	42.0	1
19.0	1	43.0	8
26.0	3	44.0	2
27.0	19	45.0	44
28.0	5	46.0	1
29.0	42	59.0	67
30.0	1	60.0	2
31.0	100	73.0	4
32.0	1	74.0	44
41.0	6	75.0	2

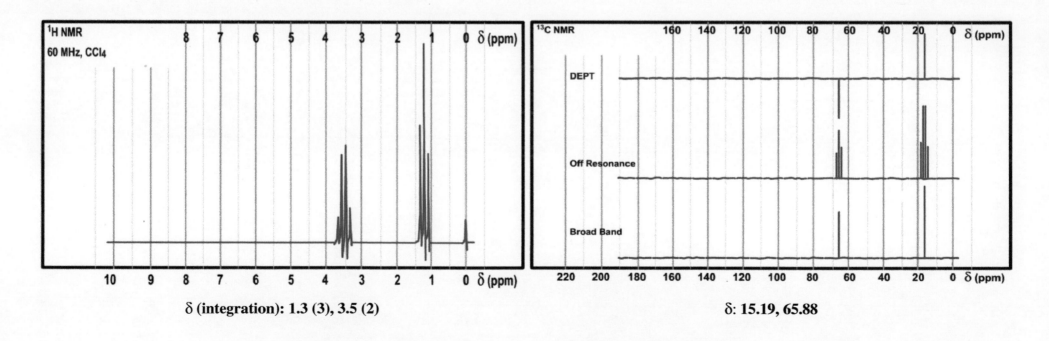

δ (integration): 1.3 (3), 3.5 (2)

δ: 15.19, 65.88

Exercise 001 (001)

Preliminary Observation

The mass spectrum can be used to get the molecular mass of the compound, to detect the presence of chlorine, bromine or other heteroatoms, and to obtain structural information from the fragmentation pattern. The IR spectrum provides evidence of functional groups and other structural features. The ^{13}C NMR spectrum gives the number of different carbon in the molecule, as well as their degree of substitution and functionality. The ^{1}H NMR spectrum can provide a count of the number of protons and information on their environment. The ^{1}H coupling patterns give indications of the connectivity of the components of the molecule.

Mass Spectrum

The isotope patterns give no indication of the presence of chlorine or bromine. The molecular ion (M^{+}) is usually the highest mass ion (with its associated isotope peaks). An odd-mass molecular ion usually indicates the presence of an odd number of nitrogen atoms. In this spectrum, the highest mass ion 74 (44%) has a ^{13}C isotope peak at 75 (2%). The natural abundance of ^{13}C is 1.1%; therefore, each carbon contributes 1.1% to the intensity of the isotope peak. The relative intensity of the M+1 peak in this compound is 4.5% ($100 \times 2/44$), indicating the presence of four carbons. The most intense peak is known as the base peak and is assigned an intensity of 100%. The ion at 31 (100%) indicates the presence of oxygen in the molecule. The highest mass hydrocarbon in that region is 29 ($CH_3CH_2^{+}$), but if oxygen is present a peak at 31 (CH_2OH^{+}) is possible. The peak at 59 corresponds to M−15, suggesting the presence of a CH_3 group.

^{13}C NMR

The simplest spectrum, the broad band (BB) decoupled spectrum in which all of the C–H coupling is removed, shows signals from only two types of carbon. In the off-resonance (OR) spectrum, the size of the direct C–H coupling is reduced to make the coupled spectrum more easily readable. In this exercise, the direct C–H coupling gives a quartet for a methyl group (CH_3), and a triplet for a methylene group (CH_2). The DEPT spectrum shows inverted signals for CH_2 carbons, and in this example has one such signal. Carbons bearing an odd number of protons show upright signals, and here it shows one such signal, that of the CH_3 group. Quaternary carbons (bearing no H) give no signal in the DEPT spectrum. The chemical shift of the lower field carbon, (66 ppm) is in the range of a methylene carbon bonded to oxygen. The molecule must have an element of symmetry, because the mass spectrum indicates that it has four carbons.

^{1}H NMR

The ^{1}H NMR spectrum shows only two types of proton, a triplet and a quartet. The small peak at 0.0 ppm is that of the tetramethylsilane reference standard. This simple coupling pattern is easily recognized as that of an ethyl group ($–CH_2CH_3$), with the methyl protons split by the neighbouring CH_2 protons into a triplet, and the methylene protons being split by the neighbouring CH_3 protons into a quartet. The integration of the spectrum indicates that the two signals have a relative intensity of 2 to 3. A molecule that has no nitrogen or halogen or other odd-valent atom, must have an even number of protons. Thus, there must be an even number of ethyl groups. The low field chemical shift (3.5 ppm) of the CH_2 group indicates that it is attached to an electron-withdrawing group, the oxygen atom in this case.

Infrared

The band at $1140 \, cm^{-1}$ is a clear indication of C–O stretching, most likely that of an ether function. Other than the bands due to the alkyl groups, there are no bands that can be assigned to other functional groups. The IR spectrum can be used very effectively to show that certain functional groups are absent. For instance, in this case one can say with certainty that there is neither a carbonyl group, nor hydroxyl present in the molecule.

Summary

The ^{1}H and ^{13}C NMR evidence indicates the presence of an ethyl group bonded to oxygen (ethoxy group). The mass spectrum gives a molecular mass of 74, and since an ethoxy group has a mass of 45, another ethyl group (mass 29) is required to complete the structure. The compound is diethyl ether, $CH_3CH_2–O–CH_2CH_3$.

Exercise 002 (003)

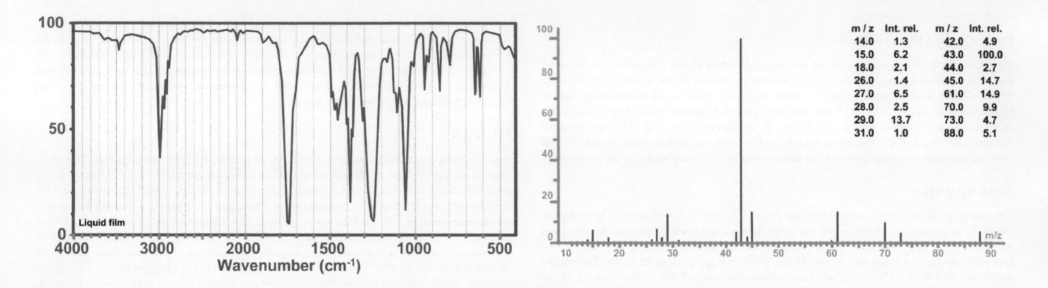

m/z	Int. rel.	m/z	Int. rel.
14.0	1.3	42.0	4.9
15.0	6.2	43.0	100.0
18.0	2.1	44.0	2.7
26.0	1.4	45.0	14.7
27.0	6.5	61.0	14.9
28.0	2.5	70.0	9.9
29.0	13.7	73.0	4.7
31.0	1.0	88.0	5.1

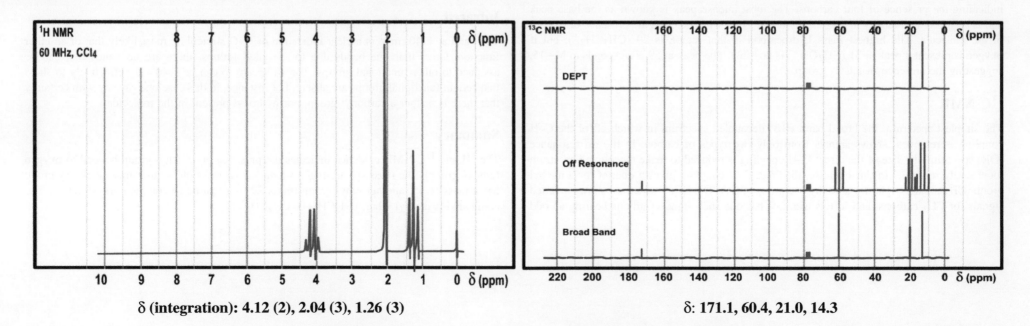

δ (integration): 4.12 (2), 2.04 (3), 1.26 (3)

δ: 171.1, 60.4, 21.0, 14.3

Exercise 002 (003)

Preliminary Observation

In a preliminary scan of the spectra there are several points that stand out. In the IR spectrum, there is a very strong peak in the carbonyl region near $1740\,cm^{-1}$. The mass spectrum appears to have a peak for the molecular ion, since the highest mass ion has an even mass. The ^{13}C NMR spectrum can be used to determine the number and type of carbons, and the ^1H NMR spectrum to get the connections between the carbons. The chemical shifts and the IR spectrum will identify the functional group. The mass spectrum can be used to get the molecular weight of the compound.

^{13}C NMR

The broad band spectrum shows that there are four types of carbon in the molecule. The off-resonance spectrum indicates that two of the carbons are methyl groups (quartets, CH_3), one is a methylene group (triplet, CH_2) and one a quaternary carbon (singlet, no attached hydrogen). The DEPT spectrum shows one inverted signal corresponding to the CH_2 carbon, and the disappearance of a peak at 171 ppm corresponding to the quaternary carbon. There are two different CH_3 groups, one CH_2 group, and a strongly deshielded carbon bearing no hydrogens (quaternary carbon). This spectrum shows that the compound has four carbons and eight protons or a multiple of these.

^1H NMR

The ^1H NMR spectrum shows the signals from three different types of proton. The integration of the spectrum shows a relative intensity of 2:3:3 (4.2, 2.0 and 1.3 ppm), for a total of eight protons, which agrees with the ^{13}C NMR spectrum that indicated a total of eight protons. One CH_3 group is not coupled to any other hydrogen (singlet at 2 ppm), and the other CH_3 (triplet at 1.3 ppm) is coupled to the CH_2 group. Thus, we have an ethyl group and an isolated methyl group. The singlet at 2 ppm is in the chemical shift region of a methyl group attached to an sp^2 carbon, while the CH_2 (4.1 ppm) is in the region of a methylene group attached to an oxygen. As the CH_2– group is coupled to a methyl group, we have a CH_3CH_2O– group.

Infrared

This infrared spectrum has two very significant bands of the functional group of the compound. The band at $1740\,cm^{-1}$ clearly shows the presence of a carbonyl group (C=O), and the band at $1250\,cm^{-1}$ indicates the presence of a C–O single bond. The frequencies of both of these bands are compatible with an ester group (C–O–C=O).

Mass Spectrum

The highest even-mass ion in the spectrum is at 88, and is probably that of the molecular ion. The most intense peak (the base peak) is found at 43, which is a fragment most commonly associated with an acylium ion ($CH_3C\equiv O^+$). This would arise from the loss of the ethoxy radical by α cleavage at the carbonyl group. The ion at 73 comes from the loss of one of the methyl groups. The ion at 70 arises from the loss of a molecule of water, a common rearrangement fragmentation in esters. The majority of the fragment ions have odd masses, which is typical of compounds without nitrogen. These odd-mass ions have an even number of electrons, and are generally more stable than the even-mass fragments, as they consist of the less stable odd-electron radical ions. Even-mass ions are generally formed when a rearrangement creates a stable small molecule, such as, in this case, the loss of water yields the ion at 70.

Summary

The structural components that have been identified are: an ethyl group, a methyl group and a quaternary carbon, the mass of these adding to 56, which is 32 less than the molar mass of 88 that was found from the mass spectrum. The missing mass is equivalent to two oxygen atoms (2×16), exactly the number required for the ester group that was identified in the IR spectrum. The NMR chemical shifts of the CH_2 group indicate that it is attached directly to the oxygen of the ester. The chemical shifts of the uncoupled methyl group indicate that it is attached to a polarized sp^2 carbon, which would place it on the carbonyl (C=O) carbon. This gives us the complete structure $CH_3(C=O)OCH_2CH_3$, ethyl acetate.

Exercise 003 (010)

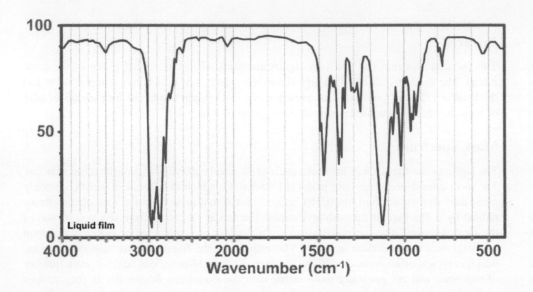

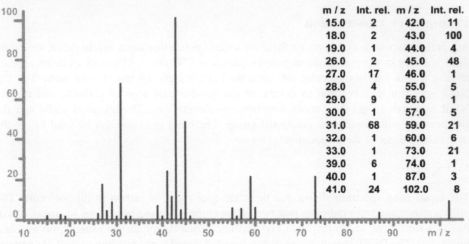

m / z	Int. rel.	m / z	Int. rel.
15.0	2	42.0	11
18.0	2	43.0	100
19.0	1	44.0	4
26.0	2	45.0	48
27.0	17	46.0	1
28.0	4	55.0	5
29.0	9	56.0	1
30.0	1	57.0	5
31.0	68	59.0	21
32.0	1	60.0	6
33.0	1	73.0	21
39.0	6	74.0	1
40.0	1	87.0	3
41.0	24	102.0	8

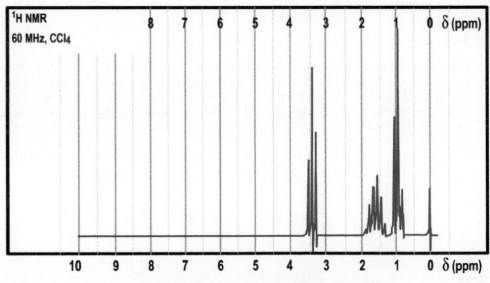

δ (integration): 3.37 (2), 1.59 (2), 0.93 (3)

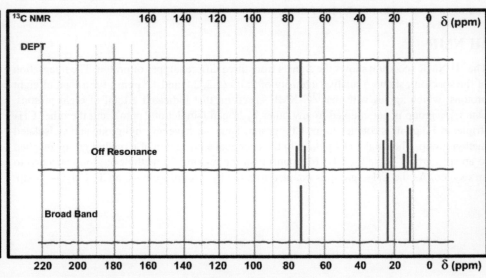

δ: 72.84, 23.34, 10.60

Exercise 003 (010)

Preliminary Observation

The infrared spectrum frequently gives the first clue as to the functional group in the molecule. When it is not obvious from the IR, then the chemical shifts in the NMR spectra may give some clues to the functional groups. The NMR spectra give the best information on the substructures within the molecule. The mass spectra may have a lot of information in the fragmentation patterns, but these must be used with caution as they arise from reactions of ions in very high energy states.

Mass Spectrum

The peak at 102 is a good candidate for the molecular ion. It has an even mass and the spectrum contains many odd-mass fragment ions: at 87 (M−15), at 73 (M−29) and at 59 (M−43). Note the fragment at 31, which usually indicates the presence of oxygen (CH_2OH^+). The molecule is possibly a mono-oxygenated saturated compound ($C_6H_{14}O$) or a dioxygenated unsaturated compound ($C_5H_{12}O_2$). Note that there is a peak for the loss of a methyl group (M−15), and for the loss of an ethyl group (M−29), so one might be tempted to conclude that the molecule contains a methyl group and an ethyl group, but this should always be verified in the ^1H and ^{13}C NMR spectra.

^1H NMR

The spectrum has three signals with an integration ratio of 2:2:3. The coupling pattern of a triplet, sextet and triplet suggests the presence of $-CH_2CH_2CH_3$, the terminal CH_2 split by two neighbours, the middle CH_2 split by five neighbours, and the terminal CH_3 split by two neighbours. In view of the chemical shift of the terminal CH_2 group at 3.4 ppm, one would expect this carbon to be attached to an oxygen. The mass spectrum of the compound requires a higher mass than an oxygen and one propyl group. As there are no peaks for other protons, there are probably two equivalent propyl groups in the molecule.

Infrared

The strong band at 1120 cm^{-1} is in the region of C–O stretching. As there is no carbonyl band or OH band, this band belongs to an ether. There is little else that one can see in the spectrum other than the bands from the alkyl chains. There is a doublet of equal intensity near 1380 cm^{-1} which normally indicates the presence of a *gem*-dimethyl group. However this pair of peaks is deceptive, as the NMR spectra shows clearly that there is no *gem*-dimethyl group in the compound.

^{13}C NMR

There are three types of carbon in the compound, two CH_2 (triplets in off-resonance and inverted in DEPT), and one CH_3 (quartet in off-resonance). The chemical shift of the carbon at 73 ppm is in the correct region for a carbon attached to the oxygen of an ether.

Summary

The molecule has a propyl group attached to oxygen and a molecular mass of 102. The mass of one propyl group ($CH_3CH_2CH_2$) is 43, and one oxygen is 16, for a total of 59. This is 43 less than the molecular mass of 102. One more propyl group is needed to complete the structure. The compound is dipropyl ether, $CH_3CH_2CH_2–O–CH_2CH_2CH_3$. Notice that the peaks of the two triplets in the ^1H NMR spectrum are not symmetrical, particularly that of the methyl group. Each one is distorted, having a larger peak on the side towards the group to which it is coupled. This is typical of systems in which the chemical shift difference is not large compared to the size of the coupling constant. If the spectrum were run in a higher field spectrometer (say, 400 MHz, instead of this one, 60 MHz), one would not expect to see such distortion, as the chemical shift difference in Hertz would be much larger, while the coupling constant would not change. The ratio of the chemical shift to coupling constant becomes much larger, consequently the distortion decreases. Compare the coupling in this compound with that of the compound in **Exercise 004**.

Exercise 004 (007)

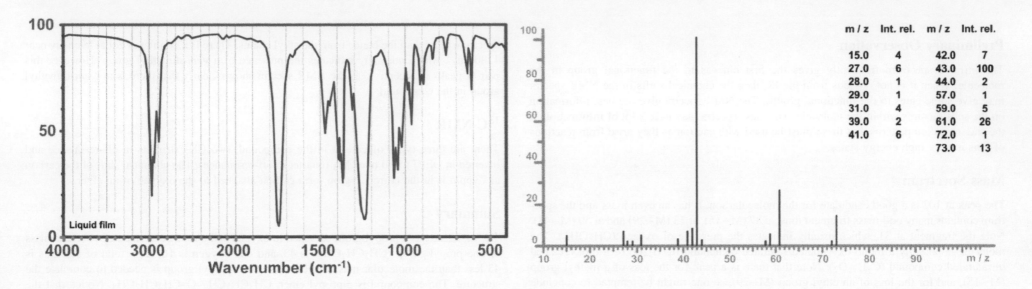

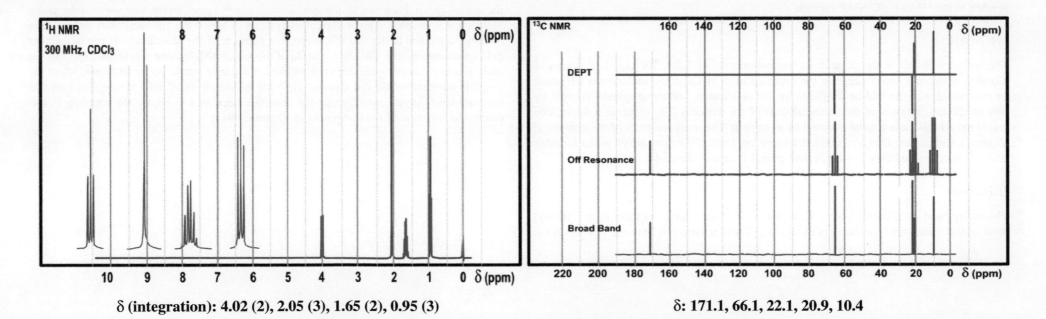

δ (integration): 4.02 (2), 2.05 (3), 1.65 (2), 0.95 (3)

δ: 171.1, 66.1, 22.1, 20.9, 10.4

Exercise 004 (007)

Preliminary Observation

A preliminary scan of the spectra shows that there are two very strong IR bands that must be accounted for in the structure. The highest mass ion in the mass spectrum is odd, so there is probably an easily fragmented functional group. The alkyl structural features stand out in the NMR spectra.

Infrared

The IR spectrum has two very significant bands of the functional group of the compound. The band at $1745\,cm^{-1}$ indicates the presence of a carbonyl. The position of a carbonyl peak can be used to determine the type of carbonyl (ketone, aldehyde, ester, amide, etc.) and whether it is conjugated. In this case it is in the range typical of a non-conjugated ester. The C–O single bond stretching at $1238\,cm^{-1}$, is in the right range for an ester, in particular an alkylacetate. The IR spectrum can also be used to eliminate possible functional groups; for instance, here one could also note that there is no alcohol, carboxylic acid, primary amine, etc.

^{13}C NMR

The DEPT spectrum indicates the presence of two CH_2 carbons (inverted signals) and one quaternary carbon (signal disappears). The chemical shift of the latter signal suggests that it is of the carbonyl carbon of an ester. The two peaks at 10 and 21 ppm are unchanged in the DEPT spectrum, and quartets in the OR spectrum indicate that they are methyl groups. The mass of these components of the compound is $5 \times 12\,(C) + 2 \times 16\,(O) + 10\,H$, for a total of 102.

Mass Spectrum

This spectrum is somewhat difficult to interpret. If indeed the peak at 73 were the molecular ion, then the compound would contain nitrogen (an odd number of nitrogens). But the main peaks have odd masses (43 and 61) which would not be expected in a compound containing nitrogen (one would expect even-mass fragments). Esters contain fragments from the associated acid and alcohol, depending on the site of the cleavage and the location of the charge (most often on the C=O). The fragments are largely derived from cleavage α to the C=O which generates an acyl cation, $RC{\equiv}O^+$. In this case it is the base peak at 43 ($CH_3C{\equiv}O^+$), typical of acetates. The ion at 61 is the relatively stable even-electron protonated acetic acid ion $(CH_3C(OH)_2)^+$, which is also typical of alkyl acetates. The peak at 73 is due to the loss of an ethyl radical by α cleavage at the alcohol oxygen.

^{1}H NMR

The ^{1}H NMR spectrum shows the presence of four signals in the ratio of: 2:3:2:3:1 for a total of 10 H. In view of the multiplicities observed for the various protons, there is a CH_2 coupled to 2 protons (triplet), a non-coupled CH_3, a CH_2 coupled to five protons (sextet), and a CH_3 coupled to two protons (triplet). The coupling pattern suggests the presence of a propyl chain, $CH_3CH_2CH_2O$, with the end CH_2 being strongly deshielded by the oxygen. The chemical shift of the CH_3 singlet suggests that it is attached to a carbonyl group.

Conclusion

The only ester that could have a propoxy group ($CH_3CH_2CH_2-O-$) and a methyl group bonded to the carbonyl would be propyl acetate, $CH_3CH_2CH_2O(C{=}O)CH_3$.

Exercise 005 (094)

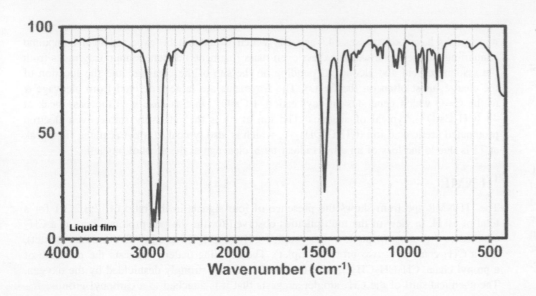

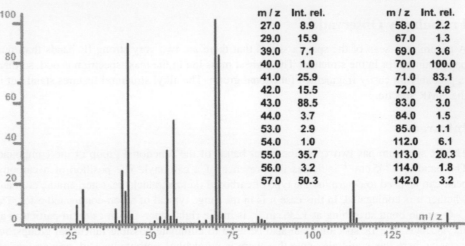

m / z	Int. rel.	m / z	Int. rel.
27.0	8.9	58.0	2.2
29.0	15.9	67.0	1.3
39.0	7.1	69.0	3.6
40.0	1.0	70.0	100.0
41.0	25.9	71.0	83.1
42.0	15.5	72.0	4.6
43.0	88.5	83.0	3.0
44.0	3.7	84.0	1.5
53.0	2.9	85.0	1.1
54.0	1.0	112.0	6.1
55.0	35.7	113.0	20.3
56.0	3.2	114.0	1.8
57.0	50.3	142.0	3.2

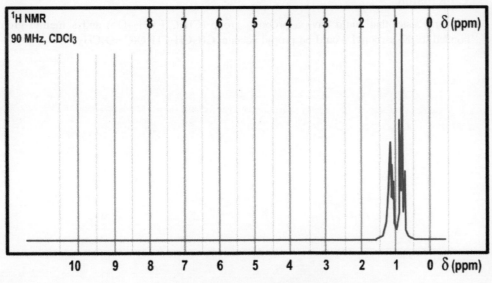

δ (integration): 1.05 (5), 0.86 (6)

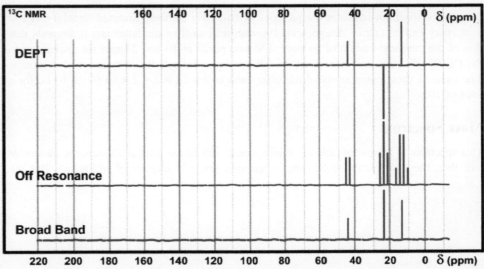

δ: 43.08, 22.88, 12.61

Exercise 005 (094)

Preliminary Observation

After a preliminary scan of the spectra to check for the presence of functional groups, one can try to determine the molecular formula from the mass spectrum. The alkyl groups can be identified by the ^{13}C and ^1H NMR spectra. From the molecular formula and the types of carbon, one can combine the groups to create a structure for the compound.

Infrared

The infrared spectrum shows no evidence of the presence of a functional group. All the peaks may be assigned to alkyl groups. The band at $1380\,cm^{-1}$ may be assigned to a deformation band of a CH_3 group. If this peak appears as a doublet, it could be taken as evidence of *gem*-dimethyl group (isopropyl $-CH(CH_3)_2$, *tert*-butyl $-C(CH_3)_3$ or $>C(CH_3)_2$).

Mass Spectrum

The highest mass ion in the spectrum (142) has an even mass, and the majority of the fragments are odd mass ions of the alkane family, 29, 43, 57, 71 and 113. This suggests that the compound has a molecular mass of 142, and is probably a hydrocarbon. Dividing the mass of 142 by 14 indicates that the molecule has a mass equivalent of 10 CH_2 units, plus a remainder of 2. This is an exact match for the saturated hydrocarbon ($C_{10}H_{22}$). The first fragment ion in the spectrum is found at 113, due to the loss of 29, CH_3CH_2; the lack of an $M-15$ peak suggests that there is no CH_3 at a branch carbon. Note that there is no peak at 31, which would have been an indication of the presence of oxygen.

^1H NMR

The proton NMR shows signals that are so close together that a precise integration is difficult. In a higher field spectrometer, the peaks would be better separated and a better integration could be obtained. The triplet near 0.9 ppm is due to methyl groups and shows the distortion typical of a methyl group split by protons with a similar chemical shift. This supports the mass spectral evidence for the presence of a CH_3CH_2 group in the molecule. The only other signals are at a slightly lower field, where one would expect to see the signals for CH_2 and CH protons.

^{13}C NMR

Although the molecular formula is $C_{10}H_{22}$, only three types of carbon are observed. There are obviously some isochronous carbons present. All the signals are found in the range for sp^3-hybridized carbons, unaffected by functional groups. The DEPT spectrum shows the absence of quaternary carbons (no signals disappear), but does show the presence of one type of secondary carbon (CH_2 signal inverted). The doublet at 43.1 ppm in the off-resonance spectrum represents CH carbons, the triplet at 22.9 ppm represents CH_2 carbons and the quartet at 12.6 ppm, the CH_3 carbons.

Structure

The accumulated evidence indicates that the molecule has ethyl groups (CH_3CH_2-) and methine carbons ($>CH-$), but no quaternary carbons. For each new branch in the chain, a methyl group is needed to terminate the new chain. A straight chain alkane has two methyl groups but no $>CH$ carbons. Insertion of a $>CH-$ adds a branch, thus requiring another CH_3. In this example the methyl groups are part of ethyl groups. Starting with a straight chain of two ethyl groups, adding a $>CH-$ requires another ethyl group giving seven carbons. Adding another $>CH-$ requires another ethyl group which gives the required 10 carbons. Thus there are four ethyl groups and two methine carbons in the molecule. The only way to arrange these to make the ethyl groups equivalent and the methines equivalent is in the molecule of 3,4-diethylhexane, $(CH_3CH_2)_2CHCH(CH_2CH_3)_2$.

Notes

The peaks at 71 and 70 arise from cleavage of the bond at the centre of the molecule, resulting in two peaks, one at 71 from the loss of a pentyl radical, and the other at 70 from the loss of a pentane molecule by a rearrangement reaction. In the 90-MHz ^1H NMR spectrum, the $>CH-$ and $>CH_2$ protons appear in an unresolved cluster below the peaks of the methyl protons. In a 400-MHz ^1H NMR spectrum, the signal would be better resolved and it may be possible to analyse the pattern. Note that the two protons of each CH_2 group are not equivalent. They are diastereotopic, being adjacent to a carbon atom bearing three different substituents. They form an AB pair ($^2J = 13\,Hz$), further coupled to the protons of the CH_3 group ($^3J = 7.2\,Hz$). In addition, they are not be equally coupled to the neighbouring methine proton, having different coupling constants ($^3J = 2$ and $^3J = 6.5\,Hz$).

Exercise 006 (009)

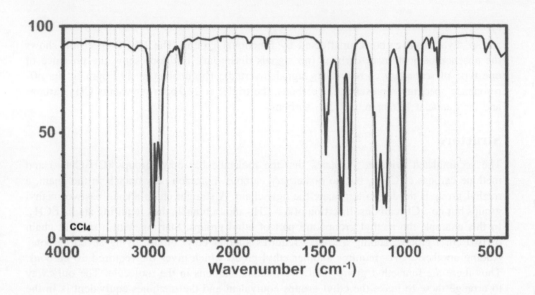

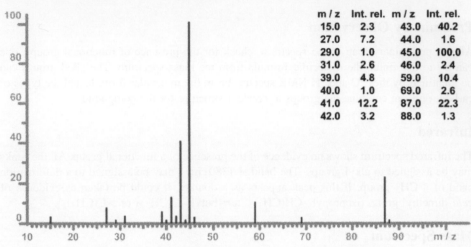

m / z	Int. rel.	m / z	Int. rel.
15.0	2.3	43.0	40.2
27.0	7.2	44.0	1.6
29.0	1.0	45.0	100.0
31.0	2.6	46.0	2.4
39.0	4.8	59.0	10.4
40.0	1.0	69.0	2.6
41.0	12.2	87.0	22.3
42.0	3.2	88.0	1.3

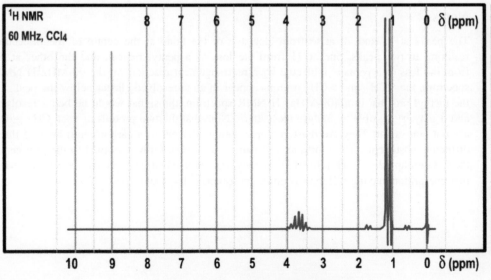

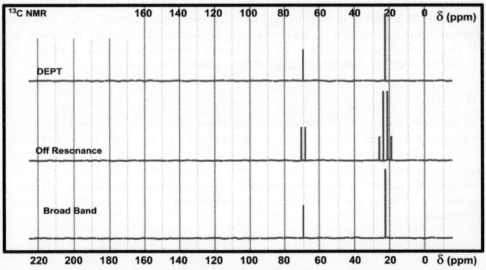

δ (integration): 1.13 (6), 3.64 (1)

δ: 68.57, 22.95

Exercise 006 (009)

Preliminary Observation

The ^{1}H NMR spectrum has small peaks on each side of the large peaks near 1 ppm. These are spurious peaks known as spinning sidebands. They are caused by gradients in the magnetic field, and are dependent on the rate of sample spinning. Ignore them in your consideration of the spectrum. Note that the band at 1380 cm^{-1} in the IR spectrum is a doublet. It may be difficult to recognize the functional group in this spectrum, but there are clues in the NMR and mass spectra.

^{1}H NMR

The ^{1}H NMR spectrum shows two types of proton, a septet and a doublet. These signals are due to an aliphatic group that is easily recognized. The high field doublet is that of a CH$_3$, which is split by a single proton. The lower field proton, at 3.7 ppm, that is responsible for this splitting, is itself split into a septet, indicating that it is coupled to six protons. This suggests the presence of an isopropyl group –CH(CH$_3$)$_2$. The integration of 1:6 supports this summary. Note that this gives an odd number of protons (seven), thus in the absence of any other monovalent atoms or nitrogen, the compound must have some symmetry element that allows the number of groups to be doubled to create an even number of protons.

Infrared

The band at 1100 cm^{-1} is the C–O stretching band of an ether group. At 1380 cm^{-1} there is a doublet of essentially equal intensities. It is characteristic of a *gem*-dimethyl group, in this case in an isopropyl group, –CH(CH$_3$)$_2$. No bands are seen that could be attributed to another functional group.

Mass Spectrum

The spectrum has a peak at the high mass end of the spectrum at 87. The peak at 88 is the isotope peak, with a relative intensity of $1.3/22.3 \times 100 = 5.8\%$, indicating there are five carbons in the ion at 87. An odd mass is characteristic of compounds containing an odd number of nitrogens; however, in this spectrum, the most intense fragments are of odd mass, whereas the fragments from nitrogen-containing compounds are usually of even mass. Thus there is some doubt that the peak at 87 represents the molecular ion! One must be careful in assuming that the highest mass ion is the molecular ion, especially if it is odd. Frequently, no molecular ions are observed in molecules that undergo easy cleavage. The peak at 87 is due to a loss of methyl group from the carbon bonded to the oxygen (α cleavage), producing a relatively stable even-electron ion. Note the peak at 31 u, which is a clue to the presence of oxygen in the molecule.

^{13}C NMR

While determining the number and types of carbon (primary, secondary, tertiary or quaternary) take note of the chemical shifts. The DEPT spectrum shows that the compound contains no CH$_2$ (no inverted signals) or quaternary carbons (no missing signals). The off-resonance spectrum shows the presence of carbons bearing one proton (>CH–), and bearing three protons (–CH$_3$). The chemical shift of the methine carbon (>CH–) is in the region for a carbon bonded to oxygen.

Summary

The structural components as identified in the various spectra can be put together so that the symmetry of the final structure generates only two types of proton. The structure is diisopropyl ether, ((CH$_3$)$_2$CH)$_2$O.

Exercise 007 (057)

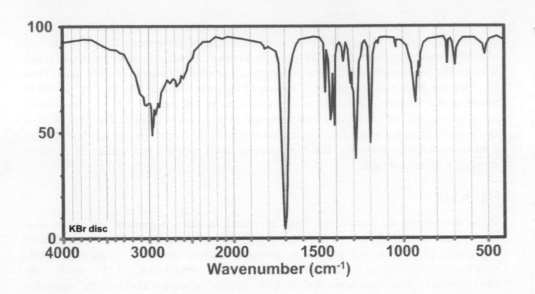

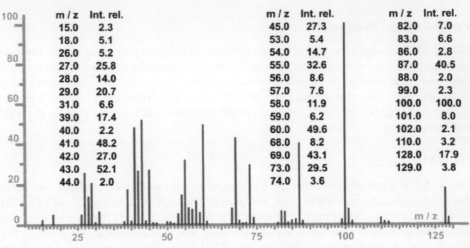

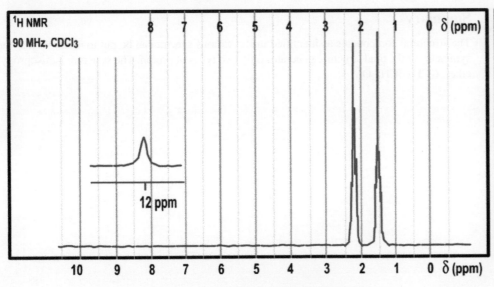

δ (integration): 12 (1), 2.21 (2), 1.51 (2)

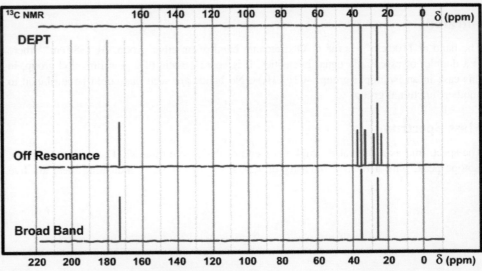

δ: 174.3, 33.4, 24.0

Exercise 007 (057)

Preliminary Observation

The IR spectrum gives a clear indication of the functional group. This can be confirmed by a quick look at the ^{13}C and ^1H NMR spectra.

Infrared

The very broad asymmetric band between 2300 and 3500 cm^{-1} is characteristicof the OH vibrations of a carboxylic acid group. That assignment is supported by the strong band at 1700 cm^{-1} due to the $>C=O$ stretching vibration, and the bandsat 1430 cm^{-1} and near 920 cm^{-1} for the OH bending vibrations.

Mass Spectrum

The peak at 128 is not the molecular ion. Indeed, if the molecular ion were at 128, then the 129 ion would be the isotope peak M+1, and the intensity relative to M of 21% (100 × 3.8/17.92) would indicate the presence of 19 or 20 carbons. This is impossible for a mass of 128 (a maximum of 10 carbons). This is a significant discrepancy in the isotope ratio which must be recognized. In mass spectrometry one rarely sees a peak for the molecular ion of carboxylic acids as they fragment readily by a variety of mechanisms. An even-mass fragment is often the product of the loss of a small neutral molecule such as water. The 128 ion could be the M−18 (M−H_2O) ion, and the 129 ion, the M−17 ion (M−OH). The base peak at 100 (M−46) corresponds to the loss of water and carbon monoxide (M−H_2O−CO).

^{13}C NMR

The spectrum shows only three types of carbon, with the one at 174 ppm being the carbonyl carbon of the carboxylic acid. The peak at 174 ppm disappears in the DEPT spectrum because it is quaternary. The DEPT spectrum also indicates that there are two types of CH_2 present (two inverted signals). In the off-resonance spectrum the two CH_2 groups appear as triplets. The peak at 33 ppm, corresponds to the α carbon of the acid, and the other at 24 ppm (less deshielded) is the carbon β to the acid group.

^1H NMR

The ^1H NMR spectrum shows three signals. with integrations of 1 (12 ppm) : 2 (2.2 ppm) : 2 (1.5 ppm) for a total of five protons. The odd number of protons is incompatible with a compound of even mass unless other monovalent atoms are present, such as halogen. However, doubling the number of the ratios would give an even number of protons: 2H : 4H : 4H, for a total of 10 protons. The signal at 12 ppm corresponds to the proton of an acid group (COOH), the one at 2.2 ppm to the CH_2 α to the acid (–CH_2–COOH), and the one at 1.5 ppm to the CH_2 β to a carboxylic acid group. The protons of these two CH_2 groups are coupled together as can be seen in the appearance of the peaks which look like distorted triplets (–CH_2CH_2COOH).

Conclusion

The carboxylic acid group is clearly identified by the very distinctive broad asymmetric band between 2300 and 3600 cm^{-1} and the strong sharp carbonyl peak near 1730 cm^{-1}. The ^{13}C NMR has the requisite signal for a carbonyl group at 194 ppm, and the ^1H NMR has the acid proton far downfield at 12 ppm. The only other groups identified are CH_2 groups. The requirements for an even number of hydrogens, and the indication of a mass larger that 128, means that the molecule has an element of symmetry, and has double the number of groups seen in the spectra. This leads to the formula of $HOOCCH_2CH_2CH_2CH_2COOH$, $C_6H_{10}O_4$, adipic acid.

Exercise 008 (011)

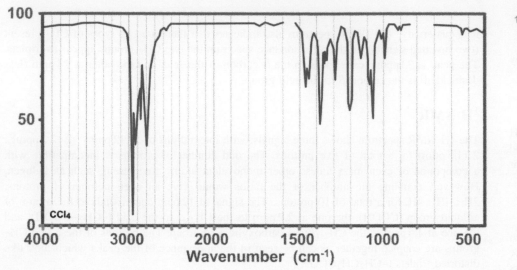

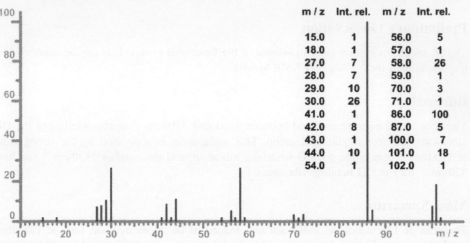

m / z	Int. rel.	m / z	Int. rel.
15.0	1	56.0	5
18.0	1	57.0	1
27.0	7	58.0	26
28.0	7	59.0	1
29.0	10	70.0	3
30.0	26	71.0	1
41.0	1	86.0	100
42.0	8	87.0	5
43.0	1	100.0	7
44.0	10	101.0	18
54.0	1	102.0	1

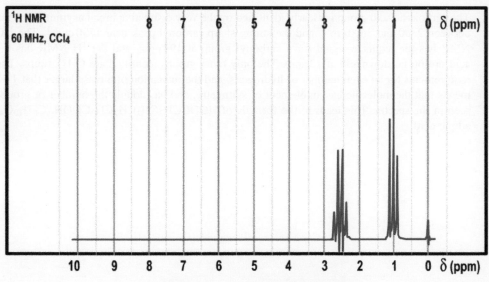

δ (integration): 2.42 (2), 0.97 (3)

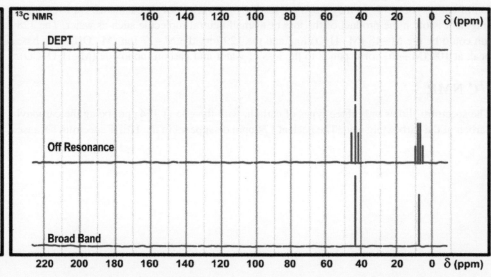

δ: 46.5, 11.8

Exercise 008 (011)

Preliminary Observation

Looking for evidence of functional groups in the IR spectrum seems somewhat unproductive in this spectrum. One might have better luck in the mass spectrum and the NMR spectra. Note that the molecular ion appears to be odd-mass, and that many of the fragment ions are even-mass.

Mass Spectrum

The molecular ion is found at 101 (18%), and has an M+1 isotope peak intensity of 1%, too weak to provide a reliable relative intensity calculation. The odd mass of the compound suggests that it contains nitrogen (or an odd number of N) and an odd number of monovalent atoms. In addition, the principal fragments have even masses, 30, 58 and 86, as might be expected for a compound containing nitrogen (normally a majority of odd fragments for non-nitrogen compounds, and even fragments for nitrogen-containing compounds). The predominant cleavage occurs by loss of a radical from the carbon bonded to the nitrogen (α cleavage), in this case yielding an ion at 86 (100%) from the loss of CH_3, and an ion at 100 (7%) from the loss of H. The greater loss of the more complex radical is a general phenomenon. The ion at 30 is due to the $H_2C=NH_2^+$ ion, which is a common fragment of amines.

^3C NMR

The DEPT spectrum indicates the presence of a CH_2 group (inverted peak), and the chemical shift (42 ppm) suggests that it is attached to a nitrogen atom. The quartet in the off-resonance spectrum indicates the presence of a methyl group (CH_3). Thus we have a CH_2 group deshielded by a heteroatom, and a CH_3 group, for a total of five protons on the two carbons. To get the required mass one needs more carbons; doubling the carbons would give 10H (an even number of monovalent atoms, and secondary amine), but still of inadequate mass. By taking three of these groups, one gets 6C and 15H (an odd number of protons atoms, and tertiary amine).

^1H NMR

The ^1H NMR spectrum shows the presence of two signals with an integration ratio of 2:3. The splitting pattern (triplet and quartet) shows that the CH_2 and CH_3 groups are coupled.

This coupling pattern and the chemical shift of the quartet suggest that the CH_3–CH_2–N chain is present in the molecule. Knowing that the molecule has some symmetry, one could consider triethylamine, as there is no indication of the presence of an NH or an NH_2 in any of the spectra. The ^1H NMR spectrum essentially confirms that the CH_2 and CH_3 groups that are observed in the ^{13}C NMR, are present as an ethyl group (as established by the observed proton–proton coupling pattern).

Infrared

The infrared spectrum does not show bands that clearly identify the functionality of the molecule. The medium intensity bands between 1000 and 1400 cm^{-1} suggest the presence of an amine, although they might be overlooked because of their proximity to the C–C stretching bands. Tertiary amines are generally difficult to detect by infrared spectroscopy because there is no N–H stretching band near 3100 cm^{-1}, and the C–N band is less intense than the C–O band (which is easier to detect). In fact, knowing the saturated nature of the compound from the molecular formula, and that the absence of bands is characteristic of primary or secondary amines, one is led by default to a tertiary amine.

Summary

The mass spectrum gives us a molecular ion peak at 101; the compound contains nitrogen and an odd number of protons atoms. The presence of nitrogen is confirmed by the principal fragments, which are at even masses 30, 58 and 86. The ^{13}C NMR spectrum shows that there are only two types of carbon, CH_2 and CH_3. The proton–proton couplings seen in the ^1H NMR spectrum establishes that there is a bond between these two carbons. The CH_3CH_2N is confirmed by the chemical shifts of the CH_2. The presence of nitrogen is difficult to determine from the infrared spectrum. Knowing the molecular mass of the compound, it is certain that the molecule is symmetric, as three ethyl groups (CH_3–CH_2) and one nitrogen atom are needed to yield the required molecular mass.

Exercise 009 (022)

Elemental analysis: C=62.07% and H=10.35%

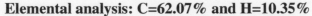

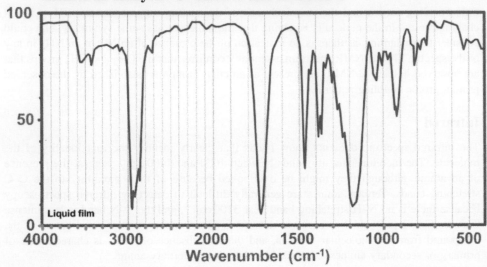

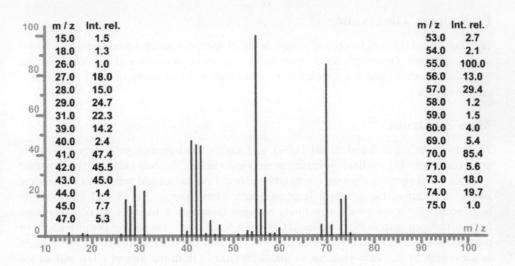

m/z	Int. rel.
15.0	1.5
18.0	1.3
26.0	1.0
27.0	18.0
28.0	15.0
29.0	24.7
31.0	22.3
39.0	14.2
40.0	2.4
41.0	47.4
42.0	45.5
43.0	45.0
44.0	1.4
45.0	7.7
47.0	5.3

m/z	Int. rel.
53.0	2.7
54.0	2.1
55.0	100.0
56.0	13.0
57.0	29.4
58.0	1.2
59.0	1.5
60.0	4.0
69.0	5.4
70.0	85.4
71.0	5.6
73.0	18.0
74.0	19.7
75.0	1.0

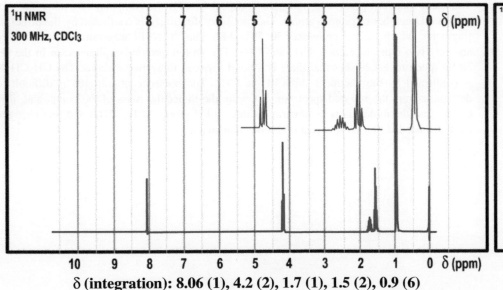

δ (integration): 8.06 (1), 4.2 (2), 1.7 (1), 1.5 (2), 0.9 (6)

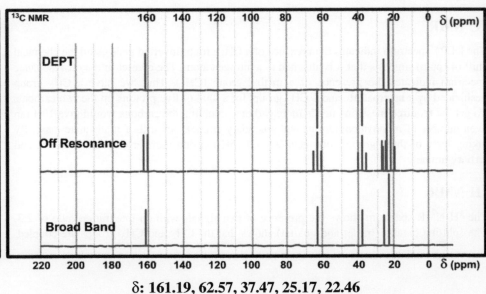

δ: 161.19, 62.57, 37.47, 25.17, 22.46

Exercise 009 (022)

Infrared

Two strong bands stand out in the IR spectrum: the band at $1730\,cm^{-1}$ representing the >C=O stretching of an ester carbonyl, and the band at $1180\,cm^{-1}$ representing the =C–O and O–C stretching vibrations of the ester. The position of the $1180\,cm^{-1}$ band is typical of the C–O stretching vibration of a formate ester. The doublet of equal intensity near $1380\,cm^{-1}$ suggests the presence of a *gem*-dimethyl group $(CH_3)_2C<$.

Mass Spectrum

The mass spectrum must be read carefully. The peak at 74 is not the molecular ion. If it were, the peak at 70 would represent a fragment from the loss of 4 mass units from the molecular ion, which is only observed by loss of two hydrogen molecules. This happens rarely, and peaks are seen for both steps. A microanalysis of the compound gives C = 62.07% and H = 10.35%, leaving a residual of 27.58% for the other atoms, probably oxygen. Calculation of the atomic ratios yields C 5.17 (62.07/12.01): H 10.35 (10.35/1.008): O 1.72 (27.58/16), which gives the simplest integer ratios of C:H:O = 3:6:1; (5.17/1.72), (10.35/1.72), (1.72/1.72). This gives an empirical formula of C_3H_6O (mass 58), which when doubled gives a molecular formula of $C_6H_{12}O_2$ (116). The peak at 74 is not the molecular ion, rather a fragment from the loss of a neutral molecule, C_3H_6.

^{13}C NMR

The spectrum shows five types of carbon atom, but one of the peaks looks as if it is twice the size of the others. There is no integration of the carbon signals, but for carbons of the same type, one would expect peaks of similar size. The peak at 22 ppm probably represents the signal from two carbon atoms, two equivalent CH_3 groups (the quartet in the OR spectrum). The other four carbons consist of two CH_2 (inverted DEPT signals) and two CH (OR doublet), but no quaternary carbons. The chemical shift and off-resonance doublet of the carbon at 161 ppm suggests that it is the carbonyl of a formate ester (HCOO–R). The chemical shift of the CH_2 peak at 63 ppm corresponds to a carbon bonded to an oxygen (O–CH_2), which agrees with the formate ester hypothesis.

^{1}H NMR

This proton NMR spectrum shows the presence of five signals: singlet, triplet, multiplet (seven or more), quartet and doublet with integration ratios of 1:2:1:2:6, respectively for the peaks at 8.1, 4.2, 1.7, 1.5 and 0.9 ppm for a total of 12 protons. The chemical shift of the triplet at 4.2 ppm is characteristic of the methylene group attached to the oxygen of an ester, RCOOCH$_2$–. The chemical shift of the signal at 8.0 ppm is characteristic of formate esters (HCOO–R). The doublet with an integration of 6 at 0.9 ppm corresponds to an isopropyl group, $(CH_3)_2CH$–. This assignment is supported by the splitting of the multiplet at 1.7 ppm. The quartet of integration 2 at 1.5 ppm corresponds to a –CH_2– split by three protons with similar couplings. This suggests a CH_2 group between a CH_2 and CH group, in the chain (–$CH_2CH_2CH<$).

Summary

Although the mass spectrum does not show a peak for the molecular ion, the elemental analysis and NMR spectra indicate that it has a molecular formula of $C_6H_{12}O_2$. The IR spectrum indicates that we have a formate ester, an observation supported by the doublet for the carbonyl in the off-resonance spectrum, and a singlet at 8 ppm in the ^{1}H NMR. The remaining portion of the molecule, C_5H_{11}, would form the alkoxy portion of the ester. The carbon attached to the oxygen is a methylene group (63 ppm, in ^{13}C, inverted signal in DEPT, triplet in OR, 4.2 ppm in ^{1}H, triplet split by an adjacent CH_2). Thus we have a sequence, H(C=O)OCH$_2$CH$_2$–. The high field doublet of integration 6 corresponds to two methyl groups $(CH_3)_2CH$–, which terminates the chain, giving a final structure of H(CO)OCH$_2$CH$_2$CH(CH$_3$)$_2$, isopentyl formate.

The ion at 70 in the mass spectrum corresponds to the loss of formic acid (HCOOH, 46) possibly by McLafferty rearrangement with charge retention on the alkene. The resulting radical ion, $[CH_2 = CH–CH(CH_3)_2]^{\cdot+}$, undergoes a typical allylic cleavage by losing a methyl radical to give the stable even-electron ion $[CH_2 = CH–CHCH_3]^+$, (base peak, 55). The ion at 74 corresponds to the loss of propylene (42) by rearrangement with a hydrogen atom being taken up by the carbonyl oxygen. The ion at 73 comes from the loss of the isopropyl radical.

Exercise 010 (005)

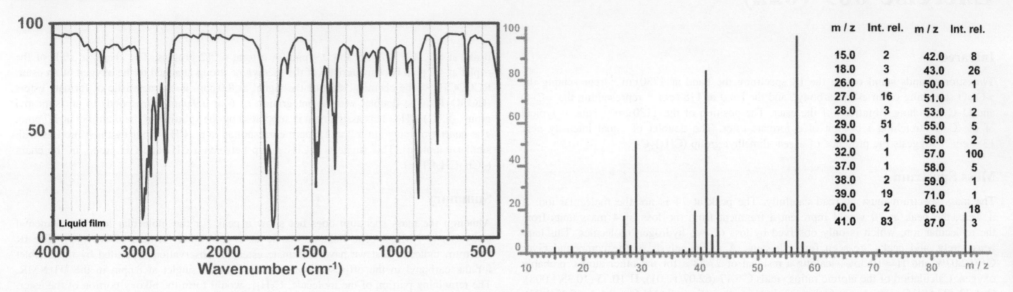

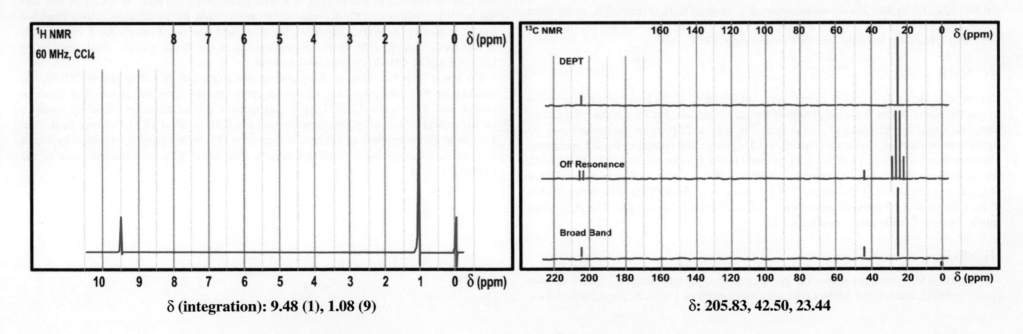

m / z	Int. rel.	m / z	Int. rel.
15.0	2	42.0	8
18.0	3	43.0	26
26.0	1	50.0	1
27.0	16	51.0	1
28.0	3	53.0	2
29.0	51	55.0	5
30.0	1	56.0	2
32.0	1	57.0	100
37.0	1	58.0	5
38.0	2	59.0	1
39.0	19	71.0	1
40.0	2	86.0	18
41.0	83	87.0	3

δ (integration): 9.48 (1), 1.08 (9)

δ: 205.83, 42.50, 23.44

Exercise 010 (005)

^{1}H NMR

This ^{1}H NMR spectrum is very simple: it has only two signals, both of which are sharp single peaks. The integration gives a relative intensity of 9 to 1, for a total of 10 protons. The signal at 1.1 ppm (9 H) is probably due to three equivalent CH_3 groups, most likely in a *tert*-butyl group. The signal at 9.5 (1 H) is strongly deshielded by an electron-withdrawing group such as a carbonyl; it is in the appropriate range for the proton of an aldehyde.

^{13}C NMR

Three types of carbon can be seen in the broadband and off-resonance spectra. The DEPT spectrum shows only two signals. The missing peak (43 ppm) corresponds to a quaternary carbon (bearing no hydrogen). The two other signals are attributed to CH (doublet in off-resonance) and CH_3 groups (quartet in off-resonance). The strongly deshielded peak at 205 ppm corresponds to the CH carbon of an aldehyde.

Infrared

The infrared spectrum has many bands, but the most significant bands are those characteristic of the functional group of the molecule. The strong band at $1720\,cm^{-1}$ represents the stretching vibration of a carbonyl group. The two bands at 2700 and $2800\,cm^{-1}$ are characteristic of an aldehyde group. The double peaks of different intensities near $1380\,cm^{-1}$ are typical of the *tert*-butyl group.

Mass Spectrum

The peak at 86 is probably the molecular ion. Its even-mass suggests that there is no nitrogen in the molecule. A mass of 86 could correspond to a C_6 alkane (C_6H_{14}) or to a compound with one oxygen and one unsaturation, or with two oxygens and two unsaturations. An accurate mass measurement of the 86 ion gave a value of 86.0732, which fits with a molecular formula of $C_5H_{10}O$ (mass 86.073165), but not with C_6H_{14} (mass 86.1096). The loss of 15 (ion at 71) is often an indication of CH_3 in the molecule. The peak at 57 is either the C_4H_9 (*tert*-butyl) or C_3H_5O ion.

Summary

Information for deriving the molecular formula ($C_5H_{10}O$) can be obtained from various spectra, such as the mass spectrum (molecular ion at 86), the ^{1}H NMR (10H, integration), the IR (oxygen of the carbonyl group), and the ^{13}C NMR (5C). The ^{13}C NMR spectra do not have integrations like proton NMR, but in some cases the information may be obtained from comparisons of peak intensities of the same types of carbon. In this case the large peak for methyl carbons represents three of them, as indicated by the 9H peak in the ^{1}H NMR. The functional group is shown to be an aldehyde by the IR carbonyl stretching band at $1720\,cm^{-1}$, and the characteristic two bands near 2700 and $2800\,cm^{-1}$. Confirmation is obtained from the chemical shift of the aldehyde H in the ^{1}H NMR, and the chemical shift and splitting of the aldehyde C in the ^{13}C NMR. The presence of a *tert*-butyl group, as indicated by the ^{1}H and ^{13}C NMR spectra, is supported by the observation of a pair of peaks of different intensities at about $1380\,cm^{-1}$ in the IR spectrum. The base peak in the mass spectrum at 57 corresponds to the *tert*-butyl cation. Adding the *tert*-butyl group to the aldehyde function gives $(CH_3)_3CCHO$, pivalaldehyde or trimethylacetaldehyde.

Exercise 011 (066)

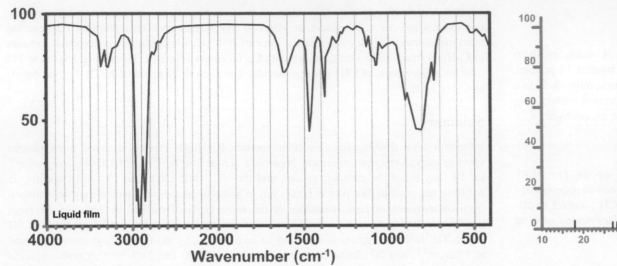

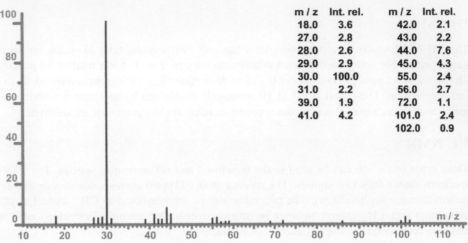

m/z	Int. rel.	m/z	Int. rel.
18.0	3.6	42.0	2.1
27.0	2.8	43.0	2.2
28.0	2.6	44.0	7.6
29.0	2.9	45.0	4.3
30.0	100.0	55.0	2.4
31.0	2.2	56.0	2.7
39.0	1.9	72.0	1.1
41.0	4.2	101.0	2.4
		102.0	0.9

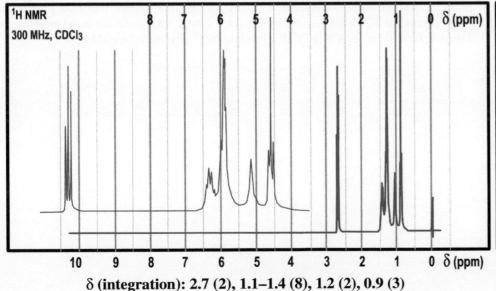

δ (integration): 2.7 (2), 1.1–1.4 (8), 1.2 (2), 0.9 (3)

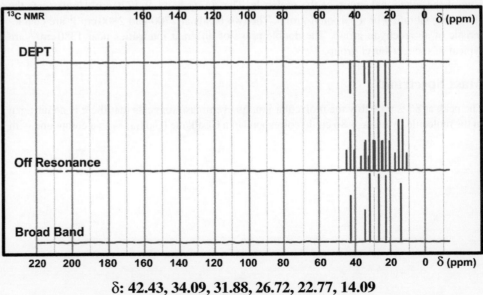

δ: 42.43, 34.09, 31.88, 26.72, 22.77, 14.09

Exercise 011 (066)

Preliminary Observation

The pair of peaks in the IR spectrum at 3300 and $3400\,cm^{-1}$ should give a clue as to the functional group. The mass spectrum provides another hint as to the nature of the functional group. Note that the molecular ion appears to be odd-mass, and that many of the fragment ions are even-mass. The ^{13}C NMR spectrum can give the number of different carbons in the molecule, and the ^{1}H NMR spectrum can give the number of different types of proton.

Mass Spectrum

The spectrum has a small peak for the molecular ion at m/z 101. The weakness of the molecular ion, M, and its isotope peak, M+1, make it impractical to use the isotope ratios to determine the number of carbons in the molecule. The presence of nitrogen in the molecule is indicated by the odd-mass molecular ion, and by the presence of several even-mass fragment ions obtained by the loss of alkyl groups (86 for M−15, 72 for M−29 and 58 for M−43). The very stable base peak at 30 ($CH_2 = NH_2^+$) is a typical fragment ion observed in alkyl amines, arising from α cleavage at the nitrogen. The molecular formula of an alkyl amine with a mass of 101 is $C_6H_{15}N$.

Infrared

The broad bands between 3300 and $3400\,cm^{-1}$ are characteristic of the symmetric and asymmetric N–H stretching bands of a primary amine. That is confirmed by the band at $1620\,cm^{-1}$ corresponding to the NH_2 scissoring movement, and the broad band at $800\,cm^{-1}$ associated with the out-of-plane bending of the NH_2. The alkyl chain bands are seen between 2980 and $2860\,cm^{-1}$ (C–H stretching), at 1470 (CH_2 scissoring) and $1370\,cm^{-1}$ (CH_3 deformation).

^{1}H NMR

In the ^{1}H NMR spectrum there are four groups of signals with some overlapping peaks. The observed integrations are 2:8:2:3 for a total of 15H, an odd number of hydrogens, as expected for an amine. The peak at 0.9 ppm (3H) corresponds to a CH_3 group of an alkyl chain, which appears as a triplet, distorted in the shape typical of linear alkyl chains (coupling to a CH_2 having a similar chemical shift). The broad peak between 1.1 and 1.4 ppm includes the CH_2 groups of the alkyl chain, all having nearly the same chemical shift. The CH_2 attached to the nitrogen is the only one that has a distinctly different chemical shift. It appears as a triplet (coupled to its neighbouring CH_2) at 2.7 ppm, (deshielded by the nitrogen). The peak at 1.2 ppm corresponds to the two hydrogens of the amino group (NH_2); note the high chemical shift of this type of proton.

^{13}C NMR

The DEPT spectrum shows the presence of five different CH_2 carbons (inverted signals). This is confirmed in the off-resonance spectrum by the presence of five triplets. One quartet in the off-resonance indicates the presence of a methyl group. The doublet at 42 ppm in the off-resonance spectrum corresponds to a CH_2 bonded to an amine nitrogen.

Summary

The presence of a nitrogen atom is confirmed by the pair of bands at $3300–3400\,cm^{-1}$ in the infrared spectrum. The ^{1}H NMR spectrum shows 15H (originating from the odd molecular mass). The ^{13}C NMR spectrum shows the presence of at least six carbons. The mass spectrum shows the presence of an odd number of nitrogen atoms (the odd mass of M). The only possible structure that could be constructed from the combination of five CH_2 groups, a CH_3 group and an NH_2 is $CH_3CH_2CH_2CH_2CH_2CH_2NH_2$, hexylamine.

Exercise 012 (004)

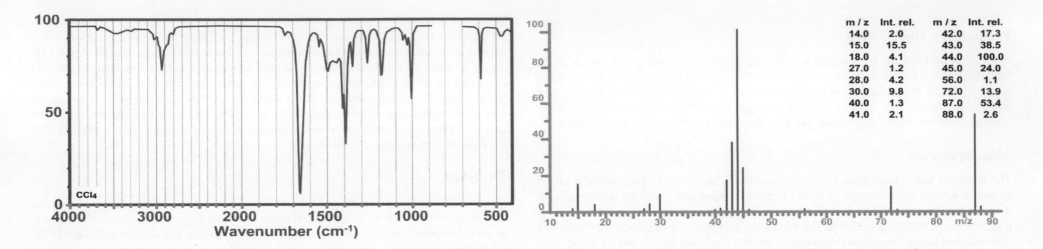

m / z	Int. rel.	m / z	Int. rel.
14.0	2.0	42.0	17.3
15.0	15.5	43.0	38.5
18.0	4.1	44.0	100.0
27.0	1.2	45.0	24.0
28.0	4.2	56.0	1.1
30.0	9.8	72.0	13.9
40.0	1.3	87.0	53.4
41.0	2.1	88.0	2.6

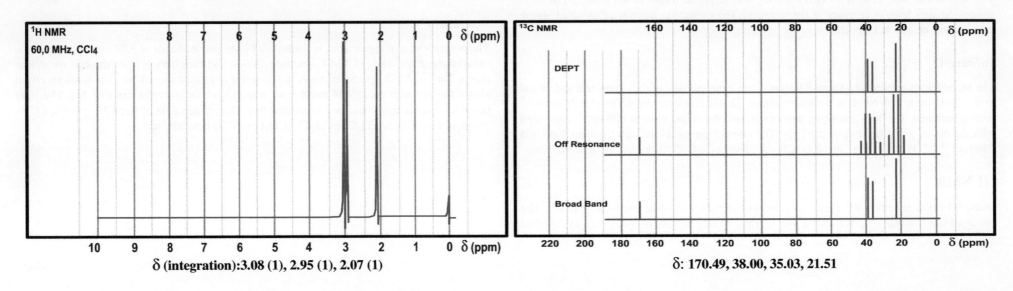

δ (integration):3.08 (1), 2.95 (1), 2.07 (1)

δ: 170.49, 38.00, 35.03, 21.51

Exercise 012 (004)

Preliminary Observation

The mass spectrum looks fairly simple; perhaps it can yield some useful information. The ^{13}C NMR spectrum can provide information on the number of different carbons in the molecule, and the ^{1}H NMR spectrum can give the number of different types of proton. The position of the very strong peak in the IR spectrum should be noted.

Mass Spectrum

The molecular ion seems to be at 87, which suggests that the compound contains a nitrogen atom. The isotope peak can be used to obtain an estimate of the number of carbons in the molecule. Each carbon contributes 1.1% to the M+1 isotope peak. In this case the relative intensity of the peak is 4.9% ($2.6/53.4 \times 100$) which suggests that the molecule has four carbons. Note that the accuracy of the relative intensity ratio can be very poor for low intensity peaks

^{13}C NMR

The broad band spectrum shows signals from four types of carbon. In the off-resonance spectrum, all of the upfield signals occur as quartets, indicating that they belong to methyl groups as each carbon is split by its three directly bonded protons (CH_3). The carbon at 170 ppm is a quaternary carbon, as it is unsplit in the off-resonance spectrum. DEPT-135 spectra give positive signals for CH and CH_3 carbon atoms, but negative signals for CH_2 carbons, and no signal for carbons with no attached hydrogen. In this compound the signal of the quaternary carbon (170 ppm) disappears, the three methyl carbons appear as upright signals, and there are no inverted peaks, indicating that there are no CH_2 groups. The chemical shift of the lower field carbon (170 ppm) is in the range of the carbonyl carbon of an amide or ester.

Infrared

The most significant band in the spectrum is the very strong band at 1660 cm^{-1}. This represents the stretching vibration of a carbonyl group, (C=O) and its strong intensity is due to the very polar nature of the carbonyl bond. The band is in the low end of the frequency range for carbonyl groups, suggesting that it is an amide.

^{1}H NMR

The ^{1}H NMR has three unsplit signals of the same integration. These peaks represent the three CH_3 groups that are observed in the ^{13}C NMR. The chemical shift of the peak at 2 ppm is in the range for a CH_3 on a carbonyl carbon, and the two peaks at 3 ppm for CH_3 groups on the nitrogen atom of an amide.

Summary

The odd mass of the molecular ion indicates that the compound contains a nitrogen atom, and the relative intensity of the M+1 peak suggests that the compound has four carbons. The Infrared spectrum indicates the presence of a carbonyl group, which is confirmed by the low-field peak in the ^{13}C NMR spectrum. The ^{13}C and ^{1}H spectra both show that the molecule has three methyl groups. Three CH_3 groups, one C=O, and a nitrogen atom add up to a formula of C_4H_9NO, which equals the observed molecular mass of 87. Construction of a molecule from these components yields only one possibility, N,N-dimethylacetamide, $CH_3(C=O)N(CH_3)_2$. The non-equivalence of the two methyl groups on the nitrogen is due to the strong conjugation of the nitrogen lone pair with the carbonyl double bond, slowing the rotation around the C–N bond so that two distinct methyl groups are seen in the NMR.

Exercise 013 (023)

Elemental analysis: C=62.60%, H=11.32% and N=12.17%

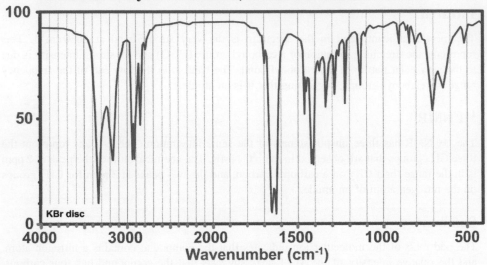

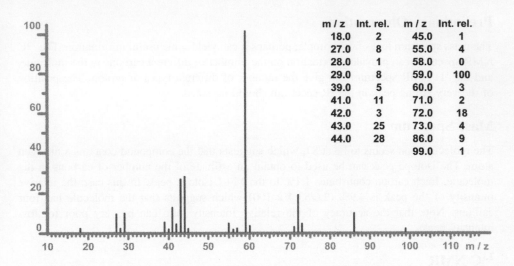

m / z	Int. rel.	m / z	Int. rel.
18.0	2	45.0	1
27.0	8	55.0	4
28.0	1	58.0	0
29.0	8	59.0	100
39.0	5	60.0	3
41.0	11	71.0	2
42.0	3	72.0	18
43.0	25	73.0	4
44.0	28	86.0	9
		99.0	1

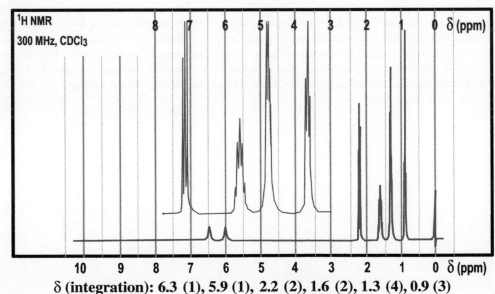

δ (integration): 6.3 (1), 5.9 (1), 2.2 (2), 1.6 (2), 1.3 (4), 0.9 (3)

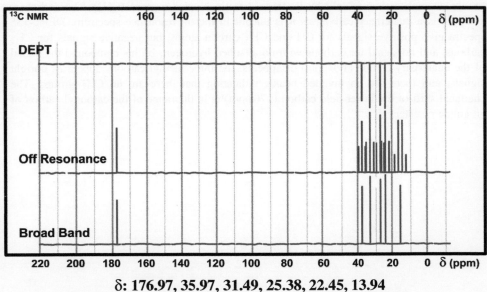

δ: 176.97, 35.97, 31.49, 25.38, 22.45, 13.94

Exercise 013 (023)

Preliminary Observations

The infrared spectrum has very distinctive bands that allow the assignment of the functional group. The mass spectrum appears to indicate that the molecule undergoes easy fragmentation.

Mass Spectrum

The peak at 99 is very weak, and does not appear to be the molecular ion. However, the series of fragment ions at even masses, 44, 72 and 86 suggest that the compound may contain a nitrogen atom. A microanalysis of the compound gives C = 62.60%, H = 11.32%, and N = 12.17, which leaves a residual of 13.91% for oxygen. The atomic ratios would be C 5.21 (62.60/12.011) : H 11.23 (11.32/1.008) : N 0.87 (12.17/14.007) : O 1.72 (27.58/16), which when divided by 0.87 gives the integer ratios of C:H:N:O = 6:13:1:2; for a molecular formula of $C_6H_{13}NO_2$. (mass 115). The fragment ion at 99 represents a loss of 16 (NH_2) which could occur by α cleavage at a carbonyl group, which would mean that the compound has a primary amide. Alpha cleavage on the other side of the carbonyl gives the ion at 44, $(O=C=NH_2)^+$. The base peak (59) comes from the loss of butene by McLafferty rearrangement (β cleavage with the carbonyl group picking up the γ hydrogen).

^{13}C NMR

The BB spectrum shows the presence of six different carbons, one CH_3 (quartet in OR), four CH_2 (inverted DEPT signals, and triplets in OR) and one quaternary carbon (signal absent in DEPT). Thus a total of 11H protons bonded to carbons are detected by the ^{13}C spectrum. The quaternary carbon at 177 ppm is in the correct range for the carbon of an amide carbonyl.

^{1}H NMR

This proton NMR spectrum shows the presence of six signals, 1H (6.3 ppm) : 1H (5.9 ppm) : 2H (2.2 ppm) : 2H (1.6 ppm) : 4H (1.3 ppm) : 3H (0.9 ppm) for a total of 13 protons.

The signals at 6.3 and 5.9 ppm correspond to two protons on the nitrogen atom. Note that they have different chemical shifts because of restricted rotation about the C–N bond, one being *cis* to the carbonyl oxygen, the other being *trans* to it. The chemical shifts of these signals suggest that they belong to a primary amide group. The signal at 2.2 ppm corresponds to a CH_2 α to the >C=O of the amide, and its splitting into a triplet indicates that it is attached to another CH_2. The peak at 0.9 ppm of integration 3, represents the terminal methyl group. The peaks at 1.6 and 1.3 ppm represent three CH_2 groups.

Infrared

The bands between 3200 and 3400 cm^{-1} show intense N–H stretching (strongly polarized) of the symmetric and asymmetric stretching bands characteristic of the NH_2 group. This is confirmed by the presence of broad bands below 750 cm^{-1}, representing >N–H deformations. The band at 1660 cm^{-1} is due to the stretching vibration of the >C=O, and that at 1630 cm^{-1}, called "band II", is due to deformations of the NH_2. These bands indicate the presence of a primary amide.

Summary

The highest mass peak in the mass spectrum is odd, and there are several even-mass fragments suggesting the presence of nitrogen, but the peak at 99 is very small and may itself be a fragment ion. The elemental analysis confirms the presence of nitrogen, yielding a molecular formula of $C_6H_{13}NO_2$ (mass 115). The ^{13}C NMR indicates that there are six different carbons in the molecule and that one of them is a carbonyl group. The carbonyl group is confirmed by the IR spectrum, which also shows bands for an NH_2 group, suggesting the presence of a primary amide. The ^{13}C NMR spectrum shows that there is only one methyl group, which would be at the end of the alkyl chain. The other carbons are all CH_2 groups which constitute the remainder of the alkyl chain. The CH_2 bonded to the carbonyl is observed in the expect positions in the ^{13}C NMR (36 ppm) and in the ^{1}H NMR (2.2 ppm). The only possible alkyl chain is a linear alkyl chain, $CH_3CH_2CH_2CH_2-$, and the compound is hexanamide, $CH_3CH_2CH_2CH_2CH_2CONH_2$.

Exercise 014 (068)

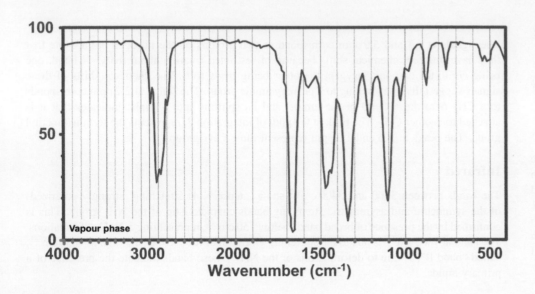

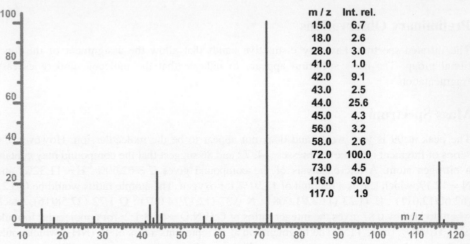

m / z	Int. rel.
15.0	6.7
18.0	2.6
28.0	3.0
41.0	1.0
42.0	9.1
43.0	2.5
44.0	25.6
45.0	4.3
56.0	3.2
58.0	2.6
72.0	100.0
73.0	4.5
116.0	30.0
117.0	1.9

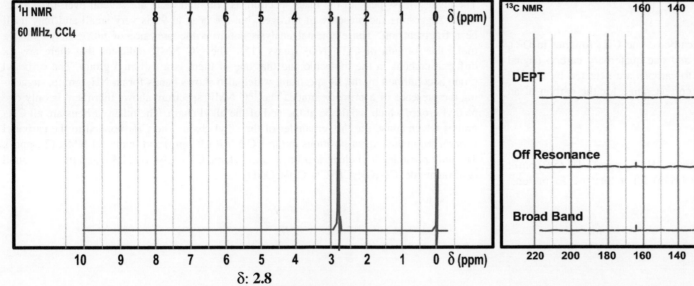

δ: 2.8

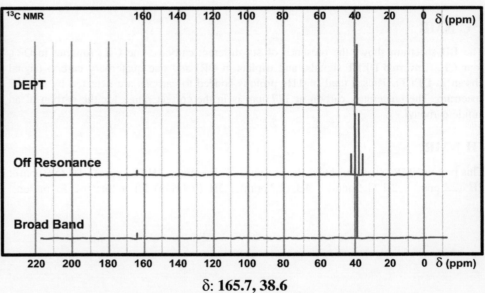

δ: 165.7, 38.6

Exercise 014 (068)

Preliminary Observations

Note the strong carbonyl band at $1685\,\mathrm{cm}^{-1}$ in the IR spectrum. This frequency is at the low end of the normal carbonyl range. The mass spectrum is interesting in that it has an even-mass molecular ion and even-mass fragment ions.

Mass Spectrum

The highest mass ion is an even-mass ion at 116, which implies that the molecule has either no nitrogen or an even number of nitrogens. The fact that the major fragment ions are also even-mass (44 and 72) indicates that the molecule has an even number of nitrogens. An even-mass compound with two nitrogen atoms can lose a fragment containing one of the N atom to generate an even-mass fragment ion containing the other nitrogen. The M+1 isotope peak at 117 may be used to estimate the number of carbons in the molecule. The relative intensity of 6.33% ($100 \times 1.9/30$) suggests the possibility of five or six carbons, based on the contribution of 1.11% for each carbon. But one must also take into account the ^{15}N contribution to the M+1, which is less than that of the ^{13}C contribution, but still significant. For five carbons and two nitrogens, one adds 1.11% per C and 0.37% per N ($5 \times 1.11 + 2 \times 0.37 = 6.29\%$), yielding a value close to the observed value. Adding $5C + 2N + 1O$ gives a mass of 102, requiring 12 H to bring the mass to 116. The molecular formula would be $C_5H_{12}N_2O$.

^{13}C NMR

There are only two types of carbon in the compound, a methyl carbon at 39 ppm (quartet in off-resonance) and a quaternary carbon at 166 ppm (signal disappears in DEPT). The peak at 39 ppm corresponds to a CH_3 bonded to a nitrogen atom and the peak at 166 corresponds to the carbonyl group of a urea functional group. The molecular formula (as determined by the mass spectrum) indicates that there must be four methyl groups in the peak at 39 ppm.

^{1}H NMR

The spectrum shows only one singlet at 2.8 ppm. It represents four isochronous CH_3 groups on the two nitrogen atoms of the urea group. Note that the assignment of 12 protons to the peak comes from the molecular formula, as a single signal cannot be integrated.

Infrared

The band at $1685\,\mathrm{cm}^{-1}$ represents a $>C{=}O$ stretching vibration which is quite a low frequency for a carbonyl group. It is in the range expected for an amide carbonyl (donor effect of the N on the $>C{=}O$, lowering the frequency of the carbonyl). The lack of N–H stretching bands between 3200 and $3400\,\mathrm{cm}^{-1}$ indicates that the functional group is neither a primary nor a secondary amide. The bands between 1000 and $1400\,\mathrm{cm}^{-1}$, particularly the strong band at $1120\,\mathrm{cm}^{-1}$, could be due to C–N stretching.

Summary

The IR spectrum has a strong carbonyl peak at the low-frequency end of the carbonyl range where one finds amides and ureas. The ^{13}C spectrum also shows a peak for the carbonyl group, as well as one other peak, belonging to isochronous methyl groups. The ^{1}H NMR shows only one type of proton, those of methyl groups bonded to nitrogen. The mass spectrum indicates that the molecular formula is $C_5H_{12}N_2O$. The only structure possible for a molecule with one carbonyl group, two nitrogen atoms and four methyl groups is tetramethyl urea, $(CH_3)_2N(C{=}O)N(CH_3)_2$. The loss of 44 from the molecular ion (base peak at 72) might suggest the loss of a neutral molecule by a rearrangement reaction; however, in this case the fragmentation is the α cleavage at the carbonyl with the loss of an even-mass radical containing nitrogen, $N(CH_3)_2$. The further loss of carbon monoxide ($C{\equiv}O$) gives the ion at 44.

Exercise 015 (006)

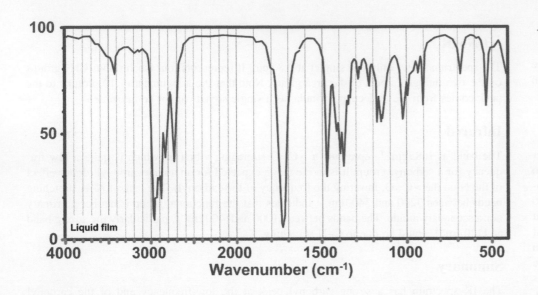

Liquid film

Wavenumber (cm⁻¹)

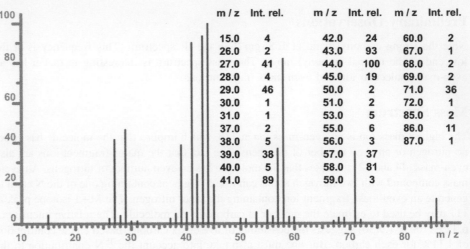

m / z	Int. rel.	m / z	Int. rel.	m / z	Int. rel.
15.0	4	42.0	24	60.0	2
26.0	2	43.0	93	67.0	1
27.0	41	44.0	100	68.0	2
28.0	4	45.0	19	69.0	2
29.0	46	50.0	2	71.0	36
30.0	1	51.0	2	72.0	1
31.0	1	53.0	5	85.0	2
37.0	2	55.0	6	86.0	11
38.0	5	56.0	3	87.0	1
39.0	38	57.0	37		
40.0	5	58.0	81		
41.0	89	59.0	3		

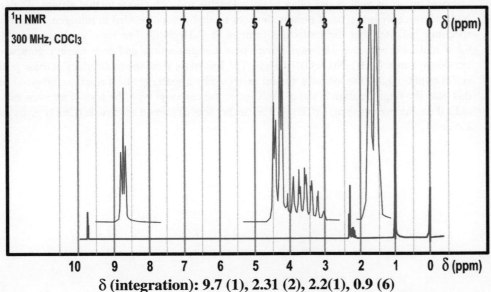

¹H NMR

300 MHz, CDCl₃

δ (integration): 9.7 (1), 2.31 (2), 2.2(1), 0.9 (6)

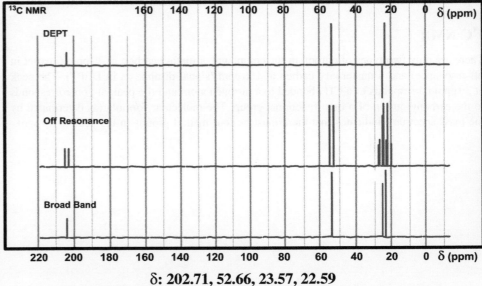

¹³C NMR

DEPT

Off Resonance

Broad Band

δ: 202.71, 52.66, 23.57, 22.59

Exercise 015 (006)

Preliminary Observation

The very strong band in the IR indicates the presence of a carbonyl group, and the bands in the C–H stretching region give further clues as to the nature of the carbonyl group. The very low-field signal in the proton NMR and its splitting pattern is very interesting.

Infrared

The strong band at $1728\,cm^{-1}$ indicates that a carbonyl group is present. The two bands near 2720 and $2820\,cm^{-1}$ indicate that it is an aldehyde group. The double peaks of similar intensities near $1380\,cm^{-1}$ suggest the presence of a *gem*-dimethyl group. The IR spectrum can be used to eliminate possible functional groups, for instance in this case the lack of OH stretching bands eliminates the possibility of an alcohol or carboxylic acid being present (3430 peak is a carbonyl overtone).

Mass Spectrum

The spectrum has a peak at m/z 86 which appears to be the molecular ion; the low intensity of the peak may offer some information about the structure. The M+1 peak at 87, of an intensity of 9% relative to the molecular ion, gives some indication of the number of carbons present; however, the very low intensity of the peak introduces a large degree of uncertainty. The M−1 peak at 85 (2%) is due to the loss of a hydrogen from the aldehyde carbonyl.

^{13}C NMR

The DEPT spectrum shows the presence of a CH_2 group (signal inverted), and indicates that there are no quaternary carbons in the compound (no signals disappear). The off-resonance spectrum shows a triplet for the CH_2, as well as a quartet for a CH_3 and doublets for two types of CH, the strongly deshielded one being the carbonyl CH and the other an aliphatic CH.

^1H NMR

The integration yields a ratio of 1:2:1:6. The signal at 9.7 ppm (a finely split triplet) corresponds to an aldehyde proton which is coupled to two other protons. The coupling is weak, which is typical of the coupling of the aldehyde proton to the α protons, thus, we have the $O=CHCH_2-$ sequence. The signal corresponding to the CH_2 group is seen at 2.3 ppm as a double doublet, with one splitting the same small size as seen in the aldehyde proton, and another larger splitting. Thus the CH_2 that is coupled to the aldehyde proton is also coupled to another CH by a larger coupling constant. This second CH must be the signal at 2.2 ppm that is seen as a poorly resolved multiplet. The peak at 0.9 ppm is a doublet representing two methyl groups split by a single proton.

Summary

The IR, ^1H NMR and ^{13}C spectra all indicate the presence of an aldehyde group. The ^{13}C NMR spectra shows signals for the aldehyde CH, as well as another CH, a CH_2 and a CH_3, but no quaternary carbons. In the ^1H NMR spectrum, the CH_2 is seen to be coupled to both the aldehyde CH and the other CH. Therefore, we have the sequence, $O=CHCH_2CH<$. This second CH is found at 2 ppm as a partly hidden multiplet, possibly nine peaks, being split by a total of eight neighbours (the protons of the CH_2 and two CH_3 groups). The two equivalent methyl groups are seen as a doublet at 0.9 ppm corresponding to six protons. This leads to the formula $(CH_3)_2CHCH_2CHO$, isovaleraldehyde.

Exercise 016 (021)

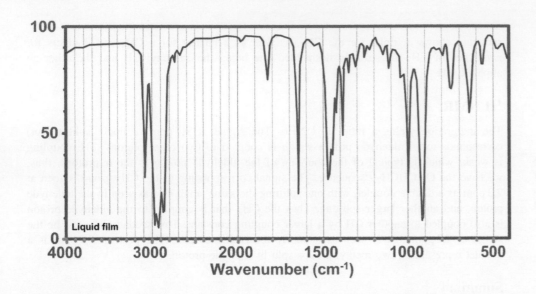

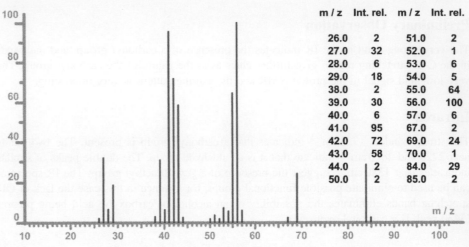

m/z	Int. rel.	m/z	Int. rel.
26.0	2	51.0	2
27.0	32	52.0	1
28.0	6	53.0	6
29.0	18	54.0	5
38.0	2	55.0	64
39.0	30	56.0	100
40.0	6	57.0	6
41.0	95	67.0	2
42.0	72	69.0	24
43.0	58	70.0	1
44.0	2	84.0	29
50.0	2	85.0	2

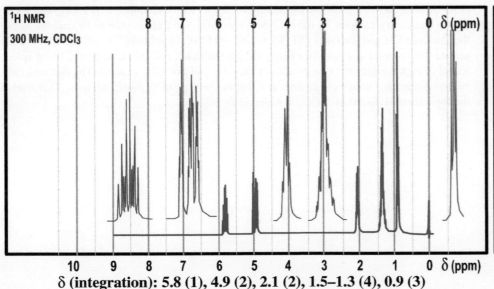

δ (integration): 5.8 (1), 4.9 (2), 2.1 (2), 1.5–1.3 (4), 0.9 (3)

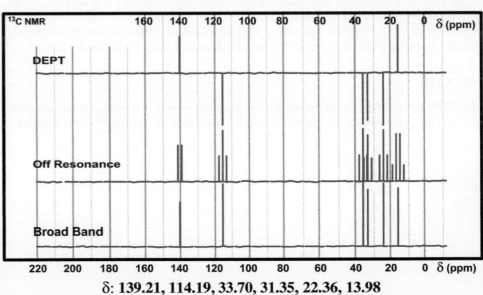

δ: 139.21, 114.19, 33.70, 31.35, 22.36, 13.98

Exercise 016 (021)

Preliminary Observation

A quick scan of the spectra gives no indication of functional groups other than an alkene group. The peak at 1810 cm^{-1} in the IR spectrum is too weak to belong to a carbonyl group, and there are no bands for OH, NH, or any heteroatoms. There are peaks on both sides of 3000 cm^{-1} indicating the presence of both sp^3 and sp^2 C–H stretching. The presence of the alkene double bonds can be confirmed by the C=C stretching in the 1600 cm^{-1} region and the out-of-plane bending bands below 1000 cm^{-1}.

Infrared

The band above 3000 cm^{-1} (3082 cm^{-1}) is characteristic of the C–H stretching on an olefinic carbon (H–C=C), indicating the presence of an alkene group. The band at 1650 cm^{-1} is characteristic of alkene >C=C< stretching. The out-of-plane bending band of the alkene hydrogens at 910 cm^{-1} is characteristic of monosubstituted alkenes. The band at 1820 cm^{-1} is a harmonic overtone of this band. This suggests that the molecule contains a vinyl group (–CH=CH$_2$).

Mass Spectrum

The molecular ion is seen at 84, with an isotope peak (M+1) at 85, of intensity 6.9%. The ^{13}C isotope peak intensity (6.9%) indicates the presence of six carbons in the molecule. The series of fragment ions at 69, 55, 41 and 27 represent simple cleavages, equivalent to the loss of methyl, ethyl, propyl and butyl radicals. The base peak at 56 arises from a rearrangement reaction, losing a neutral molecule, ethylene (28). The other intense even-mass ion at 42 arises from a similar reaction, losing a propylene molecule.

^{13}C NMR

The BB spectrum shows the presence of six different carbons, one CH$_3$ (quartet in OR), four CH$_2$ (inverted DEPT signals, and triplets in OR) and one methine carbon (doublet in OR). There are no quaternary carbons in the molecule (no signal disappears in DEPT). From the ^{13}C NMR, one can count a total of 12 protons attached to carbon. The chemical shifts of the carbons indicate that there are four sp^3 carbons and two sp^2 carbons. The signal at 139 ppm corresponds to an olefinic carbon (CH=), and the signal at 114 ppm corresponds to the other olefinic carbon (CH$_2$=). The four sp^3 carbons consist of one methyl carbon and three methylene carbons.

^{1}H NMR

The ^{1}H NMR spectrum shows the presence of five clusters of peaks, with integration ratios of 1:2:2:4:3; for a total of 12 protons. There are three olefinic protons as seen by the three multiplets, at 5.80, 4.96 and 4.92 ppm. The splitting pattern of this set of peaks is typical of those of a vinyl group. The peaks at 4.96 and 4,92 ppm correspond to the non-equivalent terminal =CH$_2$ protons, which appear as two doublets with further fine splitting. The higher field doublet has the smaller splitting, ^{3}J $\approx$ 10 Hz (cis coupling) compared to the lower doublet, ^{3}J $\approx$ 16 Hz (trans coupling). Each doublet shows the fine splitting of their mutual geminal coupling (^{2}J $\approx$ 2 Hz). The signal at 5.8 ppm corresponds to the olefinic (=CH) proton which is split by the trans proton (^{3}J $\approx$ 16 Hz), the cis proton (^{3}J $\approx$ 10 Hz), and the methylene protons (triplet ^{3}J $\approx$ 6 Hz). The chemical shift of the signal at 2.06 ppm, suggests two protons of a CH$_2$ group α to a double bond. The cluster of two overlapping signals at 1.3 to 1.5 ppm corresponds to two CH$_2$ groups. The last signal, a triplet at 0.90 ppm, corresponds to a CH$_3$ group.

Summary

The mass spectrum gives a molecular mass of 84 u, and there is no evidence of a functional group other than an alkene, thus we have a molecular formula of C$_6$H$_{12}$. Having established that the molecule contains one methyl group and one vinyl group, the remaining three methylene groups must lie in a chain between these terminal groups. This gives 1-hexene as the structure of the compound: CH$_3$CH$_2$CH$_2$CH$_2$CH=CH$_2$.

Exercise 017 (018)

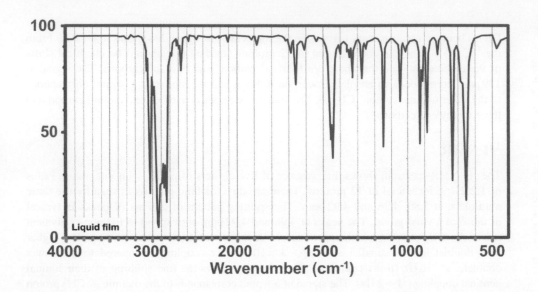

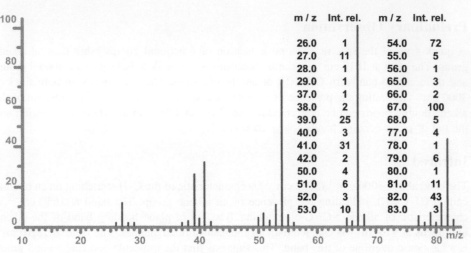

m / z	Int. rel.	m / z	Int. rel.
26.0	1	54.0	72
27.0	11	55.0	5
28.0	1	56.0	1
29.0	1	65.0	3
37.0	1	66.0	2
38.0	2	67.0	100
39.0	25	68.0	5
40.0	3	77.0	4
41.0	31	78.0	1
42.0	2	79.0	6
50.0	4	80.0	1
51.0	6	81.0	11
52.0	3	82.0	43
53.0	10	83.0	3

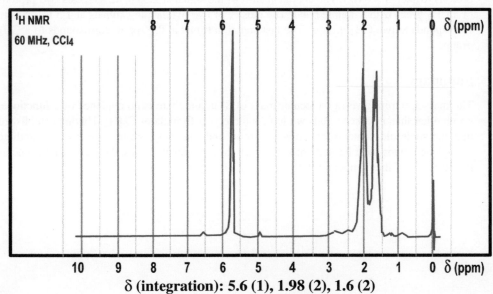

δ (integration): 5.6 (1), 1.98 (2), 1.6 (2)

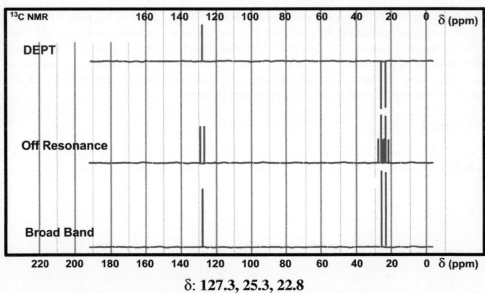

δ: 127.3, 25.3, 22.8

Exercise 017 (018)

Preliminary Observation

Close inspection of the C–H stretching region in the IR spectrum reveals that there are peaks for both sp^3 C–H and sp^2 C–H vibrations. There are also peaks in the ^{1}H and ^{13}C spectra that indicate the presence of an alkene or aromatic group.

Mass Spectrum

It appears that the peak for the molecular ion is at 82 (43%) with an isotope peak 83 (3%) for a relative intensity (3/43) of 6.9%, indicating the presence of six carbons, $6 \times 1.1\% = 6.6\%$. The molecular formula of a compound with six carbons and a molecular mass of 82, is C_6H_{10}. The ions at 27, 41, 55, indicate the presence of an unsaturation (fragments typical of alkenes). The base peak at M−15 suggests the presence of a methyl group, but you must be careful as many cyclic compounds give a peak for the loss of a methyl group. Even the simplest cyclic compounds, such as cyclohexane and cyclopentane, give a strong peak for M−15. Rearrangement of radical ions by hydrogen transfer is common in cyclic hydrocarbons. The even-mass ion at 54 is due to the loss of a molecule of ethylene, a retro Diels−Alder reaction.

^{13}C NMR

The broadband spectrum shows signals for only three types of carbon, a CH (doublet in OR), and two different CH_2 carbons (triplets in OR, and inverted signals in DEPT). The chemical shift of the signal at 127 ppm corresponds to that of an olefinic carbon. Since we see only one peak for sp^2 carbons, and we need two carbons to make the double bond ($>C=C<$), the peak must represent two equivalent carbons, and thus we have a symmetric molecule. The carbon multiplicities (two CH_2 and one CH) indicate the presence of a total of five protons, only half that in the molecular mass determination, again showing that each signal represents two carbon atoms.

^{1}H NMR

The ^{1}H NMR spectrum shows the presence of three signals, with integrations of 1:2:2 for a total of five protons. As the integration gives ratios only, the actual number of protons in the molecule is 10. The peak at 5.65 ppm represents olefinic protons, the peak 1.98 ppm represents allylic protons ($-CH_2C=C$) and the peak at 1.60 ppm represents the CH_2 β to the double bond. The coupling pattern is complex and the individual splitting cannot be discerned in this spectrum.

Infrared

The band above 3000 cm^{-1} is characteristic of olefinic $=C-H$ stretching, confirming the presence of an alkene group. The band for the $>C=C<$ stretching vibration is seen at 1660 cm^{-1}, and the out-of-plane bending band is seen at 720 cm^{-1}, in the range of the typical position for *cis*-disubstituted (Z) alkenes.

Summary

The mass spectrum gives a peak for the molecular at 82 and the M+1 indicates six carbons for a molecular formula of C_6H_{10}, two sites of unsaturation. The ^{13}C NMR shows three types of carbon, of which there are two CH_2 and one olefinic CH, for a total of three carbons and five hydrogens; clearly there is a symmetry element in the molecule. The infrared shows bands for alkene C–H stretching, the C=C stretching, as well as a band typical of the out-of-plane bending of two *cis* hydrogens. The two centres of unsaturation consist of a double bond and a ring. The compound is cyclohexene. The mass spectrum shows a peak for the loss of a methyl group, but this is due to a rearrangement reaction, common in cyclic hydrocarbons.

Exercise 018 (020)

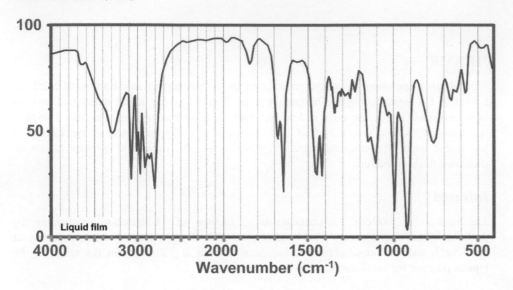

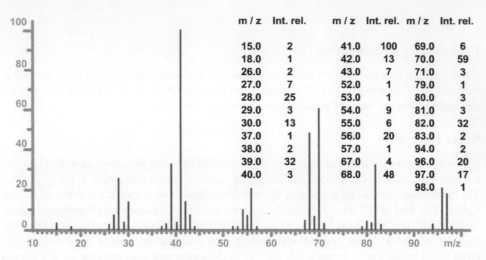

m/z	Int. rel.	m/z	Int. rel.	m/z	Int. rel.
15.0	2	41.0	100	69.0	6
18.0	1	42.0	13	70.0	59
26.0	2	43.0	7	71.0	3
27.0	7	52.0	1	79.0	1
28.0	25	53.0	1	80.0	3
29.0	3	54.0	9	81.0	3
30.0	13	55.0	6	82.0	32
37.0	1	56.0	20	83.0	2
38.0	2	57.0	1	94.0	2
39.0	32	67.0	4	96.0	20
40.0	3	68.0	48	97.0	17
				98.0	1

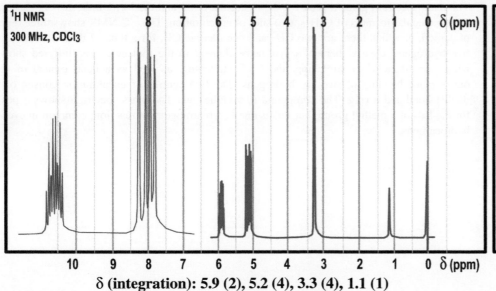

δ (integration): 5.9 (2), 5.2 (4), 3.3 (4), 1.1 (1)

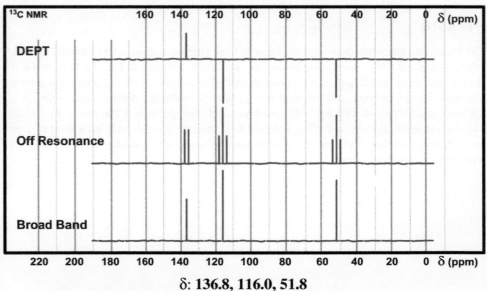

δ: 136.8, 116.0, 51.8

Exercise 018 (020)

Preliminary Observations

The C–H stretching region in the IR spectrum shows bands for alkyl and olefinic C–H stretching vibrations. There are also peaks in the ^{1}H and ^{13}C spectra that indicate the presence of an alkene group.

Infrared

The band between 3250 and 3450 cm^{-1} is typical of the N–H stretching of a secondary amine. This assignment is supported by the presence of a deformation band at 1450 cm^{-1} (scissoring movement of RR′N–H), and the out-of-plane bending band of the >N–H at 750 cm^{-1}. The band at 3100 cm^{-1} is characteristic of =C–H stretching, showing the presence of an alkene group. This is confirmed by the >C=C< stretching band near 1650 cm^{-1}, and by the deformation bands at 920 and 990 cm^{-1} characteristic of the vinyl group (–CH=CH$_2$). The band at 1840 cm^{-1} is a harmonic of the deformation band at 920 cm^{-1}.

^{13}C NMR

Three peaks are seen in the BB spectrum, a CH at 137 ppm (doublet in OR) and two different methylene groups (inverted in DEPT and triplet in OR), one at 116 ppm and one at 52 ppm. The chemical shift of one of the methylene carbons (52 ppm) suggests that it is bonded to the N of an amine. The signals at 116 ppm and 137 ppm are at the correct chemical shifts for the carbons of a vinyl group (–CH=CH$_2$).

^{1}H NMR

The ^{1}H NMR spectrum shows the presence of four signals, at 5.9, 5.2, 3.3 and 1.1 ppm, with relative integrations of 2:4:4:1 for a total of 11 protons, six of which are olefinic according to their chemical shifts. The signal at 3.3 ppm corresponds to CH$_2$ protons that are subject to deshielding by two groups, an alkene and an amine. This peak is split into a doublet (^{3}J ≈ 6 Hz) by coupling with one proton, the CH of the alkene double bond. The signals at 5.2 ppm correspond to the non-equivalent terminal =CH$_2$ protons, which appear as doublets with further fine splitting. The higher field doublet has the smaller splitting, ^{3}J ≈ 10 Hz

(*cis* coupling) compared to the lower doublet, ^{3}J ≈ 16 Hz (*trans* coupling). Each doublet shows the fine splitting of their mutual geminal coupling (^{2}J ≈ 2 Hz). The signal at 5.9 ppm corresponds to the olefinic (=CH) proton which is split by the *trans* proton (^{3}J ≈ 16 Hz), the *cis* proton (^{3}J ≈ 10 Hz) and the methylene protons (triplet ^{3}J ≈ 6 Hz). Finally, the signal at 1.1 ppm with an integration of 1 belongs to the amine NH.

Mass Spectrum

The mass spectrum indicates a molecular mass of 97, with an isotope peak intensity that is compatible with six carbons, but is not a reliable indicator because of its low intensity. The compound contains a nitrogen and an odd number of protons, and has the even-mass ions as the principal peaks. The strong M−1 peak is the result of the loss of a hydrogen from the CH$_2$ group by α cleavage (breaking a bond on the α carbon of the nitrogen). The ion at 82 corresponds to the loss of 15, even though the molecule has no CH$_3$ group. The loss of a methyl group requires rearrangement by hydrogen migration, but produces the very stable conjugated ion, $[H_2C=CHCH=NHCH=CH_2]^+$. The ion at 70 corresponds to M−27 (loss of a vinyl radical by α cleavage). The ion at 68 is formed by the loss of a mass equivalent to an ethyl group, either through loss of H$_2$ from the P-27 ion, or the loss of 29 from a rearranged molecular ion. Finally, the base peak at 41 is due to the allyl cation (H$_2$C=CH–CH$_2$$^+$), (stabilized by charge delocalization).

Summary

The mass spectrum has a peak for the molecular ion at 97, an odd-mass molecular ion, which suggests the presence of nitrogen. The M+1 peak is very weak, thus giving an imprecise estimate of the number carbons, six in this case. The ^{13}C NMR shows only three types of carbon, an olefinic CH$_2$, an olefinic CH and an aliphatic CH$_2$. The proton NMR integration indicates 11 protons, six of them being olefinic, thus indicating the presence of two equivalent vinyl groups. The IR spectrum indicates the presence of an NH, and a peak for its proton is seen in the ^{1}H NMR. Thus we have two allyl groups and an NH, which go together to complete the structure HN(CH$_2$CH=CH$_2$)$_2$, diallylamine. The other peaks in the ^{1}H NMR are all compatible with this structure.

Exercise 019 (041)

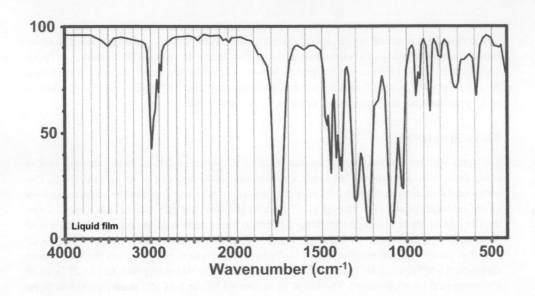

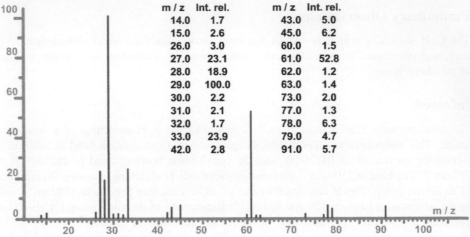

m / z	Int. rel.	m / z	Int. rel.
14.0	1.7	43.0	5.0
15.0	2.6	45.0	6.2
26.0	3.0	60.0	1.5
27.0	23.1	61.0	52.8
28.0	18.9	62.0	1.2
29.0	100.0	63.0	1.4
30.0	2.2	73.0	2.0
31.0	2.1	77.0	1.3
32.0	1.7	78.0	6.3
33.0	23.9	79.0	4.7
42.0	2.8	91.0	5.7

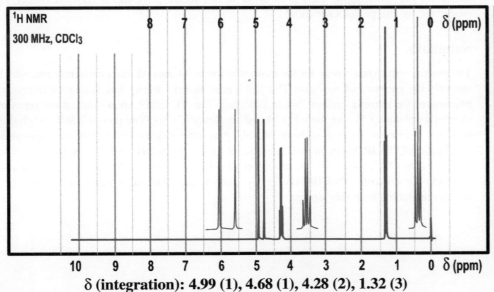

δ (integration): 4.99 (1), 4.68 (1), 4.28 (2), 1.32 (3)

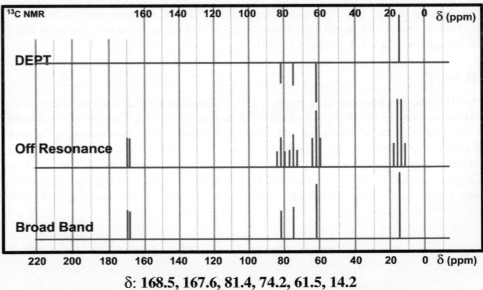

δ: 168.5, 167.6, 81.4, 74.2, 61.5, 14.2

Exercise 019 (041)

Preliminary Observations

A preliminary scan of the spectra shows some interesting features. The IR has a very strong doublet in the carbonyl region. The peak at $1744\,cm^{-1}$ is at the normal ester frequency, but one at $1767\,cm^{-1}$ is elevated beyond the normal ester range. The heaviest ion in the mass spectrum is not likely to be the molecular ion because it has an odd mass and there is no evidence of nitrogen in the molecule. The base peak at 29 (100%) is that of a normal ethyl cation, but there is an interesting unusual peak at 33 (23.9%).

Mass Spectrum

The base peak is found at 29, the mass of the common alkyl ion, $CH_3CH_2^+$. The strong peak at 33 (23.9%) is significant; any peak in the mass spectrum that is slightly greater than 29 is indicative of the presence of a heteroatom. For instance, a peak at 30 indicates the presence of nitrogen ($CH_2 = NH_2^+$) and a peak at 31 indicates the presence of oxygen ($CH_2 = OH^+$). The strong peak at 33, is due to the next element in the periodic table, fluorine. Fluorine has a mass of 19, and would form an ion at 33, ($CH_2 = F^+$). Fluorine does not have naturally occurring isotopes, so its presence cannot be detected by its isotope pattern; however, it is an NMR-active element and its presence can be detected by its coupling to carbon and to hydrogen in their respective NMR spectra.

Infrared

This infrared spectrum has a very strong band in the carbonyl region, in fact it appears that there are two peaks here, at 1767 and $1744\,cm^{-1}$. The $1744\,cm^{-1}$ band is within the normal range of an ester, but the $1767\,cm^{-1}$ peak is beyond the range of a normal ester. It is known that halogens on the α position of a carbonyl group cause an increase in the carbonyl stretching frequency. This suggests that there is a fluorine atom on the α carbon of the ester. The ester should have C–O stretching bands in the $1000–1250\,cm^{-1}$ region and the C–F stretching band should also be seen in the $1000–1350\,cm^{-1}$ region. There are in fact several strong bands in that region.

^1H NMR

The ^1H NMR spectrum has four signals: two singlets at 4.99 and 4.68, a quartet at 4.28 and a triplet at 1.32 ppm. The respective integration ratios are 1:1:2:3 for a total of seven protons. The addition of one fluorine atom gives an even number of monovalent atoms. The quartet–triplet system, with a splitting of 7.2 Hz, indicates the presence of an ethyl group. The chemical shift of the CH_2 group (4.28 ppm) indicates that it is attached to the oxygen of an ester. The other two peaks at 4.99 and 4.68 ppm are actually two parts of the signal from a pair of equivalent protons of a CH_2 group. The signal (centred at 4.84 ppm) is split by a fluorine on the same carbon into a doublet with a coupling, $^2J_{HF}$ of 47 Hz. These are the protons of the FCH_2 group attached to the ester carbonyl ($FCH_2(C{=}O)OR$). Estimation of the chemical shift using Shoolery's rules gives a value of 4.99 ppm ($0.23 + 4.32 + 1.55$), very close to the observed value of 4.84 ppm.

^{13}C NMR

The proton-decoupled spectrum shows four types of carbon in the molecule. The downfield pair of peaks at 167.65 and 168.52 ppm represent the signal of a carbon of an ester carbonyl (quaternary C=O, disappears in DEPT) split by coupling to a fluorine on an adjacent carbon ($^2J_{CF} = 22$ Hz). The pair of peaks at 74.22 and 81.44 ppm represent the signal of a CH_2 (inverted in DEPT, triplet in OR) split by coupling to a fluorine atom ($^1J_{CF} = 181$ Hz). The triplet in the off-resonance spectrum at 61.5 ppm represents the CH_2 attached to the oxygen of an ester. The quartet in the off-resonance spectrum at 14.2 ppm represents the CH_3 of the ethyl group.

Summary

The structural components have been identified as: an ethoxy group ($CH_3CH_2O–$), an ester carbonyl (C=O) and a fluoromethylene group ($–CH_2F$). Attaching these components gives the molecule $CH_3CH_2O–(C{=}O)–CH_2F$, ethyl fluoroacetate. The molecular mass is 106, and the main peaks in the mass spectrum are due to: loss of CH_3 at 91, loss of CH_3CH_2O at 61, and loss of ethylene by a McLafferty rearrangement at 78. The increase in frequency of the carbonyl group is due to the inductive withdrawal effect of the α fluorine, which increases the bond order of the C=O bond. The occurrence of the split peak is due to the existence of conformations with different relative orientations of the fluorine and carbonyl group which affect the inductive contribution.

Exercise 020 (012)

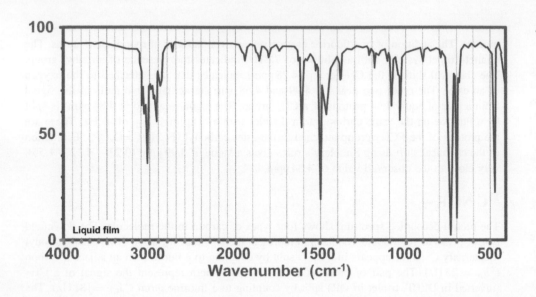

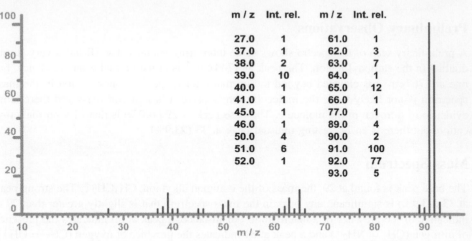

m / z	Int. rel.	m / z	Int. rel.
27.0	1	61.0	1
37.0	1	62.0	3
38.0	2	63.0	7
39.0	10	64.0	1
40.0	1	65.0	12
41.0	1	66.0	1
45.0	1	77.0	1
46.0	1	89.0	4
50.0	4	90.0	2
51.0	6	91.0	100
52.0	1	92.0	77
		93.0	5

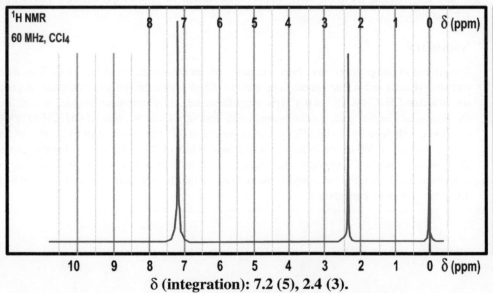

δ (integration): 7.2 (5), 2.4 (3).

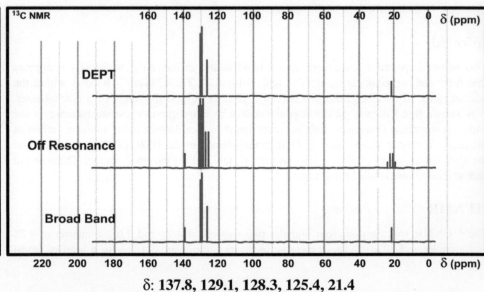

δ: 137.8, 129.1, 128.3, 125.4, 21.4

Exercise 020 (012)

Preliminary Observations

Scanning the IR spectrum for evidence of a functional group gives a clear indication of an aromatic ring. There are bands in all four regions that one would search for signs of an aromatic ring: CH stretching bands between 3000 and 3100 cm^{-1}, small overtone–combination bands between 2000 and 1600 cm^{-1}, strong peaks near 1600 and 500 cm^{-1}, and finally out-of-plane bending bands in the region of 700–900 cm^{-1}.

Mass Spectrum

This spectrum appears to have a particularly strong peak for the molecular ion at 92 (77%) and a strong M−1 peak at 91 (100%). The peak at 91 cannot be the molecular ion because the M+1 isotope peak would be too high. The principal peaks of the fragments ions have odd masses at 39, 63 and 65, which are the typical ions of a phenyl ring.

^{13}C NMR

There are five signals in the spectrum, one methyl (quartet in OR), one quaternary carbon in the aromatic region (disappears in DEPT) and three aromatic CH carbons (inverted in DEPT, doublets in OR). The three signals of the CH carbons are not of the same intensity. Two of them are about twice the intensity of the third, suggesting that they each represent two carbons. This is a peak pattern that is typical of a monosubstituted benzene ring.

^{1}H NMR

The ^{1}H NMR spectrum shows two types of protons with an intensity ratio of 5:3. The peak at 2.4 ppm represents a CH_3 bonded to an sp^2 carbon, and the signal at 7.2 ppm is due to the five aromatic protons (monosubstitution). Although there are three different types of aromatic proton (o, m, p), the chemical shifts are very close and are not seen as separate peaks in this low-resolution spectrum (60 MHz). The differences in the three positions are seen clearly in the ^{13}C spectrum, which shows three types of tertiary sp^2 carbons (H–C=), one para to the CH_3 group and, because of the symmetry of the molecule, two equivalent ortho carbons and two equivalent meta carbons.

Infrared

The bands between 3000 and 3100 cm^{-1} show the presence of hydrogen on an sp^2 carbon (C–H stretching), either the C–H of an alkene or an aromatic ring. The presence of bands at 1500 and 1600 cm^{-1} for the ring >C=C< stretching, and the bands at 700 and 750 cm^{-1} for the out-of-plane C–H bending, are typical of a monosubstituted benzene ring.

Summary

The mass spectrum gives us a peak for the molecular ion at 92, and a large M−1 ion at 91, the tropylium ion, which is commonly observed for alkyl benzenes. The ^{13}C NMR indicates the presence of five types of carbon, one quaternary, three types of CH and one CH_3. The ^{1}H NMR confirms the presence of a CH_3 group, and five aromatic five protons. Therefore there is a benzene ring with a CH_3 group on the ring, $C_6H_5CH_3$, toluene.

Exercise 021 (013)

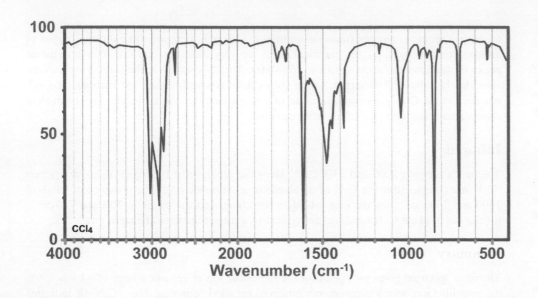

CCl₄

100

50

0

Wavenumber (cm⁻¹)
4000   3000   2000   1500   1000   500

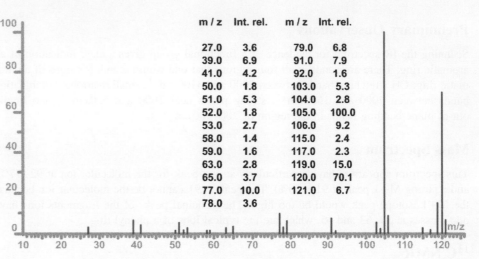

m / z	Int. rel.	m / z	Int. rel.
27.0	3.6	79.0	6.8
39.0	6.9	91.0	7.9
41.0	4.2	92.0	1.6
50.0	1.8	103.0	5.3
51.0	5.3	104.0	2.8
52.0	1.8	105.0	100.0
53.0	2.7	106.0	9.2
58.0	1.4	115.0	2.4
59.0	1.6	117.0	2.3
63.0	2.8	119.0	15.0
65.0	3.7	120.0	70.1
77.0	10.0	121.0	6.7
78.0	3.6		

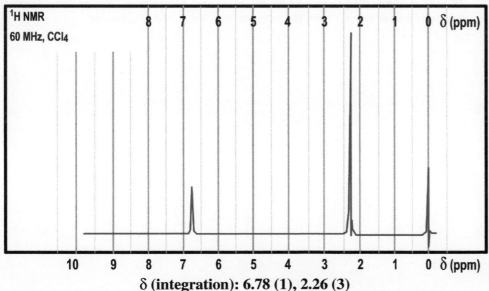

¹H NMR
60 MHz, CCl₄

δ (integration): 6.78 (1), 2.26 (3)

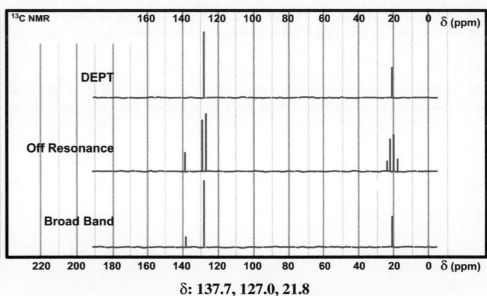

¹³C NMR

DEPT

Off Resonance

Broad Band

δ: 137.7, 127.0, 21.8

Exercise 021 (013)

Preliminary Observation

There are no indications of a functional group other than an aromatic ring in the IR spectrum. The simplicity of the NMR spectra indicates that the molecule has a great deal of symmetry.

Mass Spectrum

The mass spectrum indicates that the molecular mass of the compound is 120. The relative intensity of the M+1 peak, 9.56% ($100 \times 6.7/70.1$) indicates that there are nine carbons in the molecule. The expected value of the M+1 isotope peak for nine carbons is actually 9.9%. In fact, a more accurate calculation of the relative ratio of M+1/M would give 9.8% ($100 \times 6.7/68.4$) if one reduced the size of M by taking out the contribution of the isotope peak of the M−1 ion. The peaks at 77 ($C_6H_5^+$), at 91 ($C_7H_7^+$) and at 105 ($C_8H_9^+$) suggest the presence of alkyl groups on a benzene ring.

^{13}C NMR

There are three peaks in the broad band spectrum: a quaternary carbon (signal disappears in DEPT) in the aromatic region (137.7 ppm); a CH carbon (doublet in OR spectrum), also in the aromatic region (127.7 ppm); and a methyl carbon (quartet in OR spectrum) in the aliphatic region (21.8 ppm). The compound has only three types of carbon although the molecular formula has nine, therefore there must be some three-fold symmetry element within the compound.

Infrared

The peak at 3017 cm^{-1} is due to C–H stretching of hydrogen on an sp^2 carbon. It could be the C–H on either an alkene or an aromatic ring. The bands at 1500 and 1600 cm^{-1} represent the ring >C=C< stretching, and the band at 835 cm^{-1}, an out-of-plane C–H bending typical of an isolated hydrogen on an aromatic ring. This indicates that we have a trisubstituted benzene ring, the substituents being meta to one another.

^{1}H NMR

The ^{1}H NMR spectrum shows two signals, with no perceptible coupling and an integration of 1:3. The peak at 2.3 ppm corresponds to a CH$_3$ group on an sp^2 carbon, and the peak at 6.7 ppm to protons on an aromatic ring. The chemical shift of the aromatic protons is upfield from the benzene position by 0.5 ppm, which agrees with the effect calculated for two ortho and one para methyl groups.

Summary

The mass spectrum has a molecular peak at 120 (C_9H_{12}), and the ^{13}C NMR shows three types of carbon – sp^2 quaternary, sp^2 tertiary (CH) and primary (CH$_3$). The ^{1}H NMR has only signals for the methyl group and an aromatic proton. The IR spectrum shows =C–H and >C=C< stretching bands, and an out-of-plane bending band for an isolated H (between two substituents) on a benzene ring. Therefore the compound has three methyl groups on a benzene ring. The ^{13}C NMR confirms the substitution pattern, as there are only three signals for the nine carbons, therefore there is a three-fold element of symmetry in the compound, a 1,3,5 trisubstituted benzene ring. The compound is mesitylene, or 1,3,5-trimethylbenzene.

Exercise 022 (025)

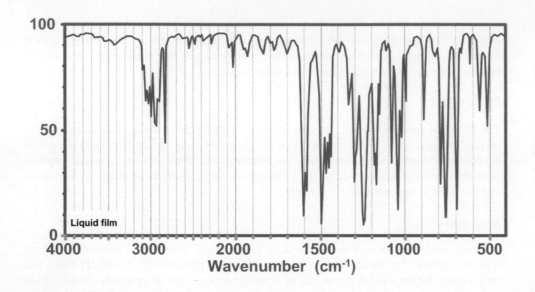

Liquid film

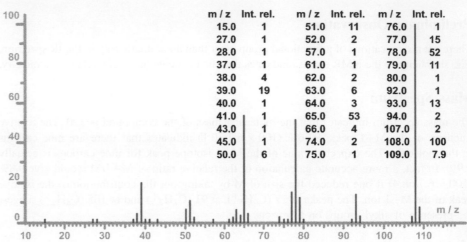

m / z	Int. rel.	m / z	Int. rel.	m / z	Int. rel.
15.0	1	51.0	11	76.0	1
27.0	1	52.0	2	77.0	15
28.0	1	57.0	1	78.0	52
37.0	1	61.0	1	79.0	11
38.0	4	62.0	2	80.0	1
39.0	19	63.0	6	92.0	1
40.0	1	64.0	3	93.0	13
41.0	1	65.0	53	94.0	2
43.0	1	66.0	4	107.0	2
45.0	1	74.0	2	108.0	100
50.0	6	75.0	1	109.0	7.9

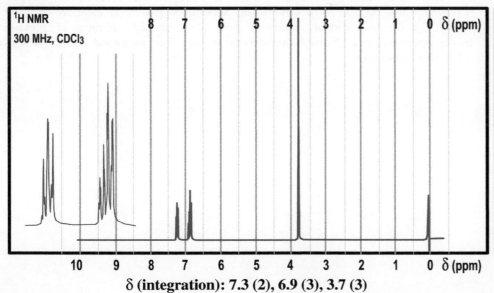

δ (integration): 7.3 (2), 6.9 (3), 3.7 (3)

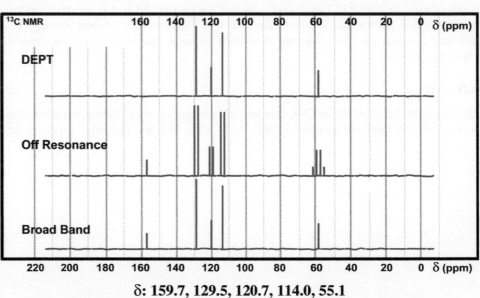

δ: 159.7, 129.5, 120.7, 114.0, 55.1

Exercise 022 (025)

Preliminary Observation

A scan of the spectra tells one that there is an aromatic system, and that a substituent has a shielding effect on some of the protons in the ^{1}H NMR. The compound has a significant peak for the molecular ion, probably because of the aromatic π system.

Infrared

The typical peaks of an aromatic ring stand out in this spectrum. The C–H stretching bands are seen between 3000 and 3100 cm^{-1}, small bands are obvious in the fingerprint region (2000–1600 cm^{-1}), C=C stretching bands are seen at 1600 and 1500 cm^{-1}, and there are bands in the out-of-plane region of 675–870 cm^{-1}. The fingerprint pattern and the out-of-plane bending bands at 700 and 750 cm^{-1} are typical of five adjacent protons on an aromatic ring (monosubstituted benzene ring). The band at 2836 cm^{-1} corresponds to C–H stretching in a methoxy group, and the band at 1250 cm^{-1} is characteristic of =C–OC stretching.

^{13}C NMR

There are five signals in the BB spectrum, two of them being much more intense, indicating that they each may represent more than one carbon. The DEPT spectrum indicates the presence of one quaternary carbon (loss of signal at 160), quite deshielded by being directly bonded to an oxygen (ipso carbon). The off-resonance spectrum indicates the presence of three sp^2 carbons as doublets (CH) and one quartet (CH$_3$), and confirms the presence of a quaternary carbon. Of the four types of aromatic carbon, the two more intense peaks belong to the ortho and meta isochronous pairs, with the higher field signal belonging to the more highly shielded ortho carbons. The lower intensity CH signal belongs to the para carbon. The chemical shifts are explained by the resonance donor effect shielding the ortho and para carbons but having little effect on the meta carbons. The carbons directly bonded to the oxygen are strongly deshielded; OCH$_3$ carbon at 55 ppm, and ipso carbon of the ring at 160 ppm.

^{1}H NMR

The proton NMR shows the presence of three groups of signals, with integration ratios of 2:3:3 at 7.3, 6.9 and 3.7 ppm respectively. The singlet at 3.7 ppm is at the characteristic shift of a methoxy group on a benzene ring. The signals situated between 6.9 and 7.3 ppm represent five aromatic protons. The triplet at 7.3 ppm represents the meta protons coupled to an ortho and para proton. Unlike the para and ortho protons, the meta protons are not significantly shielded by the resonance-donating effect of the oxygen substituent. The signals for the para and ortho protons are found near 6.9 ppm, which is upfield from the normal benzene position of 7.3 ppm. They are shielded by the resonance effect of the oxygen at the ortho and para positions. The signals consist of a triplet for the para proton and a doublet for the ortho protons.

Mass Spectrum

The mass spectrum has a peak for the molecular ion at 108 (100%), and the M+1 isotope peak of 7.9%, indicating that the compound has seven carbons. The ion at 93 corresponds to the loss of a methyl group, the ion at 78 to the loss of formaldehyde, and the ion at 77 is the phenyl ion (C$_6$H$_5^+$) corresponding to the loss of CH$_3$O. In this case, the mass spectrum alone gives strong evidence that the compound is anisole: C$_6$H$_5$OCH$_3$.

Summary

In each spectrum there are strong indications of the presence of a phenyl ring. The ^{1}H NMR spectrum has two signals in the aromatic region, one near the chemical shift of benzene and the second at a higher field, which indicates that it is shielded by an electron-donating substituent. Shielding of aromatic protons occurs by resonance donation from a substituent with non-bonding electrons, usually oxygen or nitrogen, and is most effective at the ortho and para positions. The methoxy group is an example of such a group. The methoxy group is found as a singlet at the expected chemical shift (3.8 ppm). The compound is clearly anisole, C$_6$H$_5$OCH$_3$.

Exercise 023 (024)

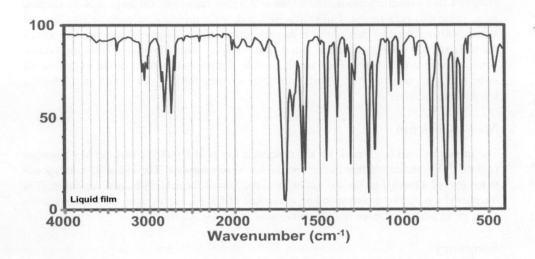

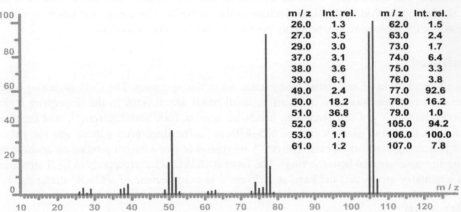

m/z	Int. rel.	m/z	Int. rel.
26.0	1.3	62.0	1.5
27.0	3.5	63.0	2.4
29.0	3.0	73.0	1.7
37.0	3.1	74.0	6.4
38.0	3.6	75.0	3.3
39.0	6.1	76.0	3.8
49.0	2.4	77.0	92.6
50.0	18.2	78.0	16.2
51.0	36.8	79.0	1.0
52.0	9.9	105.0	94.2
53.0	1.1	106.0	100.0
61.0	1.2	107.0	7.8

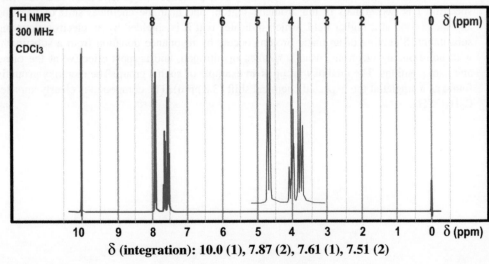

δ (integration): 10.0 (1), 7.87 (2), 7.61 (1), 7.51 (2)

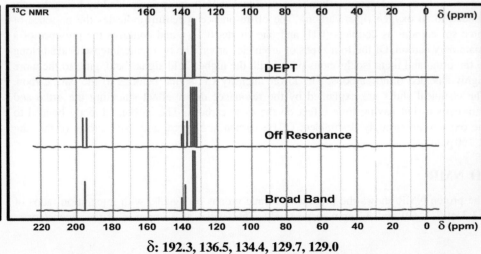

δ: 192.3, 136.5, 134.4, 129.7, 129.0

Exercise 023 (024)

Preliminary Observation

The typical peaks of an aromatic ring can be seen in the IR and NMR spectra. In the IR spectrum the peaks at 2720 and 2820 cm^{-1}, as well the strong carbonyl band at 1700 cm^{-1}, are particularly significant. The peak in the ^{1}H NMR at 10 ppm is very revealing.

Mass Spectrum

The peak at 106 is that of the molecular ion, and the M+1 isotope peak of 7.8% indicates that there are seven carbons in the molecule. There is a major peak at 105 (94%) for the loss of a hydrogen radical, followed by another very strong peak at 77 (93%) for the loss of an additional 28. The IR spectrum indicates the presence of a carbonyl, thus the intense peak at 105 comes from α cleavage at the carbonyl group. The loss of a hydrogen atom is common in aldehydes. The ion at 77 – the phenyl ion ($C_6H_5^+$) – results from the loss of carbon monoxide from the 105 ion. The ion at 51 is typical of benzene rings, resulting from the loss of acetylene from the phenyl cation.

^{13}C NMR

It is particularly noteworthy that there are no peaks in the aliphatic region. There are four peaks in the aromatic region and one in the carbonyl region. The DEPT spectrum shows that there is one quaternary aromatic carbon and all the others are tertiary (CH) carbons. As there are seven carbons in the molecule and only five peaks, there has to be an element of symmetry leading to some carbons being isochronous. Looking at the intensities of the CH carbons, it appears that two of the peaks are about twice the intensity of the others. Comparisons like this may be made for similar types of carbon in ^{13}C spectra, but not for carbon of different types. For example, quaternary carbons appear less intense than carbons bearing hydrogens. The two larger peaks would each represent a pair of isochronous carbons, so that the five peaks account for all seven carbons. Only four types of aromatic carbon are seen because the two ortho carbons are equivalent, as are the two meta carbons. The peak at 192 ppm is a carbonyl, and the doublet in the OR spectrum indicates that it is of an aldehyde, HC=O.

^{1}H NMR

There are four signals in the ^{1}H NMR spectrum. The singlet at 10.0 ppm is that of the aldehyde proton. The most deshielded aromatic protons, the doublet at 7.87 ppm (integration 2), belong to the protons ortho to the carbonyl group. The para proton is seen as a triplet (integration 1) at 7.61 ppm. The meta protons are seen as a distorted triplet (integration 2) at 7.51 ppm.

Infrared

There are four regions where one looks for absorptions due to aromatic compounds. The C–H stretching bands are found between 3000 and 3100 cm^{-1}, weak overtone-combination bands are found between 2000 and 1600 cm^{-1} (fingerprint region), C=C ring stretching band at 1600, and out-of-plane C–H bending vibrations between 675 and 870 cm^{-1}. Bands are found in all those regions in this IR spectrum. The pattern in the fingerprint region, as well as the out-of-plane bending bands, can be used to determine the substitution pattern. The bands at 695 and 750 cm^{-1} indicate that the ring is monosubstituted, and the pattern in the fingerprint region is in agreement. The strong band at 1700 cm^{-1} corresponds to a carbonyl group, and the pair of bands at 2820 and 2740 cm^{-1} are characteristic of aldehydes (C–H stretch); the deformation band (H–C=O) at 1400 cm^{-1} is also characteristic of aldehydes.

Summary

There are several indicators that point to the presence of an aldehyde in the compound. The strong carbonyl band stands out in the IR as well as the typical pair of C–H stretching bands between 2720 and 2850 cm^{-1}. The sharp singlet at 10 ppm in the proton NMR, and the doublet in the OR ^{13}C spectrum at 192 ppm, both indicate an aldehyde. The loss of a hydrogen atom followed by the loss of carbon monoxide in the mass spectrum is also an indication of an aldehyde. There are also several indicators of a benzene ring. Both the ^{13}C and ^{1}H NMR spectra have peaks in the aromatic regions. The IR spectrum has bands in all four of the typical aromatic regions. The mass spectrum has a strong peak for the phenyl ion at 77 and its typical product ion at 51. A monosubstituted benzene ring and an aldehyde group together make benzaldehyde, C_6H_5CHO.

Exercise 024 (060)

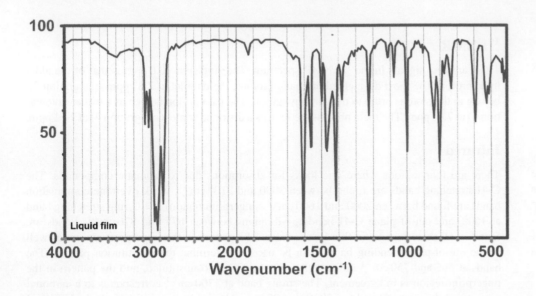

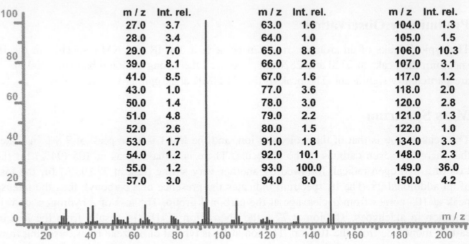

m / z	Int. rel.	m / z	Int. rel.	m / z	Int. rel.
27.0	3.7	63.0	1.6	104.0	1.3
28.0	3.4	64.0	1.0	105.0	1.5
29.0	7.0	65.0	8.8	106.0	10.3
39.0	8.1	66.0	2.6	107.0	3.1
41.0	8.5	67.0	1.6	117.0	1.2
43.0	1.0	77.0	3.6	118.0	2.0
50.0	1.4	78.0	3.0	120.0	2.8
51.0	4.8	79.0	2.2	121.0	1.3
52.0	2.6	80.0	1.5	122.0	1.0
53.0	1.7	91.0	1.8	134.0	3.3
54.0	1.2	92.0	10.1	148.0	2.3
55.0	1.2	93.0	100.0	149.0	36.0
57.0	3.0	94.0	8.0	150.0	4.2

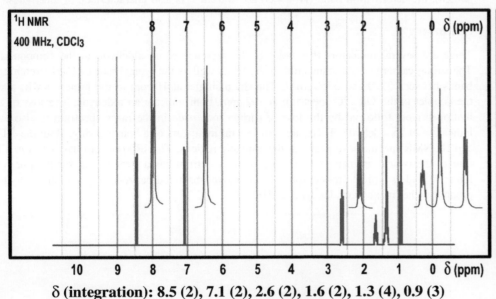

δ (integration): 8.5 (2), 7.1 (2), 2.6 (2), 1.6 (2), 1.3 (4), 0.9 (3)

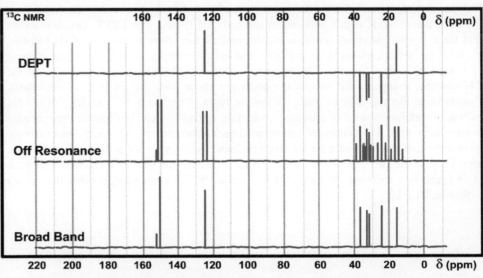

δ: 151.8, 149.6, 123.9, 35.2, 31.4, 30.0, 22.4, 14.0

Exercise 024 (060)

Preliminary Observation

Take note of the peaks above $3000 \, cm^{-1}$ in the IR spectrum, and the signal for the strongly deshielded aromatic protons in the NMR.

Mass Spectrum

The compound has a molecular mass of 149, the odd mass indicating that the compound contains a nitrogen atom. The M+1 isotope peak at 150 has a relative intensity of 11.6%, which gives an estimate of 10 carbons in the molecule. The molecular ion peak is relatively strong (36%), suggesting the presence of readily ionized functional group (such as a cyclic π system). The series of weak peaks, 14 units apart at 134 (M−15), 120 (M−29), 106 (M−43) and 92 (M−57) is characteristic of a linear alkyl chain. The strong peak at 93 (100%) is due to a rearrangement in which butene is lost (M−56). This suggests the presence of a butyl chain on the β position of a carbonyl or similar polar group, from which it can be lost by a McLafferty-type rearrangement.

^{13}C NMR

The spectrum shows eight types of carbon in the compound: three in the aromatic region, and five in the aliphatic region. In the DEPT spectrum only one signal disappears, a quaternary carbon in the aromatic region (152 ppm). The other two aromatic peaks (149.6 and 123.9 ppm) are doublets (CH) in the off-resonance spectrum. These peaks represent pairs of isochronous carbons; there are only eight signals for a compound which appears to have 10 carbons. The aromatic ring can be completed by adding a nitrogen to the five aromatic carbons already identified. The doublet at 149.6 can be assigned to the α carbons adjacent to the nitrogen (more deshielded) and the doublet at 123.9 to the carbons β to the nitrogen. The final ring carbon, which is a quaternary deshielded by the para nitrogen and the alkyl substituent, is found at 151.8 ppm. The aliphatic carbons consist of four CH_2 carbons (inverted in DEPT) and a CH_3 carbon (quartet in off-resonance).

^{1}H NMR

The ^{1}H NMR spectrum has six signals, with integration ratios of 2:2:2:2:4:3 for a total of 15 protons. The doublet (integration 2) at 8.5 ppm corresponds to two isochronous aromatic protons situated ortho to the nitrogen of a pyridine ring (α hydrogens), deshielded by the inductive withdrawal effect of the nitrogen. The doublet (integration 2) at 7.1 ppm corresponds to two isochronous aromatic protons meta to the nitrogen of a pyridine ring (β hydrogens), shielded by the resonance-donor effect of the nitrogen. The ring protons form an AA′XX′ coupling system, with $J_{AX} = 4 \, Hz$. The ^3J coupling of the α and β protons (4–6 Hz) in pyridine rings is weaker than the ortho coupling in benzene rings (6–9 Hz). The triplet (integration 2) at 2.6 ppm represents the CH_2 bonded to the pyridine ring. The multiplet (integration 2) 1.6 ppm represents the CH_2 β to the pyridine ring. The multiplet (integration 4) at 1.3 ppm corresponds to two aliphatic CH_2 groups. The triplet (integration 3) at 0.9 ppm represents the terminal methyl group.

Infrared

There are bands in each of the four regions where one finds absorptions due to aromatic compounds: the C–H stretching bands between 3000 and $3100 \, cm^{-1}$; weak overtone-combination bands between 2000 and $1600 \, cm^{-1}$ (fingerprint region); C=C ring-stretching bands at 1600 and 1500 (weak); and out-of-plane C–H bending vibrations between 675 and $870 \, cm^{-1}$. The intensity of the ring-stretching bands indicates that the ring is highly polarized, suggesting the presence of a heteroatom. The band near $800 \, cm^{-1}$ represents the bending bands corresponding to two adjacent hydrogens. The aliphatic chain is seen in the bands between 2850 and $2980 \, cm^{-1}$, and the CH_2 rocking band at $720 \, cm^{-1}$ (can be difficult to identify in the presence of aromatic bands).

Summary

The molecule has a molecular mass of 149, indicating that the compound contains a nitrogen atom ($C_{10}H_{15}N$). The peak for the molecular ion is relatively strong (36%), indicating the presence of a readily ionized functional group such as a cyclic π system. An aromatic ring is also indicated by the IR and NMR spectra. The ring has a quaternary carbon, four CH carbons and one nitrogen, a pyridine ring. The symmetry of the ring indicates that it has a substituent in the para position. The NMR spectra indicate that the substituent is a pentyl group. The compound is 4-pentylpyridine. The peaks at 78, 65, 51 and 39 in the mass spectrum represent products of further fragmentations of the pyridine ring system (primarily by the loss of HC≡CH or HC≡N).

Exercise 025 (040)

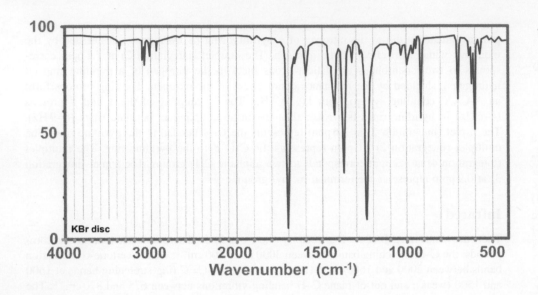

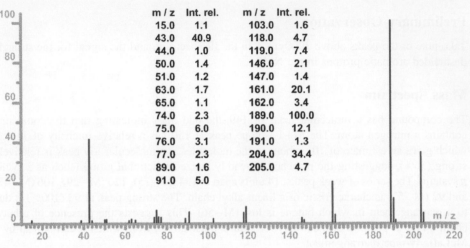

m / z	Int. rel.	m / z	Int. rel.
15.0	1.1	103.0	1.6
43.0	40.9	118.0	4.7
44.0	1.0	119.0	7.4
50.0	1.4	146.0	2.1
51.0	1.2	147.0	1.4
63.0	1.7	161.0	20.1
65.0	1.1	162.0	3.4
74.0	2.3	189.0	100.0
75.0	6.0	190.0	12.1
76.0	3.1	191.0	1.3
77.0	2.3	204.0	34.4
89.0	1.6	205.0	4.7
91.0	5.0		

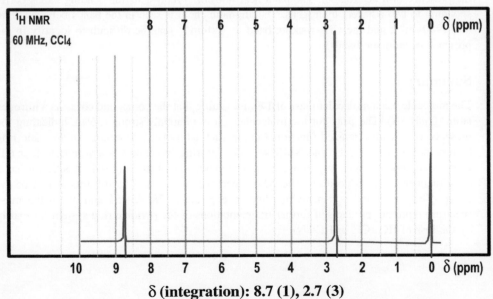

δ (integration): 8.7 (1), 2.7 (3)

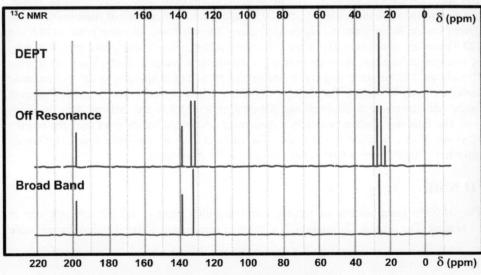

δ: 196.7, 137.8, 131.8, 26.9

Exercise 025 (040)

Preliminary Observation

A preliminary scan of the spectra shows that the presence of a ketone is indicated by the peak at 198 ppm in the ^{13}C spectrum, and the strong peak at 1695 cm^{-1} in the IR spectrum. There is also clear evidence of an aromatic ring in the IR and NMR spectra.

Mass Spectrum

The peak at 204 (34.4%) seems to be a good candidate for the molecular ion; it has an even mass and there is a very large peak at 189 (M−CH$_3$). The M+1 isotope peak at 205 (4.7%) has a relative intensity of 13.66%, suggesting that there are 12 carbons in the molecule (13.66/1.1 = 12.4). A similar calculation using the ion at 189 (100%) and its isotope peak at 190 (12.1%) yields a result of 11 carbons (12.1/1.1 = 11). Adding the lost methyl group to this fragment gives 12 carbons for the complete molecule. The fragment at 161 (20.1%), a loss of 28 from 189, is likely to be due to the loss of CO, considering the large carbonyl peak in the IR spectrum. The loss of CH$_3$, and of CO, and the appearance of a large peak at 43 (40.9%), provide strong evidence for the presence of an acetyl group, CH$_3$C=O.

^{13}C NMR

There are signals for four types of carbon, two quaternary carbons (signals disappear in DEPT), one CH carbon (doublet in OR) and one methyl group (quartet in OR). The methyl carbon at 27 ppm is at the correct chemical shift for a methyl ketone. The quaternary carbon at 197 ppm corresponds to a carbon in the carbonyl group of a ketone (>C=O) and the one at 1378 ppm corresponds to an aromatic carbon bearing a carbonyl group. The peak at 132 ppm corresponds to an aromatic ring CH carbon. Thus we have two types of aromatic carbon: a CH and a quaternary carbon bearing an acetyl group.

^{1}H NMR

There are singlets at 8.7 and 2.7 ppm with an integration ratio of 1:3. The signal at 8.7 ppm is far removed from the normal chemical shift (7.3 ppm) of benzene protons. This strong deshielding of 1.4 ppm represents the effect of more than one acetyl group, as the ortho effect of one >C=O group is only 0.60 ppm. The signal of integration 3 at 2.7 ppm has the characteristic chemical shift of the CH$_3$ group of a methyl ketone.

Infrared

The bands between 3000 and 3100 cm^{-1} are due to aromatic (=C–H) stretching. This agrees with the presence of the band near 1600 cm^{-1}, representing the stretching of the ring bonds (>C=C<) and the band at 900 cm^{-1} representing the deformation of an isolated aromatic C–H between two substituents. Finally, the band at 1695 cm^{-1} corresponds to ketone carbonyl stretching. The frequency is lower than the normal position of about 1720 cm^{-1} because of conjugation with the aromatic ring. The band at 1230 cm^{-1}, which is often observed for aromatic rings bearing a ketone group, is assigned to the polar C–C stretching in the C–C(=O) group. The methyl groups are associated with bands at 3005 and 2910 cm^{-1} respectively of the C–H asymmetric and symmetric stretching and the typical deformation of the CH$_3$ at 1360 cm^{-1}.

Summary

The mass spectrum gives a strong peak for the molecular ion, and provides good evidence for the presence of an acetyl group (CH$_3$C=O). The position of the carbonyl peak (1695 cm^{-1}) in the IR spectrum indicates that the carbonyl group is conjugated. The normal position of an unconjugated methyl ketone would be expected to be near 1720 cm^{-1}. The molecular mass indicates that we have three acetyl groups on the benzene ring, and the ^{13}C and ^{1}H NMR spectra indicate that we have only one type of aromatic CH, and a quaternary carbon. With three acetyl groups, the only arrangement that provides the required symmetry is 1,3,5-substitution. The compound is 1,3,5-triacetylbenzene.

Exercise 026 (084)

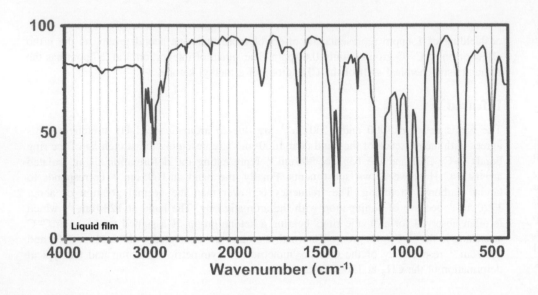

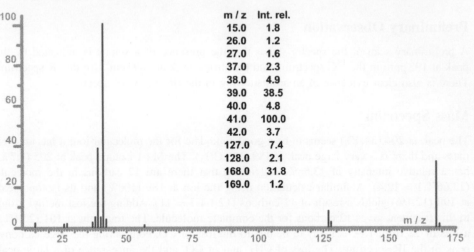

m/z	Int. rel.
15.0	1.8
26.0	1.2
27.0	1.6
37.0	2.3
38.0	4.9
39.0	38.5
40.0	4.8
41.0	100.0
42.0	3.7
127.0	7.4
128.0	2.1
168.0	31.8
169.0	1.2

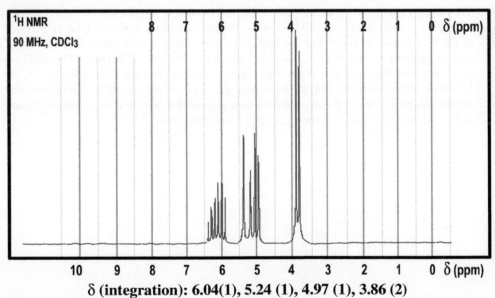

δ (integration): 6.04(1), 5.24 (1), 4.97 (1), 3.86 (2)

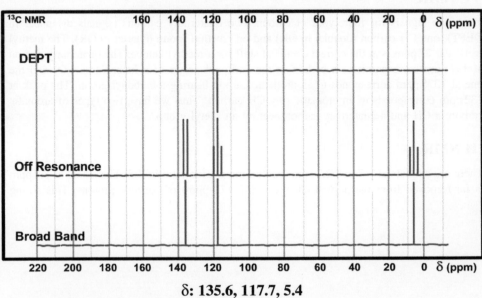

δ: 135.6, 117.7, 5.4

Exercise 026 (084)

Preliminary Observation

Note that there are C–H stretching bands above and below $3000\,cm^{-1}$.

Mass Spectrum

The molecular ion has a mass of 168 (31.8%) and an M+1 isotope at 169 (1.2%) for a relative intensity of 3.8% indicating that the compound has three carbons. Because of the low intensity of the isotope peak, this is not a very accurate determination, but the number is confirmed by the ^{13}C NMR spectrum. Three carbons contribute only 36 to the molecular mass, leaving another 132 to be found. The peak at 127 (7.4%) provides the clue as to the source of the extra mass; iodine has a mass of 127. The I^+ ion at 127 is accompanied by a peak at 128 with a relative intensity of 28%, but this cannot be an isotope peak because iodine has no naturally occurring isotopes. This 128 ion is due to the presence of an HI ion, produced in a rearrangement reaction. The three carbons are found in the ion at 41 ($C_3H_5^+$). This peak represents the stable allyl cation ($CH_2 = CHCH_2^+$) and is accompanied by the ion at 39 ($C_3H_3^+$) arising from the loss of a molecule of hydrogen. The presence of a peak at 127 and a peak for the loss of 127 provide clear evidence of the presence of one iodine atom.

^{13}C NMR

There are only three types of carbon showing in the spectrum. In the DEPT spectrum there are two inverted signals indicating that there are two CH_2 groups. In the off-resonance spectrum the doublet indicates the presence of a >CH– carbon, and the chemical shift of 135.6 ppm indicates that it is an olefinic CH. The other peaks in the off-resonance spectrum are triplets (CH_2), the one at 117.7 ppm being in the olefinic region, and the other triplet at 5.4 ppm, being very strongly shielded, is that of a CH_2 bonded to an iodine atom. The very strong shielding of this carbon is due to the 'heavy atom effect', caused by the electron cloud of the many electrons of the iodine atom, which counteracts the weaker deshielding from the polarization effect.

^{1}H NMR

This ^{1}H NMR spectrum has four signals with integration ratios of 1:1:1:2 at 6.04, 5.24, 4.97 and 3.86 ppm respectively. The doublet of integration 2 at 3.86 ppm corresponds to a CH_2 situated α to an alkene group, and bearing an iodine atom. This CH_2 is coupled by ^3J vicinal coupling to an olefinic proton represented by the multiplet at 6.04 ppm, and also has a weak coupling (^4J) with the two other olefinic protons. The two protons on the terminal carbon of the double bond can be easily identified by the size of their couplings with the third olefinic proton. The doublet at 4.97 ppm has a smaller coupling than the doublet at 5.24 ppm, therefore it is cis to the proton on the other carbon of the double bond. The size of the cis and $trans$ couplings on alkenes varies somewhat depending on the substituents, but for alkyl substituents typical values are $^3J_{cis} = 10\,Hz$, and $^3J_{trans} = 17\,Hz$. The measured values in this case are 16.2 and 9.7 Hz. The smaller splitting on the doublets includes a $^2J_{gem}$ coupling of about 1.5 Hz. The group of signals correspond to the specific profile of a terminal allyl group: $CH_2=CHCH_2–$.

Infrared

The band at $3100\,cm^{-1}$ is characteristic of =C–H stretching. This band is at the high end of the range which is characteristic of a terminal =CH_2, more precisely a CH_2 of a vinyl or allyl group. This is confirmed by the presence of the band at $1850\,cm^{-1}$, which is the harmonic of the band at $920\,cm^{-1}$, representing the typical deformation of a monosubstituted alkene. The band at $1630\,cm^{-1}$ is characteristic of >C=C< stretching, the lower frequency than that normally seen for a similar alkene ($1643\,cm^{-1}$ in 1-pentene) is probably due to the interaction with the heavy heteroatom because of its mass (in the harmonic oscillator). The iodine atom does not have an absorption band in the range of this IR spectrum.

Summary

There are three carbons in the ^{13}C NMR and five hydrogens in the ^{1}H NMR, together forming an allyl group, $CH_2=CHCH_2–$. Adding an iodine atom to this group gives the complete structure, allyl iodide, $CH_2=CHCH_2I$.

Exercise 027 (016)

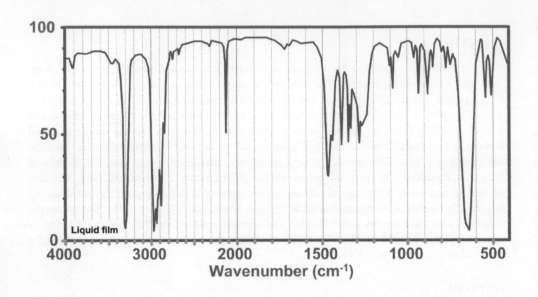

Liquid film

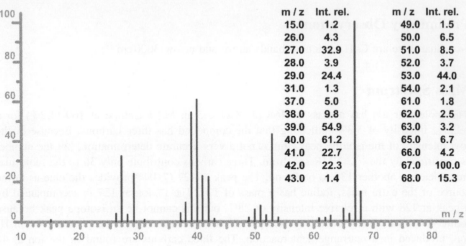

m / z	Int. rel.	m / z	Int. rel.
15.0	1.2	49.0	1.5
26.0	4.3	50.0	6.5
27.0	32.9	51.0	8.5
28.0	3.9	52.0	3.7
29.0	24.4	53.0	44.0
31.0	1.3	54.0	2.1
37.0	5.0	61.0	1.8
38.0	9.8	62.0	2.5
39.0	54.9	63.0	3.2
40.0	61.2	65.0	6.7
41.0	22.7	66.0	4.2
42.0	22.3	67.0	100.0
43.0	1.4	68.0	15.3

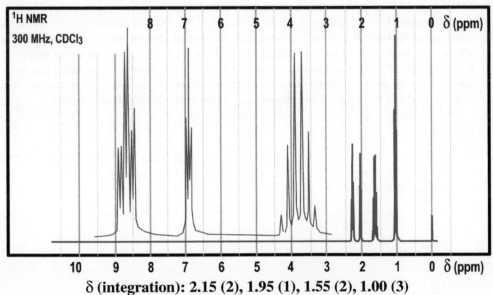

¹H NMR
300 MHz, CDCl₃

δ (integration): 2.15 (2), 1.95 (1), 1.55 (2), 1.00 (3)

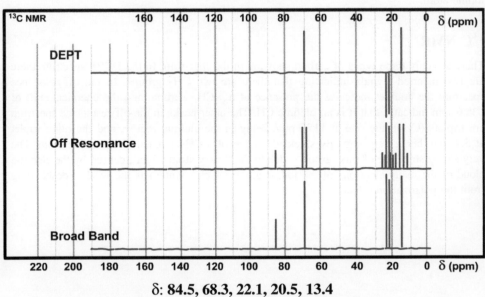

¹³C NMR

DEPT

Off Resonance

Broad Band

δ: 84.5, 68.3, 22.1, 20.5, 13.4

Exercise 027 (016)

Preliminary Observation

The combination of sharp bands at 3300 and 2100 cm^{-1} is a strong indication of a particular functional group.

Infrared

The band at 3307 cm^{-1} represents the stretching of a C–H on an sp carbon, indicating that the compound is a terminal alkyne. This is confirmed by the position of the C≡C triple bond stretching band at 2120 cm^{-1} (specific for terminal alkynes). The broad band between 600 and 700 cm^{-1} is due to the deformation of the ≡C–H bond.

Mass Spectrum

The large peak at 67 is a candidate for the molecular ion, but there are a few factors that conspire against this possibility. First is the size of the M+1 peak at 68; if this is the isotope peak, it would indicate that there are 14 carbons present, which is impossible for this mass. Second, the molecular ion would be an odd-mass ion and thus require that a nitrogen be present, for which there is no evidence. The peak at 68 is more likely to be the molecular ion, and the peak at 67 is probably the result of the loss of a hydrogen atom. This would give a molecular formula of C_5H_8, a hydrocarbon with two unsaturations. The strong peaks at 39 and 53 are the analogues of the saturated alkyl ions 43 and 57(minus four hydrogens).

^{13}C NMR

There are peaks present for five carbons, a CH_3 (quartet in OR), two CH_2 carbons (inverted peaks in DEPT, and triplets in OR), a CH carbon (doublet in OR) and a quaternary carbon (peak disappears in DEPT). The chemical shift (85 ppm) of the quaternary carbon suggests that it is an sp carbon of an alkyne. The other carbon of the triple bond is found at 68 ppm, which is the correct position for the CH of a terminal alkyne, –C≡CH.

^{1}H NMR

The ^{1}H NMR spectrum has four signals with integrations of 2:1:2:3 for a total of eight protons. The splitting pattern is a little difficult to observe, but it can be used to help to determine the attachment of the groups identified from the ^{13}C spectrum. The lone methyl group is at the highest field (1.0 ppm) and appears as a triplet, thus it is attached to one of the CH_2 groups. That CH_2 group (1.55 ppm) appears as a sextet, thus it is also coupled to the second CH_2 group as well as the methyl group, split by a total of five protons. This splitting of the second CH_2 (2.15 ppm) is more complex as it is also split by the acetylenic CH with a finer splitting constant (^{4}J coupling). The peak at 1.95 ppm corresponds to a terminal alkyne proton, and appears as a very fine triplet (J ≈ 2.5 Hz) because of the long-range ^{4}J coupling (–CH_2C≡C–H) between the CH_2 and the acetylenic hydrogen.

Summary

The mass spectrum gives a peak for the molecular ion at 68, and a base peak at 67, indicating the presence of a very labile hydrogen atom. The ^{13}C NMR shows five types of carbon, which doesn't allow for the presence of any heteroatoms as its molecular mass is only 68. The OR and DEPT spectra show that the five carbons consist of one CH_3, two CH_2, one CH and one quaternary carbon. The proton coupling constants show that the atoms are arranged in the sequence, $CH_3CH_2CH_2$C≡CH, thus the compound is 1-pentyne.

Exercise 028 (017)

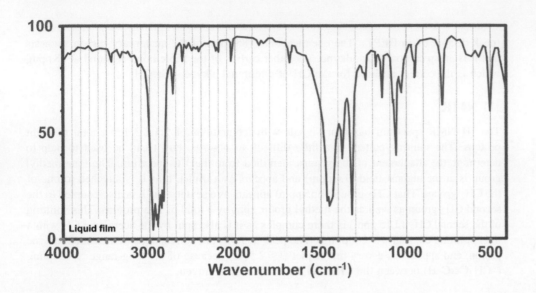

Liquid film

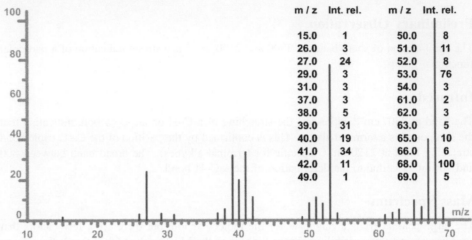

m/z	Int. rel.	m/z	Int. rel.
15.0	1	50.0	8
26.0	3	51.0	11
27.0	24	52.0	8
29.0	3	53.0	76
31.0	3	54.0	3
37.0	3	61.0	2
38.0	5	62.0	3
39.0	31	63.0	5
40.0	19	65.0	10
41.0	34	66.0	6
42.0	11	68.0	100
49.0	1	69.0	5

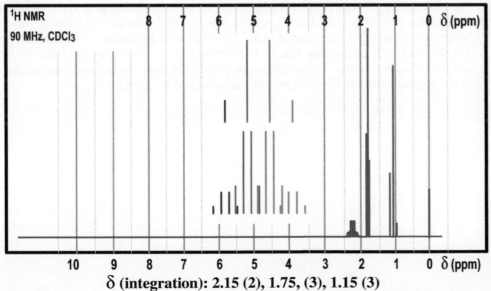

^{1}H NMR

90 MHz, CDCl$_3$

δ (integration): 2.15 (2), 1.75, (3), 1.15 (3)

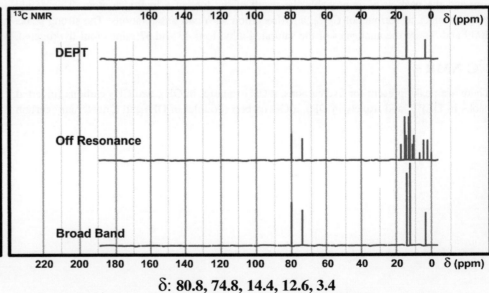

^{13}C NMR

DEPT

Off Resonance

Broad Band

δ: 80.8, 74.8, 14.4, 12.6, 3.4

Exercise 028 (017)

Preliminary Observation

Although the infrared spectrum offers little information on the functional group of the compound, the strong peak for the mass spectrum indicates that the molecule has a readily ionized functional group.

^{13}C NMR

There are peaks for five carbons showing in the broadband spectrum. The OR and DEPT spectra show that there are two different CH_3 groups (quartets in OR), one CH_2 (inverted peak in DEPT, triplet in OR) and two quaternary carbons (peaks disappear in DEPT). The chemical shift of the quaternary carbons indicates that they are sp carbons. To summarize: CH_3, CH_3, CH_2, C, C.

Mass Spectrum

The mass spectrum gives a peak for the molecular ion at 68, and a base peak at 67. The molecular mass of 68 and five carbons gives a formula of C_5H_8, indicating the presence of two unsaturations. There is a significant peak for the loss of a methyl group (M−15), as well as the loss of a hydrogen (M−1).

^{1}H NMR

The proton NMR spectrum shows the presence of three signals with integration ratios of 2 (2.15 ppm) : 3 (1.75 ppm) : 3 (1.15 ppm) for a total of eight protons. The CH_3 at 1.15 ppm is coupled with a CH_2 (2.15 ppm) by a vicinal coupling ($^3J = 7$ Hz), producing a triplet.

The chemical shift of the CH_2 shows that it is bonded to an sp carbon, as is also the CH_3 at 1.75 ppm. The methylene (CH_2), which is coupled by the vicinal coupling ($^3J = 7$ Hz) to one of the CH_3 groups (at 1.15 ppm), appears as a quartet further split into finer quartets by long-distance coupling to the other CH_3 ($^4J = 2.5$ Hz). This yields a poorly resolved cluster of 16 lines. The other CH_3 attached to the triple bond also exhibits the results of the long-range coupling, appearing as a finely coupled triplet ($^4J = 2.5$ Hz). This establishes the sequence: $CH_3CH_2C{\equiv}CCH_3$.

Infrared

This spectrum doesn't show any bands that are particularly useful in determining the functional group of the compound. Because the triple bond $-C{\equiv}C-$ has two very similar substituents, which minimize the dipolar moment, only a very weak peak for the triple bond stretching is seen at 2055 cm^{-1}. The absence of a ${\equiv}C-H$ band near 3300 cm^{-1} indicates that the compound does not have a terminal alkyne group.

Summary

The mass spectrum gives a molecular mass of 68 which equates to C_5H_8, a hydrocarbon with double unsaturation. The ^{13}C NMR shows peaks for all the five carbons, and they can be assigned to two quaternary carbons, 1 CH_2 and 2 CH_3, The ^{1}H NMR confirms the presence of a total of eight hydrogens in one CH_3 and one CH_3CH_2 group. The chemical shifts indicate that the two groups are on the triple bond of an alkyne. The absence of a ${\equiv}C-H$ stretching, and the lack of significant extremely weak $C{\equiv}C$ stretching in the IR spectrum, confirms that it is a disubstituted alkyne.

Exercise 029 (027)

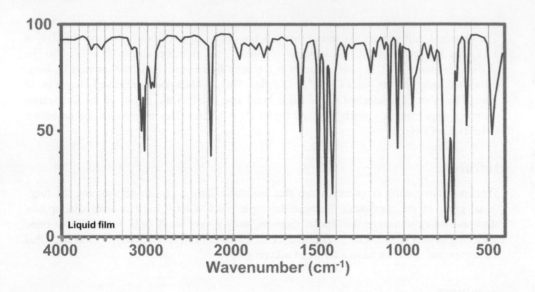

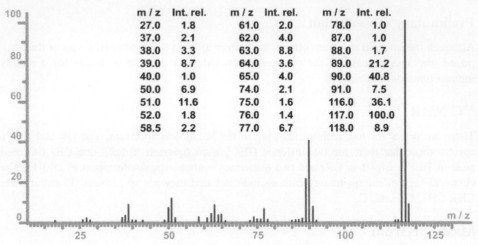

m / z	Int. rel.	m / z	Int. rel.	m / z	Int. rel.
27.0	1.8	61.0	2.0	78.0	1.0
37.0	2.1	62.0	4.0	87.0	1.0
38.0	3.3	63.0	8.8	88.0	1.7
39.0	8.7	64.0	3.6	89.0	21.2
40.0	1.0	65.0	4.0	90.0	40.8
50.0	6.9	74.0	2.1	91.0	7.5
51.0	11.6	75.0	1.6	116.0	36.1
52.0	1.8	76.0	1.4	117.0	100.0
58.5	2.2	77.0	6.7	118.0	8.9

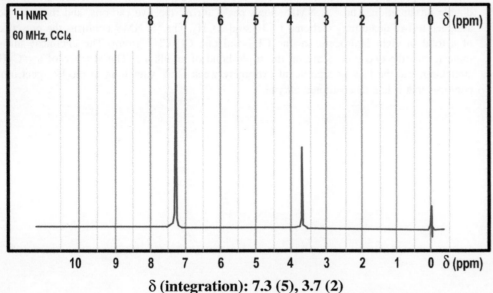

δ (integration): 7.3 (5), 3.7 (2)

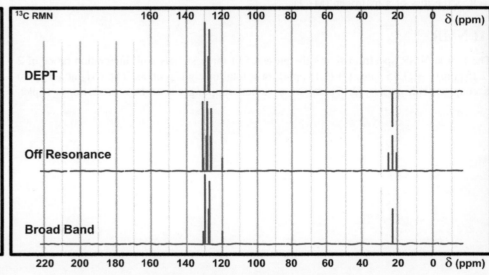

δ: 131, 130, 128, 127, 120, 25

Exercise 029 (027)

Preliminary Observation

In looking for evidence of a functional group, the infrared spectrum gives a strong indication of a functional group, as does the mass spectrum.

Infrared

The very sharp band at $2260\,cm^{-1}$ stands out in a fairly exclusive region of the spectrum – the triple bond stretching region. Alkyne triple bonds usually have very weak bands unless they are terminal bonds, in which case they would also show a band at $3300\,cm^{-1}$ for the $\equiv$C–H stretch, which is absent in this case. The most likely possibility is that this band belongs to a nitrile group, –C$\equiv$N. An aromatic ring is indicated by the sp^2 C–H stretching bands between 3000 and $3100\,cm^{-1}$, the weak overtone-combination bands between 2000 and $1600\,cm^{-1}$, the ring bond stretching at 1600 and $1500\,cm^{-1}$ and the out-of-plane bending bands at 700 and $750\,cm^{-1}$.

^{13}C NMR

There are six peaks in the BB spectrum, with only one being in the sp^3 region, a methylene group at 25 ppm (peak inverted in DEPT, and triplet in OR). The lower field peaks are close together and have to be examined carefully. Two of them (120 and 131 ppm) are identified as quaternary carbons by the disappearance of their signals in the DEPT spectrum. The other three peaks are tertiary carbons (>CH), and are unchanged in the DEPT spectrum. Only CH and CH$_3$ carbons are unchanged in DEPT spectra, and no methyl group would occur in this region. Also, if you look carefully in the OR spectrum, you will see that each is split into a doublet (>CH). In summary, we have the quaternary carbon of the nitrile at 120 ppm, the quaternary ring carbon at 131 ppm, and peaks for three types of ring CH carbon (two ortho, two meta, and a para carbon). The carbon of a C$\equiv$N is frequently missed in the ^{13}C NMR, as it is in the same region as sp^2 carbons, but can be easily detected in the IR spectrum. Finally, the signal at 25 ppm represents a CH$_2$ subsitituted with two deshielding groups, an aromatic ring and a nitrile group.

Mass Spectrum

This mass spectrum is not difficult to interpret. The strong M-1 at 116 arises from the loss of a hydrogen from the position α to both the benzene ring and the nitrile. The molecular ion is the base peak, thus the intensity of the M$+1$ at 118 is known with some precision, allowing one to make a reliable determination of the number of carbons. The M$+1$ at 118 of intensity 8.9% indicates the presence of eight carbons (1.1% for each carbon). The ion at 90 results from the loss of a molecule of HC$\equiv$N (27), often observed in nitriles. The lower intensity ions (77, 51 and 65) are the typical fragment ions of a benzene ring.

^{1}H NMR

The proton NMR shows only two signals, one for five protons at 7.3 ppm and one for two protons at 3.7 ppm. The peak at 3.7 ppm represents the CH$_2$ and is substituted with two deshielding substituents, an aromatic ring and a nitrile group. The signal at 7.3 ppm for 5 H shows the presence of five aromatic H in one unresolved peak, a situation commonly found in benzene rings substituted with a CH$_2$ group, (C$_6$H$_5$CH$_2$–).

Summary

The nitrile group can be recognized readily in the IR spectrum, but it does not stand out in the ^{13}C NMR because it occurs in the same region as the aromatic carbons. However, the presence of the nitrile can be confirmed by the signal for the quaternary carbon near 120 ppm. In the mass spectrum, one frequently observes peaks for the loss of HC$\equiv$N. The aromatic ring is easily detected in the IR spectrum, the ^{1}H NMR and ^{13}C NMR spectra. The only sp^3 carbon in the molecule is identified by the singlet in the aliphatic region of the ^{1}H NMR and the triplet in the off-resonance ^{13}C spectrum. Joining the components gives the complete structure, C$_6$H$_5$CH$_2$C$\equiv$N, benzylnitrile.

Exercise 030 (002)

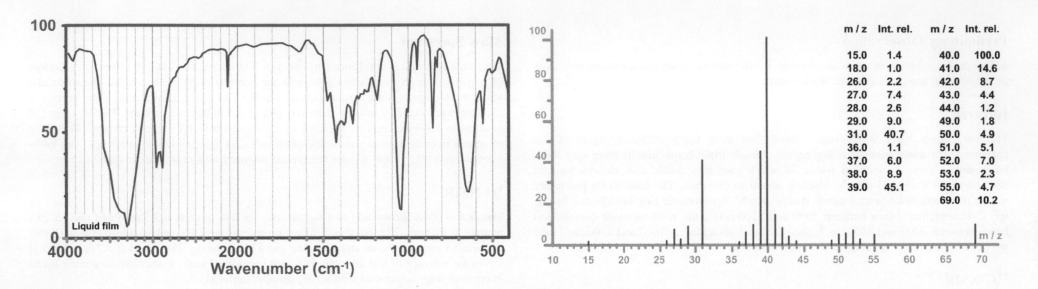

m / z	Int. rel.	m / z	Int. rel.
15.0	1.4	40.0	100.0
18.0	1.0	41.0	14.6
26.0	2.2	42.0	8.7
27.0	7.4	43.0	4.4
28.0	2.6	44.0	1.2
29.0	9.0	49.0	1.8
31.0	40.7	50.0	4.9
36.0	1.1	51.0	5.1
37.0	6.0	52.0	7.0
38.0	8.9	53.0	2.3
39.0	45.1	55.0	4.7
		69.0	10.2

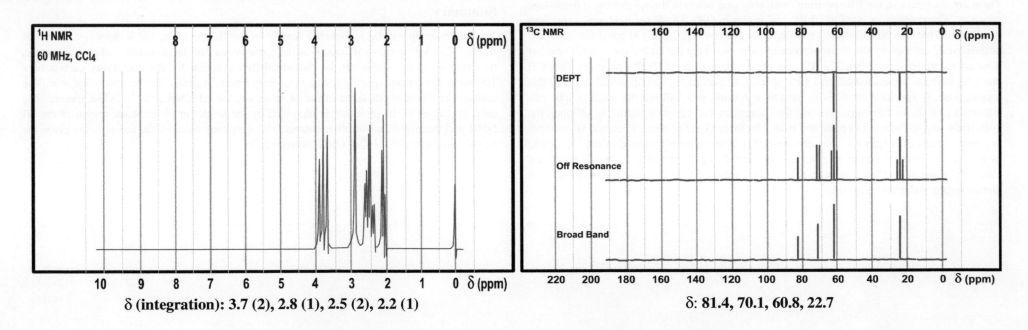

δ (integration): 3.7 (2), 2.8 (1), 2.5 (2), 2.2 (1)

δ: 81.4, 70.1, 60.8, 22.7

Exercise 030 (002)

Preliminary Observation

The very broad band centred near 3400 cm^{-1} indicates the presence of an alcohol. There is no evidence of an aromatic ring, but there is a significant peak at 2100 cm^{-1}.

^{13}C NMR

The broad band spectrum shows signals from four types of carbon. The splitting in the off-resonance spectrum shows that there are two methylene carbons (CH$_2$ triplets), a methine carbon (CH doublet) and a quaternary carbon (C singlet). The DEPT spectrum has two inverted signals, confirming the presence of the CH$_2$ groups, and the peak for the quaternary carbon (81 ppm) has disappeared. The upright signal at 70 ppm indicates a carbon bearing an odd number of protons (CH). The chemical shifts of the lower field carbons (70 and 81 ppm) are in the characteristic region for the two carbons of a terminal alkyne group, $-$C≡CH.

^{1}H NMR

The spectrum has four signals: two types of CH$_2$ and one CH, exactly as expected from the ^{13}C NMR spectrum, as well as one additional hydrogen attached to a heteroatom. The integration ratios of 2:1:2:1, correspond to a CH$_2$ (3.7 ppm), a CH (2.2 ppm), a CH$_2$ (2.5 ppm) and the additional proton (2.8 ppm). The splitting pattern can help to establish the arrangement of these carbons. The triplet at 3.7 ppm is split by two protons, and the same size splitting (coupling size is the same between mutually coupled atoms) is found in the protons at 2.5 ppm. Additional splitting is seen in the signal at 2.5 ppm due to a smaller coupling to the proton at 2.2 ppm. The strong coupling between the protons of the two CH$_2$ carbons indicates that they are attached to one another. The weak coupling between the CH and one of the CH$_2$ groups will also be helpful in developing the full structure.

Infrared

A predominant band in the infrared spectrum is found at 3300 cm^{-1}. The very broad nearly symmetrical shape of the band is typical of the O–H stretching vibration of an alcohol. The other strong band associated with the alcohol is found at 1050 cm^{-1} due to C–O stretching.

The proton that was detected in the NMR as being bonded to a heteroatom is therefore an OH. The typical band for the C≡C stretching of a terminal alkyne is seen at 2100 cm^{-1}, but the ≡C–H stretching band, which is expected around 3300 cm^{-1}, is barely detectable because it is masked by the very broad OH stretching band. The terminal alkyne is also confirmed by the CH bending band observed 650 cm^{-1}.

Mass Spectrum

The spectrum appears to indicate that the compound contains nitrogen, because the highest mass ion is odd (69) and has an even-mass fragment (40) as the base peak. However, it is also possible that the high-mass ion is a fragment ion and that the base peak results from the loss of a molecule to give an even-mass fragment. In this case the peak at 69 is due to the loss of a hydrogen atom from the molecular ion. The molecule has two hydrogens that are in positions susceptile to loss, the terminal alkyne hydrogen, and the hydrogen on the α carbon of the alcohol. The base peak at 40 comes from the loss of a neutral molecule, formaldehyde, CH$_2$=O. Note the peak at 31, which is usually taken as a sign of oxygen in the molecule.

Summary

With the molecular formula of C$_4$H$_6$O – in which the carbons are identified as CH$_2$, CH$_2$, CH and C, and the functional groups as a terminal alkyne and alcohol – the compound has to be HC≡CCH$_2$CH$_2$OH, but-3-yn-1-ol. The ^{1}H NMR shows that the protons of the two CH$_2$ groups are coupled, thus we have –CH$_2$CH$_2$–. The low field triplet at 3.7 ppm belongs to the CH$_2$ bearing the OH, and is coupled to only two other protons, those of the neighbouring CH$_2$ (which should have coupling of the same size). The signal at 2.5 ppm shows a triplet with the same size coupling, but each of the peaks is split into a doublet by a weaker coupling interaction with a single proton. The coupling sizes are different, the first corresponding to the classical ^{3}J aliphatic coupling and the second to a much weaker ^{4}J coupling over a longer distance. The signal at 2.1 ppm is attributed to the acetylenic≡C–H, split by a weak four–bond coupling to the CH$_2$ protons. This coupling pattern fits the sequence – CH$_2$CH$_2$C≡C–H. The singlet at 2.9 ppm is attributed to OH proton, which shows no splitting because it is rapidly exchanging.

Exercise 031 (045)

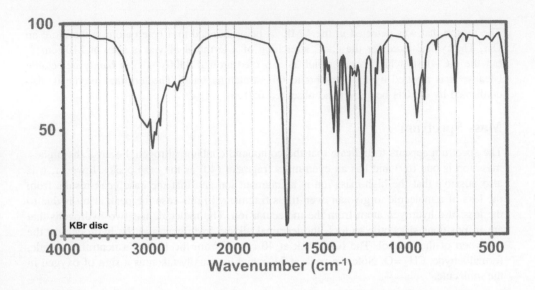

KBr disc

100 ... 50 ... 0
4000   3000   2000   1500   1000   500
Wavenumber (cm⁻¹)

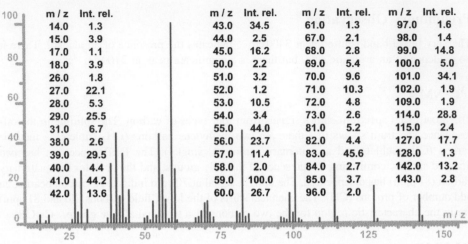

m / z	Int. rel.	m / z	Int. rel.	m / z	Int. rel.	m / z	Int. rel.
14.0	1.3	43.0	34.5	61.0	1.3	97.0	1.6
15.0	3.9	44.0	2.5	67.0	2.1	98.0	1.4
17.0	1.1	45.0	16.2	68.0	2.8	99.0	14.8
18.0	3.9	50.0	2.2	69.0	5.4	100.0	5.0
26.0	1.8	51.0	3.2	70.0	9.6	101.0	34.1
27.0	22.1	52.0	1.2	71.0	10.3	102.0	1.9
28.0	5.3	53.0	10.5	72.0	4.8	109.0	1.9
29.0	25.5	54.0	3.4	73.0	2.6	114.0	28.8
31.0	6.7	55.0	44.0	81.0	5.2	115.0	2.4
38.0	2.6	56.0	23.7	82.0	4.4	127.0	17.7
39.0	29.5	57.0	11.4	83.0	45.6	128.0	1.3
40.0	4.4	58.0	2.0	84.0	2.7	142.0	13.2
41.0	44.2	59.0	100.0	85.0	3.7	143.0	2.8
42.0	13.6	60.0	26.7	96.0	2.0		

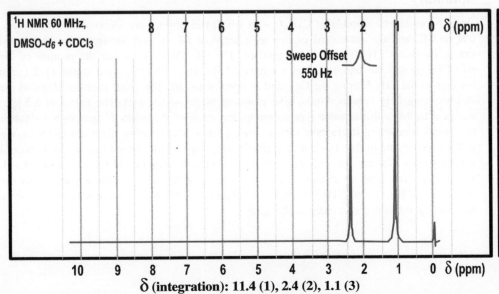

¹H NMR 60 MHz, DMSO-d₆ + CDCl₃

Sweep Offset 550 Hz

δ (integration): 11.4 (1), 2.4 (2), 1.1 (3)

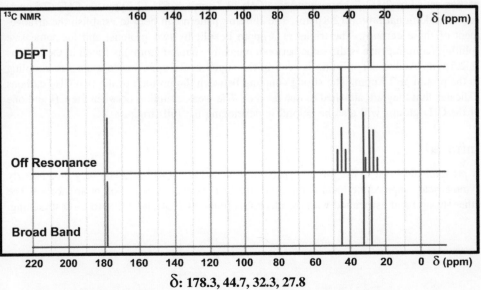

¹³C NMR

DEPT

Off Resonance

Broad Band

δ: 178.3, 44.7, 32.3, 27.8

Exercise 031 (045)

Preliminary Observation

The IR spectrum has very distinctive peaks for a particular functional group. The mass spectrum appears to have a peak for the molecular ion. However the functional group of the molecule may undergo rearrangement to lose a neutral molecule, giving an even-mass fragment ion that could be mistaken for the molecular ion. Note that a peak in the 60-MHz ^{1}H NMR spectrum is offset by 550 Hz, which is equivalent to being offset by 9.17 ppm, (550/60).

Infrared

The IR spectrum has bands that are very characteristic of a carboxylic acid. The broad O–H stretching bands between 3500 and 2300 cm^{-1} and the strong sharp band at 1700 cm^{-1} are very typical of a carboxylic acid. The other bands that would be expected for a carboxylic acid are bands near 1430 and 1260 cm^{-1} for the C–O stretching, and a band near 940 cm^{-1} representing OH deformation.

^{13}C NMR

The compound contains four types of carbon atom: a quaternary carbon at 178.3 ppm (disappears in DEPT), a quaternary carbon at 32.3 ppm (disappears in DEPT), a methylene carbon at 44.7 ppm (inverted in DEPT, triplet in OR) and a methyl carbon (quartet in OR). The downfield quaternary carbon is that of the carboxylic acid group, and the high-field quaternary carbon is an aliphatic carbon.

^{1}H NMR

The signals at 11.4, 2.4 and 1.1 ppm have an integration ratio of 1:2:3, for a minimum of six hydrogens. The peak at 11.4 ppm is in the appropriate region for protons of carboxylic acid groups. The fact that there is no splitting of any of the protons shows that the methyl group and methylene groups are not bonded to one another. The higher field singlet at 1.1 ppm is at the correct position for CH_3 groups on a quaternary carbon. The singlet at 2.4 ppm has a chemical shift appropriate for a methylene group ($-CH_2-$) in the α position of a carboxylic acid.

Mass Spectrum

The peak at 142 could be the molecular ion as it is the highest mass peak and is even. However, it is known that some carboxylic acids lose water, particularly when another labile hydrogen is present in the molecule. If this is so, the 142 peak may be the M−18 ion, and the molecular mass of the compound is 160 (142 + 18). The fragment at 127 corresponds to the loss of a methyl from the 142 fragment ion. The base peak at 59 corresponds to the $[CH_2COOH]^+$ ion, and the peak at 60 to acetic acid, $[CH_3COOH]^+$.

Summary

The spectral evidence indicates that the molecule has these groups: $-CH_2COOH$, $-CH_3$ and $>C<$. The structure can be made by adding two of each of the groups to the quaternary carbon. The compound is therefore $(CH_3)_2C(CH_2COOH)_2$, 3,3-dimethylglutaric acid, molecular mass 160. In the IR spectrum, one might expect to see a doublet of equal intensity near 1380 cm^{-1} corresponding to a *gem*-dimethyl group $(CH_3)_2C<$, but it is not discernible.

Exercise 032 (026)

Elemental analysis: C=60.76% and H=8.91%

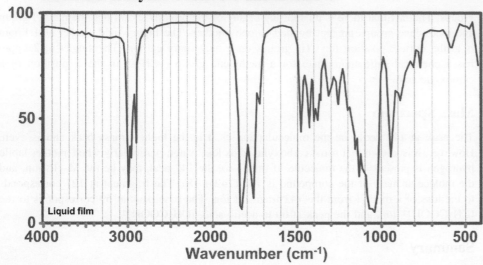

Liquid film

m / z	Int. rel.	m / z	Int. rel.
26.0	1.3	44.0	1.4
27.0	13.1	45.0	1.1
28.0	1.1	55.0	5.3
29.0	2.3	60.0	8.3
39.0	5.6	70.0	3.9
40.0	1.1	71.0	100.0
41.0	13.4	72.0	4.7
42.0	4.4	73.0	2.7
43.0	45.8		

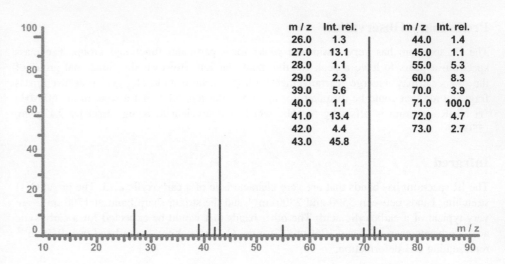

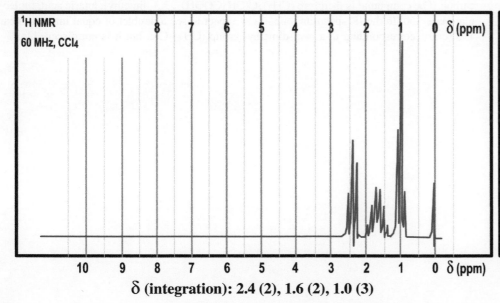

1H NMR
60 MHz, CCl4

δ (integration): 2.4 (2), 1.6 (2), 1.0 (3)

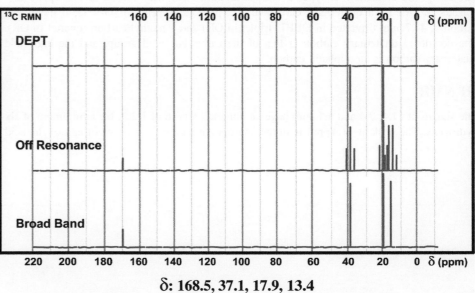

13C RMN
DEPT
Off Resonance
Broad Band

δ: 168.5, 37.1, 17.9, 13.4

Exercise 032 (026)

Preliminary Observation

The infrared spectrum has two strong peaks in a very distinctive carbonyl region. There is no evidence of an aromatic ring or any hydroxyl group.

Infrared

The bands near 1820 and 1760 cm^{-1} are typical of the carbonyl stretching of a carboxylic acid anhydride. The other band for the C–O stretching is found as an intense broad band at 1030 cm^{-1}. The band at 1820 cm^{-1} (the higher frequency band relative to the second C=O band) is more intense, indicating that the anhydride is acyclic.

^{1}H NMR

This ^{1}H NMR spectrum shows three signals in the aliphatic region, a triplet, a sextet and a triplet. The methyl protons (1.0 ppm) and the lower field methylene protons (2.4 ppm) are each split by the central methylene protons (1.6 ppm) into triplets. The central methylene protons are split by their five neighbours into a sextet. The chemical shift of the CH$_2$ at 2.4 ppm is as expected for a CH$_2$ attached to a carbonyl group. We therefore have a sequence: CH$_3$CH$_2$CH$_2$C=O.

^{13}C NMR

The BB spectrum shows three signals in the aliphatic region and one in the carbonyl region. The OR and DEPT spectra indicate the types of carbon to be one methyl group (quartet in OR spectrum), two CH$_2$ groups (inverted signal in DEPT) and a quaternary carbon (no signal in DEPT) in the carbonyl region. The signal at 169 ppm is at a higher field than that of a ketone, which would be expected near 200 ppm, but is in the correct range for an ester or anhydride. One of the CH$_2$ groups, which is at a lower field, is attached to the carbonyl group. Note that the number of hydrogens can be obtained from the C–H coupling, as expressed in the DEPT and OR spectra. In this case the total number of protons observed is seven, in agreement with the number seen in the ^{1}H NMR spectrum.

Mass Spectrum

The mass spectrum has a base peak at 71, but this is unlikely to be the molecular ion as there is no other evidence of the presence of nitrogen. The principal peaks are odd-mass peaks, 27, 43 and 55, which would not be expected for a compound containing nitrogen. The ^{13}C isotope peak (72) of the base peak (71) has a relative intensity of 4.7%, indicating there are four carbons in that ion. The peak at 73(2.7%) is too intense to be an M+2 isotope peak: it must be a fragment from a higher precursor. Microanalysis gives C = 60.76% and H = 8.91%, which leaves a residual of 30.33% for oxygen. The calculated atomic ratios are C:H:O = 5.059:8.84:1.90 from (60.76/12.011), (8.91/1.008) and (30.33/16). Attempts to obtain integer ratios using values of 1 or 2 for oxygen fail, but a trial with 3 oxygens gives integer atomic ratios. Multiplying the calculated fractional atomic ratios by 1.58 (3/1.90) converts them to integer ratios of C:H:O = 8:14:3. This gives an empirical formula of C$_8$H$_{14}$O$_3$. for a mass of 158. The base peak at 71 ion is due to α cleavage at the carbonyl with the loss of O(C=O)C$_3$H$_7$ (87), generating the acyl cation ion C$_3$H$_7$C≡O$^+$.

Summary

The NMR spectra tell little about the compound except the fact that a propyl group is present and it is attached to the quaternary carbon of a carbonyl, possibly an ester. The chemical shift of the CH$_2$ protons is not far enough downfield to be attached to the oxygen, so the alkyl chain must be attached to the carbonyl: CH$_3$CH$_2$CH$_2$C=O. This group corresponds to the peak at 71 in the mass spectrum. As no other carbon atoms are present, two of these units must be joined together via a common oxygen atom. This produces an anhydride which gives the typical double peaks for the carbonyl stretching. The compound is therefore butyric anhydride, (CH$_3$CH$_2$CH$_2$C=O)$_2$O.

Exercise 033 (038)

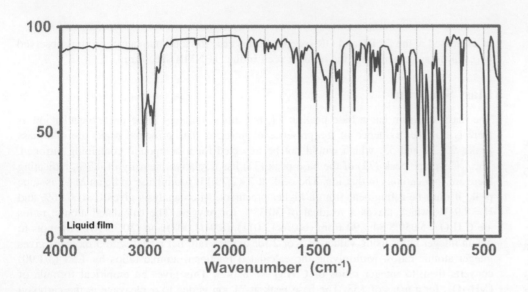

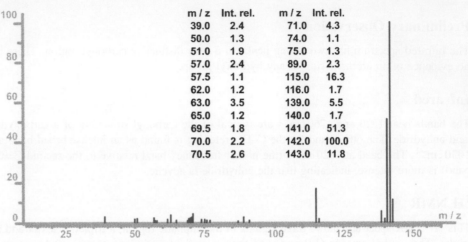

m/z	Int. rel.	m/z	Int. rel.
39.0	2.4	71.0	4.3
50.0	1.3	74.0	1.1
51.0	1.9	75.0	1.3
57.0	2.4	89.0	2.3
57.5	1.1	115.0	16.3
62.0	1.2	116.0	1.7
63.0	3.5	139.0	5.5
65.0	1.2	140.0	1.7
69.5	1.8	141.0	51.3
70.0	2.2	142.0	100.0
70.5	2.6	143.0	11.8

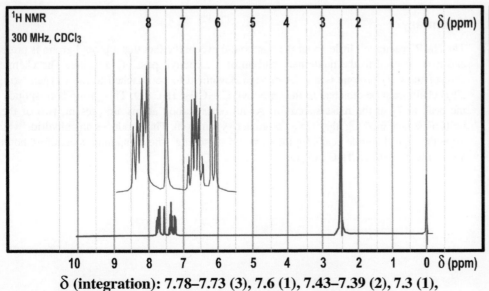

δ (integration): 7.78–7.73 (3), 7.6 (1), 7.43–7.39 (2), 7.3 (1), 2.5 (3)

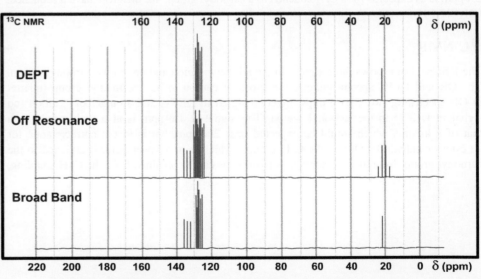

δ: 135.3, 133.7, 131.8, 128.1, 127.7, 127.6, 127.2, 126.8, 125.8, 124.9, 21.6

Exercise 033 (038)

Preliminary Observation

The compound presents evidence for an aromatic system and a methyl group, but no evidence for any heteroatoms or other functional groups.

Mass Spectrum

The mass spectrum shows a peak for the molecular ion at 142. Calculation of the relative intensity of M+1/M must be done carefully in this example. Calculation of the number of carbons from the isotope ratio gives 10.7 carbons (11.8/1.1). It should be noted that an M−1 peak has an isotope peak that contributes to the intensity of the M peak which has an effect on the intensity ratio. The most likely molecular formula is $C_{11}H_{10}$, which equates to the presence of seven centres of unsaturation. The parent peak is also the base peak, indicating a very stable molecular ion (probably a highly unsaturated cyclic structure). This is compatible with the observation of a very strong M−1 peak, possibly resulting from the loss of a proton from the α carbon of an aromatic ring. The other ions are weak, except for the M−27 ion at 115 corresponding to the loss of acetylene (26) from the M−1 ion.

^{13}C NMR

There are 11 carbon signals in the BB spectrum, one CH_3 carbon at 21.6 ppm (quartet in OR), seven CH carbons between 125 and 128 ppm (unchanged in DEPT, doublets in OR) and three quaternary carbons between 132 and 135 ppm (peaks disappear in DEPT). The chemical shifts of all of the lower field carbons are in the aromatic region, and that of the methyl group is correct for a methyl on an aromatic ring.

^{1}H NMR

The single peak at 2.5 ppm represents the methyl group, while the peaks between 7.3 and 7.8 ppm represent seven aromatic protons. A potential aromatic system with seven protons in a monosubstituted system would be naphthalene. In naphthalene there are only two types of ring position, α and β. The protons in the two positions are different in their chemical shifts. The α protons are closer to the second ring and thus more deshielded. In naphthalene, the α protons resonate near 7.8 ppm, while the β protons are found near 7.4 ppm. In this spectrum there are four protons occupying positions with a larger chemical shift (lower field) than 7.5 (the α protons), and only three protons are at lower shifts (the β protons). There is one β hydrogen missing, the one replaced by the CH_3 group. The protons ortho to the methyl group are slightly shielded, with the H_α appearing as a singlet (slightly broadened by meta coupling) at 7.6 ppm, and the H_β as a double doublet at 7.3 ppm.

Infrared

The bands between 3000 and 3100 cm^{-1} show the presence of an aromatic ring (=C–H stretching). This is confirmed by the presence of bands at 1510 and 1600 cm^{-1}, representing the >C=C< bond stretching. The band at 740 cm^{-1} represents out-of-plane bending, corresponding to four adjacent hydrogens on the ring. The band at 820 cm^{-1} represents the out-of-plane bending of two adjacent hydrogens on the ring, and the band at 850 cm^{-1} represents the bending at an isolated hydrogen. This confirms the position of the CH_3 in the β position of the naphthalene.

Summary

The mass spectrum shows a strong peak for the molecular ion at 142. There is little fragmentation other than the loss of a hydrogen radical and a molecule of acetylene. The lack of fragmentation suggests the presence of a stable π system. The occurrence of several obviously doubly charged ions (57.5, 59.5 and 70.5) testifies to the fact that the molecule has such a system. The relative intensity of the isotope peak indicates the presence of 11 carbons, an observation that is confirmed by the ^{13}C NMR spectrum. The proton NMR integration gives a total of 10 hydrogens. The molecular formula of the compound is therefore $C_{11}H_{10}$. As there is only one methyl group on an aromatic ring with seven protons, this suggests a naphthalene skeleton. There are only two possible choices for the position of the substituent: α or β. The ^{1}H NMR and IR spectra give conclusive evidence that it is β-methylnaphthalene. Another ring system that might have come to mind with the same formula as naphthalene ($C_{10}H_8$) is azulene, a 7–5 ring system. However, its proton and carbon chemical shifts are outside the range of values observed for this compound.

Exercise 034 (093)

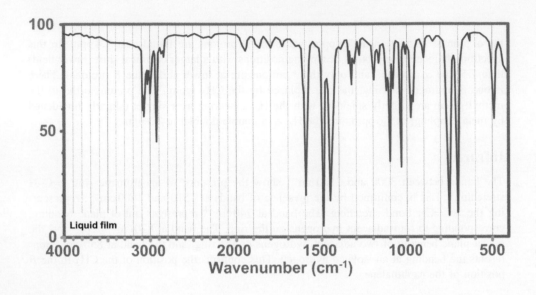

Liquid film

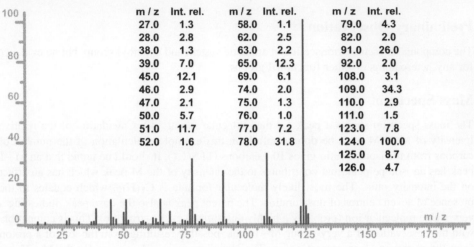

m / z	Int. rel.	m / z	Int. rel.	m / z	Int. rel.
27.0	1.3	58.0	1.1	79.0	4.3
28.0	2.8	62.0	2.5	82.0	2.0
38.0	1.3	63.0	2.2	91.0	26.0
39.0	7.0	65.0	12.3	92.0	2.0
45.0	12.1	69.0	6.1	108.0	3.1
46.0	2.9	74.0	2.0	109.0	34.3
47.0	2.1	75.0	1.3	110.0	2.9
50.0	5.7	76.0	1.0	111.0	1.5
51.0	11.7	77.0	7.2	123.0	7.8
52.0	1.6	78.0	31.8	124.0	100.0
				125.0	8.7
				126.0	4.7

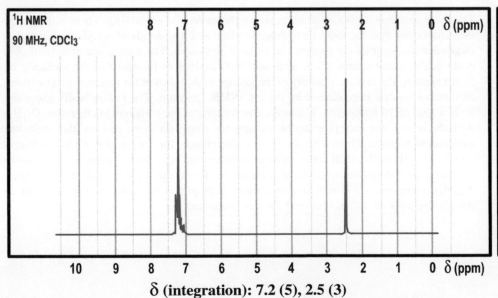

¹H NMR

90 MHz, CDCl₃

δ (integration): 7.2 (5), 2.5 (3)

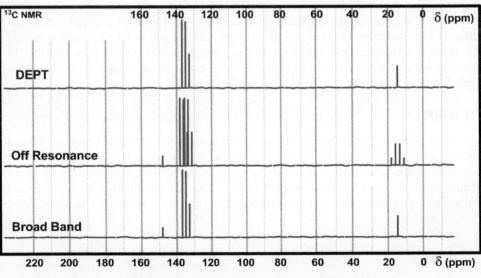

¹³C NMR

DEPT

Off Resonance

Broad Band

δ: 138.5, 128.8, 126.6, 124.9, 15.7

Exercise 034 (093)

Preliminary Observation

The IR spectrum shows clear evidence for an aromatic ring, and this is supported by the ^{1}H and ^{13}C NMR spectra. There seems to be a strong molecular ion in the mass spectrum; take note of the isotope peaks.

Mass Spectrum

If the peak at 124 (100%) is the molecular ion, then the peaks for M+1 at 125 (8.7%) and for M+2 at 126 (4.7%) are significant isotope peaks. The large M+2 could be due to the presence of sulfur or silicon in the molecule (Cl and Br would give a much stronger M+2 peak). Silicon has an isotope abundance of 5.07 for M+1, and 3.36 for M+2, while sulfur has an isotope abundance of 0.80 for M+1 and 4.44 for M+2. The size of the observed M+2 peak is larger than expected for a silicon atom, but fits well for the presence of one sulfur atom. The size of the M+1 isotope peak is compatible with a formula containing one sulfur atom and seven carbons: M+1%: $0.80 + 7 \times 1.1 = 8.5$. This gives a mass of 116, which requires the addition of eight hydrogens to bring the mass to 124, resulting in a formula of C_7H_8S. The molecular ion is very stable (strong conjugation and the presence of a heteroatom). The ion at 109 corresponds to the M−15 ion (note the isotope ion at 111, sulfur is still present, not surprising with the loss of only 15). The ion at 91 is a tropylium ion resulting from a rearrangement in which an SH radical is lost. The ion of mass 45 (isotope peaks 46 and 47) belongs to the $HC\equiv S^+$ ion. The ions at 78, 77, 65 and 39 are the typical series of a benzene ring.

^{13}C NMR

Of the five carbon signals, the one at 138.5 ppm, represents a quaternary aromatic carbon (signal disappears in DEPT), the one at 15.7 ppm represents a methyl group (quartet in off-resonance), and the other three represent aromatic CH carbons (unchanged in DEPT, doublets in OR). The two doublets at 126.6 and 128.8 ppm correspond to two pairs of isochronous CH carbons of a phenyl ring. Therefore, we have a monosubstituted benzene ring with the quaternary carbon bearing a sulfur atom. The calculation of chemical shifts from tables of models gives values matching the observed values: 138.5 ppm (for the ipso carbon bearing the SCH_3), 126.6 ppm (for the ortho carbons), 128.8 ppm (for the meta carbons) and 124.9 ppm (for the para carbon).

^{1}H NMR

The ^{1}H NMR spectrum has two signals with an integration ratio of 3:5 at 2.5 and 7.2 ppm respectively. The singlet at 2.5 ppm corresponds to a CH_3 bonded to a sulfur atom. The five aromatic hydrogens are only slightly different, and appear as a broadened singlet at 7.2 ppm. With a higher field spectrometer, it is possible to see a separate signal for the para proton (most shielded, highest field) near 7.1 ppm.

Infrared

The bands above 3000 cm^{-1} correspond to aromatic CH stretching, and the aromatic ring is confirmed by the bands near 1600 and 1500 cm^{-1} (>C=C< stretching). The band at 700 cm^{-1} represents the out-of-plane deformations of five adjacent aromatic hydrogens. This band is often accompanied by another band near 760 cm^{-1} that is also characteristic of four adjacent hydrogens.

Summary

The evidence indicates that there is a monosubstituted benzene ring, a sulfur atom, and a methyl group. These can be put together in one way only, with the benzene ring and the methyl group both attached to the sulfur atom. This gives thioanisole as the structure of the compound, $C_6H_5SCH_3$.

Exercise 035 (082)

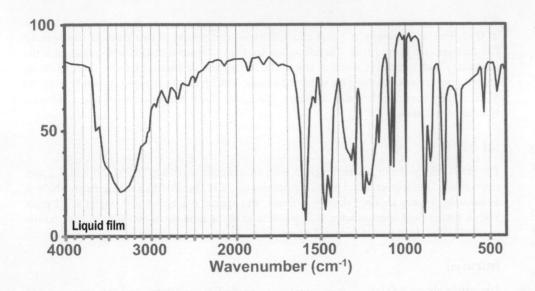

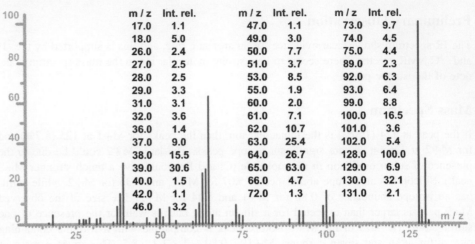

m / z	Int. rel.	m / z	Int. rel.	m / z	Int. rel.
17.0	1.1	47.0	1.1	73.0	9.7
18.0	5.0	49.0	3.0	74.0	4.5
26.0	2.4	50.0	7.7	75.0	4.4
27.0	2.5	51.0	3.7	89.0	2.3
28.0	2.5	53.0	8.5	92.0	6.3
29.0	3.3	55.0	1.9	93.0	6.4
31.0	3.1	60.0	2.0	99.0	8.8
32.0	3.6	61.0	7.1	100.0	16.5
36.0	1.4	62.0	10.7	101.0	3.6
37.0	9.0	63.0	25.4	102.0	5.4
38.0	15.5	64.0	26.7	128.0	100.0
39.0	30.6	65.0	63.0	129.0	6.9
40.0	1.5	66.0	4.7	130.0	32.1
42.0	1.1	72.0	1.3	131.0	2.1
46.0	3.2				

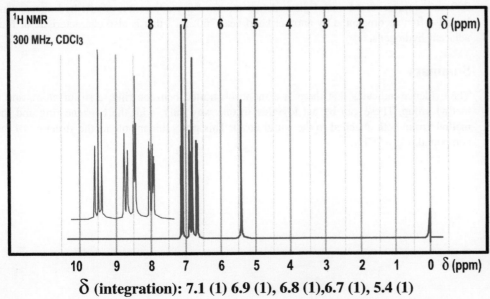

δ (integration): 7.1 (1) 6.9 (1), 6.8 (1),6.7 (1), 5.4 (1)

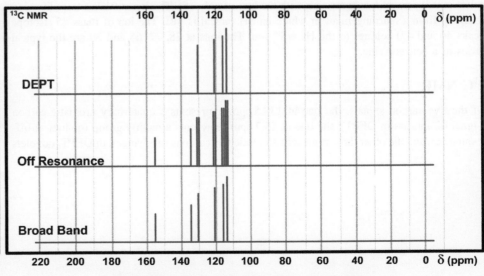

δ: 155.8, 135.0, 130.6, 121.5, 116.0, 113.9

Exercise 035 (082)

Preliminary Observation

Be careful in the interpretation of the weak bands near 2700 and 2800 cm^{-1}; they are possibly overtone bands.

Infrared

The very broad band centred near 3350 cm^{-1} represents the O–H stretching of strongly hydrogen bonded OH. Only a small peak on its shoulder at about 3650 cm^{-1} can be seen for the free OH stretching vibration. The associated intense band at 1220 cm^{-1}, suggests an enol or a phenol. Note that the intensity of the stretching band of the O–H of phenols is medium to strong, somewhat weaker than that of alcohols. The bands near 1580 and 1480 cm^{-1} are representative of >C=C< stretching vibrations of the aromatic ring, the strength of the band at 1580 cm^{-1} indicating strong polarization. The band near 780 cm^{-1} is characteristic of three adjacent hydrogens on an aromatic ring, while the band at 850 cm^{-1} is characteristic of a single hydrogen between two substituents.

Mass Spectrum

There is a strong peak for the molecular ion at m/z 128 (100%). The M+1 isotope peak 129 (6.9%) indicates that the compound has six carbons (6 × 1.1%). The M+2 isotope peak 130 (32.1%) indicates that the compound contains one chlorine atom. The IR spectrum has clear evidence of an OH group. The six carbons, one oxygen and one chlorine atom have a mass of 123 (6 × 12+6+35). As the molecular mass of the compound is 128, five hydrogens are needed to complete the molecular formula, C_6H_5ClO. The molecular ion is very stable, having a large electron source in the ring π system, and the non-bonding electrons of the oxygen and chlorine. The loss of a chlorine atom provides the fragment ion at 93 (6.4%), and loss of HCl yields the peak at 92 (6/3%). The peak at 100 (16.5%), and its associated isotope peak at 102 (5.4%) are due to the loss of C≡O. The fragment at 65 (53%) for $C_5H_5^+$, and its fragmentation product at 39 (30%) for $C_3H_3^+$ (from the loss of acetylene), are classic fragments of benzene rings.

^{13}C NMR

Signals are seen for the six carbons of the benzene ring, two quaternary carbons (signals disappear in DEPT), and four methine carbons (doublets in OR). The lack of symmetry in the ring indicates that it is not para substituted. The quaternary carbon at 156 ppm is furthest downfield, being attached to the strongly electronegative oxygen atom. The other quaternary carbon at 135 ppm corresponds to the one having the chlorine substituent. The methine carbons (four CH doublets in OR) at 114, 116, 121 and 131 ppm represent the four unsubstituted carbons.

^{1}H NMR

The ^{1}H NMR spectrum has signals for five different protons, four of them in the aromatic region and one at a chemical shift (5.4 ppm) typical of a phenol in a CDCl$_3$ solution. The ring protons that are shifted significantly upfield compared to benzene (7.27 ppm), would be in positions ortho and para to the oxygen, a strong resonance donor to the ortho position (−0.50 ppm) and para position (−0.40 ppm). The protons meta to the oxygen are much less strongly shielded (−0.14 ppm). The effect of the chlorine is insignificant compared to the effect of oxygen (ortho, +0.03; meta, −0.06; para, −0.04). One of the upfield protons (6.84 ppm) shows no ortho coupling; it appears as a very narrowly split triplet (meta coupling). This proton must have the two substituents on either side of it, resulting in a meta-substituted ring. The proton with an ortho Cl and ortho OH has a calculated chemical shift of 6.80 ppm (7.27 − 0.50 + 0.03), very close to the observed value of 6.84 ppm. The triplet with the larger splitting would be between two ortho protons. The calculated shift of this proton is 7.07 ppm (7.27 − 0.14 − 0.06), very close to the observed value of 7.11 ppm. The doublet at 6.70 ppm corresponds to the proton para to the Cl (7.27 − 0.50 − 0.04 = 6.73), and the doublet at 6.91 ppm corresponds to the proton para to the OH (7.27 − 0.40 + 0.03 = 6.90). Each doublet also shows fine meta coupling.

Conclusion

The compound is *m*-chlorophenol, ClC_6H_4OH.

Exercise 036 (091)

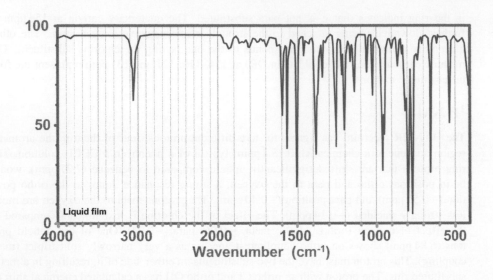

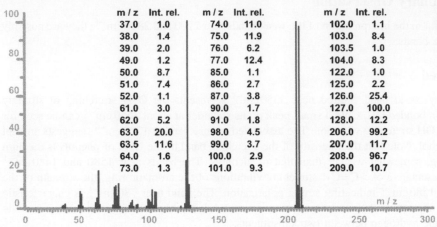

m/z	Int. rel.	m/z	Int. rel.	m/z	Int. rel.
37.0	1.0	74.0	11.0	102.0	1.1
38.0	1.4	75.0	11.9	103.0	8.4
39.0	2.0	76.0	6.2	103.5	1.0
49.0	1.2	77.0	12.4	104.0	8.3
50.0	8.7	85.0	1.1	122.0	1.0
51.0	7.4	86.0	2.7	125.0	2.2
52.0	1.1	87.0	3.8	126.0	25.4
61.0	3.0	90.0	1.7	127.0	100.0
62.0	5.2	91.0	1.8	128.0	12.2
63.0	20.0	98.0	4.5	206.0	99.2
63.5	11.6	99.0	3.7	207.0	11.7
64.0	1.6	100.0	2.9	208.0	96.7
73.0	1.3	101.0	9.3	209.0	10.7

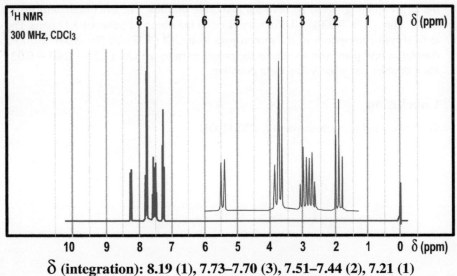

δ (integration): 8.19 (1), 7.73–7.70 (3), 7.51–7.44 (2), 7.21 (1)

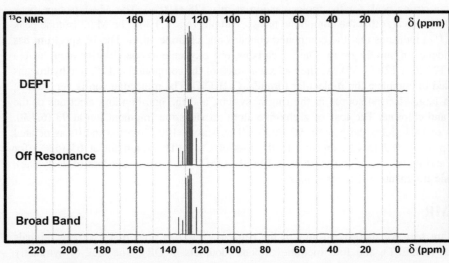

δ: 134.5, 131.9, 129.8, 128.2, 127.8, 127.2, 127.0, 126.6, 126.0, 122.8

Exercise 036 (091)

Preliminary Observation

The IR spectrum has a peak above $3000\,cm^{-1}$ for aromatic protons, but none below $3000\,cm^{-1}$ for aliphatic protons. The isotope peaks in the mass spectrum are interesting.

Mass Spectrum

The mass spectrum has a molecular ion at 206 (99.2%), with an isotope peak at 208 (96.7%). The natural abundance isotope ratio of ^{79}Br to ^{81}Br is 100 to 97.2. The observed isotope ratio in this spectrum is 97.5% ($100 \times 96.7/99.2$), indicating that the molecule contains a bromine atom. The base peak at 127 (100%) represents the fragment derived from the loss of the bromine atom. The ^{13}C isotope peaks at 207 (11.7%) and 209 (10.7%) indicate the presence of 10 carbons. Ten carbons and one bromine have a mass of 199 ($10 \times 12 + 79$); seven hydrogens are needed to make up the rest of the mass of 206 ($C_{10}H_7Br$). The other weak fragments are those of the 'aromatic family'; for example, the peak at 101 is $127 - 26$ (loss of acetylene), typical of aromatic rings. Note also the typical ions of phenyl ring fragmentation at 77, 51 and 39.

^{13}C NMR

There are peaks for 10 types of carbon, all aromatic. The DEPT spectrum has three peaks disappear, thus there are three types of quaternary carbon. The 10-carbon aromatic system formed from two benzene rings (naphthalene) has two quaternary carbons at the ring juncture; a substituent would account for a third quaternary carbon. The α and β carbons of naphthalene have chemical shifts of 127.8 and 125.6 ppm respectively, and the ring junction carbons 134.5 ppm. Bromine has a shielding effect on the ipso carbon (-5.5, heavy atom effect) and para carbon (-1.3), and a deshielding effect on the ortho ($+3.4$) and meta carbons ($+1.7$). The substituent effect on the junction carbons of naphthalene is less than that of unsubstituted carbons. In this spectrum, the junction carbons are found at 131.9 and 134.5 ppm, and the quaternary carbon bearing the bromine at 122.8 ppm. The other carbons are found between 129.8 and 126.0 ppm.

^1H NMR

The spectrum has a doublet at 8.19, three overlapping doublets between 7.73 and 7.70, two triplets between 7.51 and 7.44 and a triplet at 7.21 ppm. Naphthalene itself has only two different protons, the α and β protons, seen at 7.84 (α) and 7.47 (β). The effect of bromine on the chemical shifts in a benzene ring are: ortho ($+0.22$), meta (-0.13) and para (-0.03). The other significant effect in a naphthalene ring is the peri effect ($\alpha - \alpha$ on adjacent rings), which is about $+0.30$ for bromine. The major couplings in naphthalene are the ortho couplings (7–8 Hz). The meta couplings (about 1 Hz) and para couplings (<1 Hz) are significantly smaller. The α and β protons can be easily distinguished by their splitting pattern. The α protons are doublets, and the β protons are triplets (each with the additional small meta coupling). The doublet ($^3J = 8\,Hz$) at 8.19 ppm is an α proton shifted downfield by the peri effect of the bromine in the α position of the other ring. The two other α protons are found in the group of three doublets near 7.7 ppm which also includes the doublet of the β proton ortho to the bromine. The triplet at 7.21 ppm is the β proton meta to the bromine. The two β protons on the unsubstituted ring are seen as triplets at 7.51 and 7.44 ppm.

Infrared

The spectrum has peaks in the aromatic regions: between 3000 and $3100\,cm^{-1}$ for sp^2 CH stretching, between 2000 and 1600 for weak overtone-combination bands, at 1505 and $1600\,cm^{-1}$ for aromatic $>C=C<$ stretching, at $760\,cm^{-1}$ for out-of-plane deformation of four adjacent ring hydrogens, and at $800\,cm^{-1}$ for three adjacent hydrogens. The infrared spectrum supports substitution at the α position. The frequency of the C–Br stretching is below $500\,cm^{-1}$, thus it is not observed.

Summary

The compound is 1-bromonaphthalene $C_{10}H_7Br$. The mass spectrum is critical for the detection of bromine. The molecular mass and the doublet from the isotope pattern indicate the presence of bromine. The evidence from the ^{13}C NMR, ^1H NMR and IR spectra supports the assignment of substitution at the α position. In the ^1H NMR, the major couplings produce three triplets and four doublets as expected for an α-substituted naphthalene. The bands at $760\,cm^{-1}$ and $800\,cm^{-1}$ represent out-of-plane deformations of four adjacent hydrogens and of three adjacent hydrogens on the ring.

Exercise 037 (044)

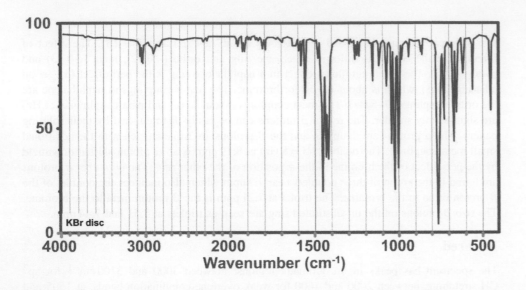

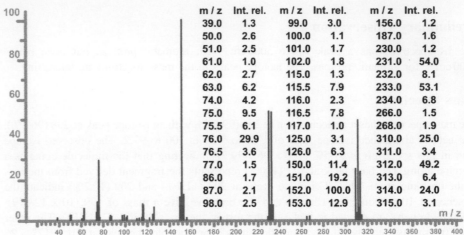

m / z	Int. rel.	m / z	Int. rel.	m / z	Int. rel.
39.0	1.3	99.0	3.0	156.0	1.2
50.0	2.6	100.0	1.1	187.0	1.6
51.0	2.5	101.0	1.7	230.0	1.2
61.0	1.0	102.0	1.8	231.0	54.0
62.0	2.7	115.0	1.3	232.0	8.1
63.0	6.2	115.5	7.9	233.0	53.1
74.0	4.2	116.0	2.3	234.0	6.8
75.0	9.5	116.5	7.8	266.0	1.5
75.5	6.1	117.0	1.0	268.0	1.9
76.0	29.9	125.0	3.1	310.0	25.0
76.5	3.6	126.0	6.3	311.0	3.4
77.0	1.5	150.0	11.4	312.0	49.2
86.0	1.7	151.0	19.2	313.0	6.4
87.0	2.1	152.0	100.0	314.0	24.0
98.0	2.5	153.0	12.9	315.0	3.1

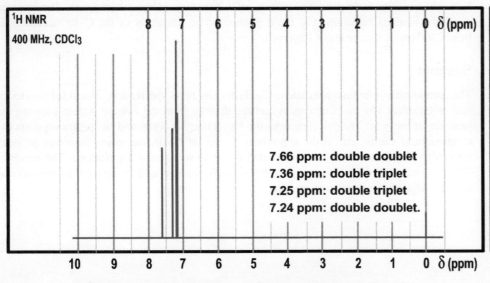

7.66 ppm: double doublet
7.36 ppm: double triplet
7.25 ppm: double triplet
7.24 ppm: double doublet.

δ (integration): 7.66 (1) 7.36 (1), 7.25 (1), 7.24 (1)

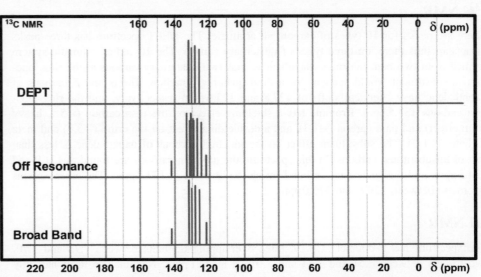

δ: 142.1, 132.6, 131.0, 129.4, 127.1, 123.5

Exercise 037 (044)

Preliminary Observation

The patterns of the isotope peaks in the mass spectrum seem particularly interesting.

Mass Spectrum

The mass spectrum shows a parent ion at 310. The peaks at M, M+2 and M+4 in a ratio of 1:2:1 are characteristic of the presence of two atoms of bromine. The peaks at 231 and 233 arise from the loss of a bromine atom. The 231 ion (contains ^{79}Br) would come from the 310 ion by loss of ^{79}Br, or from the 312 ion by loss of ^{81}Br, while the 233 ion (contains ^{81}Br) would come from the 312 ion by loss of ^{79}Br or from the 314 ion by loss of ^{81}Br. The base peak at 152 is due to the loss of both bromine atoms. One could look for the carbon skeleton of the molecule by replacing the two monovalent atoms (two Br in this case) with two H; thus giving a mass of 154 (152+2 = 154). Dividing this mass of 154 by 12 (mass of 1 C) gives us 12 C and a residue of 10 (10 H), suggesting a possible formula of $C_{12}H_{10}$. Reintroducing the two Br atoms leads to a formula of $C_{12}H_8Br_2$. In the mass spectrum, the ion obtained from the loss of the two bromine atoms (152) is very stable, as are many aromatic ions. The ion at 76 could be a doubly positively charged version of the 152 ion. Remember that the x-axis on the mass spectral chart is not a 'mass axis', but an axis of the mass to charge ratio (m/z). When z equals the charge of 1 electron (singly charged ions), one can read the mass directly, but for doubly charged ions (z = 2) the number read on the axis is actually m/2. Thus, the peak at 76 could correspond to a doubly charged ion of mass 152 (76 × 2 = 152).

^{13}C NMR

The compound has signals for six types of carbon atom, all in the aromatic region. Since the mass spectrum indicates a molecular formula with 12 carbons, each of these signals must represent two carbons. The DEPT spectrum shows that there are two different quaternary carbons, the high-field one at 124 ppm being shifted by a bromine atom (*ipso* effect of large atoms), and the peak at 142 having a carbon substituent. The off-resonance spectrum indicates the presence of four aromatic CH type carbons, and confirms the presence of the quaternary carbons. Each of these signals represents two isochronous carbons.

^{1}H NMR

The ^{1}H NMR spectrum has four signals between 7.24 and 7.66 ppm of equal integration ratios. Each signal corresponds to two isochronous protons. The downfield peak at 7.66 is deshielded by the bromine atom in the ortho position. A bromine substituent has a deshielding effect in the ortho position (+0.22), but is shielding to the meta position (−0.03) and the para position (−0.13). There is only one proton of the ring that has been shifted downfield; therefore, there is only one ortho to the bromine atom (thus, the ring must be ortho-disubstituted). The coupling pattern of the aromatic protons also shows that the four hydrogen atoms are adjacent. The presence of two ^{3}J doublets and two ^{3}J triplets is typical of ortho-disubstituted benzene rings. The protons next to the substituent see only one ortho neighbour and one meta neighbour, thus having a doublet from ^{3}J coupling with additional weak ^{4}J coupling. The two other protons see two ortho neighbours, thus having a triple from ^{3}J coupling with additional weak ^{4}J coupling from the meta proton.

Infrared

The bands between 3000 and 3100 cm^{-1} indicate the presence of an aromatic ring (=C–H stretching). This is supported by the presence of bands near 1600 and 1500 cm^{-1} representing stretching of the >C=C< bonds. The band at 770 cm^{-1} is a deformation band characteristic of a benzene ring bearing four adjacent hydrogens. Thus, we have an ortho-disubstituted aromatic ring. The C–Br bands are not observed (due to the mass effect of bromine which lowers the frequency below the range of the spectrometer).

Summary

With a molecular formula of $C_{12}H_8Br_2$ and a symmetrical molecule having only six types of carbon, the two benzene rings must be equivalently substituted. Each of the benzene rings has a bromine atom, and it is simply a matter of the location of the bromine. The ^{1}H NMR has only one proton deshielded by the bromine, thus the substitution of the rings is ortho. This is confirmed by the proton coupling pattern and out-of-plane bending bands in the IR spectrum. The compound is 2,2′-dibromobiphenyl.

Exercise 038 (046)

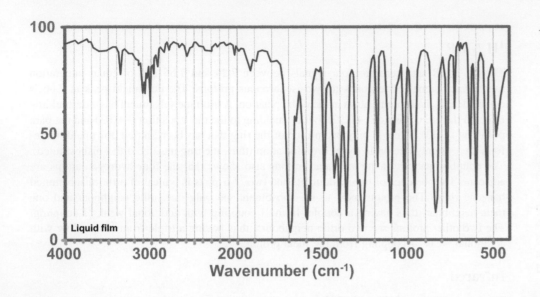

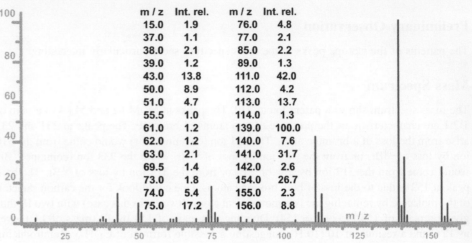

m / z	Int. rel.	m / z	Int. rel.
15.0	1.9	76.0	4.8
37.0	1.1	77.0	2.1
38.0	2.1	85.0	2.2
39.0	1.2	89.0	1.3
43.0	13.8	111.0	42.0
50.0	8.9	112.0	4.2
51.0	4.7	113.0	13.7
55.5	1.0	114.0	1.3
61.0	1.2	139.0	100.0
62.0	1.2	140.0	7.6
63.0	2.1	141.0	31.7
69.5	1.4	142.0	2.6
73.0	1.8	154.0	26.7
74.0	5.4	155.0	2.3
75.0	17.2	156.0	8.8

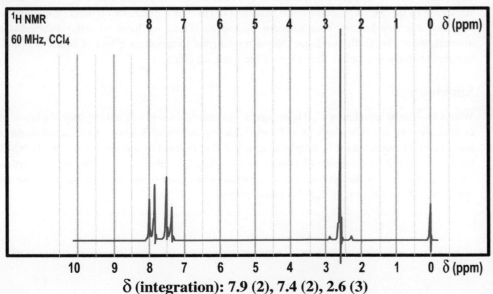

δ (integration): 7.9 (2), 7.4 (2), 2.6 (3)

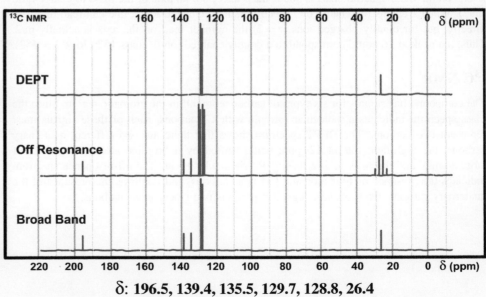

δ: 196.5, 139.4, 135.5, 129.7, 128.8, 26.4

Exercise 038 (046)

Preliminary Observation

The compound appears to have an aromatic ring, a carbonyl and a chlorine atom.

Mass Spectrum

The relative intensities of the isotope peaks give very important information about the presence of halogens in this compound. The peak for the molecular ion at 154 (26.7%) and the ions for the loss of 15 at 139 (100%), and for the loss of 43 at 111 (42%), each have an isotope 2 mass units higher with a relative abundance of 32%. This intensity of the M+2 isotope peak is characteristic of the presence of chlorine. Each of these ions has a ^{35}Cl atom and an isotope peak with a ^{37}Cl atom in a ratio of 100:32. The loss of 15 is equal to the loss of a methyl group, and a further loss of 28 is equivalent to the loss of carbon monoxide. Subsequent loss of HCl from this ion (111 − 36), a typical fragmentation of chlorinated aromatics, leads to the peak at 75. The compound is almost identified solely by the mass spectrum; only the orientation of the substituents has to be determined.

^{13}C NMR

The DEPT spectrum shows that there are three quaternary carbons in the molecule. The chemical shift of the carbon at 197 ppm corresponds to that of the carbonyl of a ketone. The two other quaternary carbons at 139 and 136 ppm correspond to aromatic carbons, one bonded to the Cl, the other bonded to the carbonyl group. The off-resonance spectrum indicates the presence of two aromatic CH carbons and confirms the presence of the quaternary carbons. The CH doublet at 129.7 ppm and the CH doublet at 128.8 ppm each represent two isochronous carbons. This is the classical situation of a para-disubstituted benzene ring. Finally, the signal at 26 ppm is in the range of a methyl group (CH_3) on a carbonyl or other sp^2 carbon.

^1H NMR

The ^1H NMR spectrum has three signals with integration ratios of 2:2:3 at 7.9, 7.4 and 2.6 ppm. The 'doublet' at 7.9 ppm corresponds to two isochronous aromatic protons ortho to a carbonyl group, and the 'doublet' at 7.4 ppm corresponds to two isochronous aromatic protons ortho to a chlorine. The four protons actually form an AA′XX′ system, which may show additional smaller peaks. The singlet at 2.6 ppm corresponds to a CH_3 bonded to a carbonyl group.

Infrared

The bands between 3000 and 3150 cm^{-1} indicate the presence of double bonds (=C–H stretching). This is supported by the presence of bands near 1600 and 1500 cm^{-1} representing stretching of the >C=C< bonds of a strongly polarized aromatic ring. The band at 830 cm^{-1} is a deformation band characteristic of a para-disubstituted benzene ring. In effect the band between 800 and 855 cm^{-1} is a deformation band characteristic of a ring with two adjacent protons. The strong band at 1690 cm^{-1} corresponds to a conjugated carbonyl group, which has a harmonic band at 3470 cm^{-1}.

Conclusion

The evidence indicates that there is a benzene ring with a chlorine and acetyl group as substituents. The ^{13}C NMR spectrum shows only four different carbons, and the ^1H NMR has a symmetrical AA′XX′ pattern for the ring protons, proving that the ring is para-substituted. This gives the structure: p-chloroacetophenone, $CH_3(CO)C_6H_4Cl$.

Exercise 039 (087)

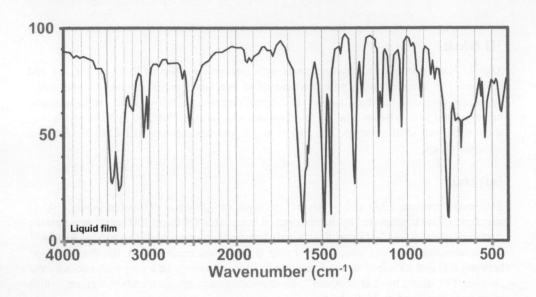

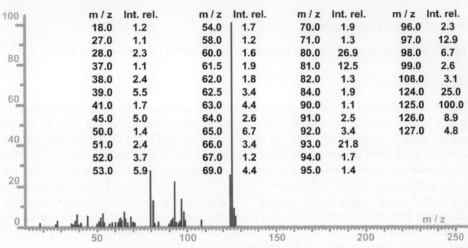

m/z	Int. rel.	m/z	Int. rel.	m/z	Int. rel.	m/z	Int. rel.
18.0	1.2	54.0	1.7	70.0	1.9	96.0	2.3
27.0	1.1	58.0	1.2	71.0	1.3	97.0	12.9
28.0	2.3	60.0	1.6	80.0	26.9	98.0	6.7
37.0	1.1	61.5	1.9	81.0	12.5	99.0	2.6
38.0	2.4	62.0	1.8	82.0	1.3	108.0	3.1
39.0	5.5	62.5	3.4	84.0	1.9	124.0	25.0
41.0	1.7	63.0	4.4	90.0	1.1	125.0	100.0
45.0	5.0	64.0	2.6	91.0	2.5	126.0	8.9
50.0	1.4	65.0	6.7	92.0	3.4	127.0	4.8
51.0	2.4	66.0	3.4	93.0	21.8		
52.0	3.7	67.0	1.2	94.0	1.7		
53.0	5.9	69.0	4.4	95.0	1.4		

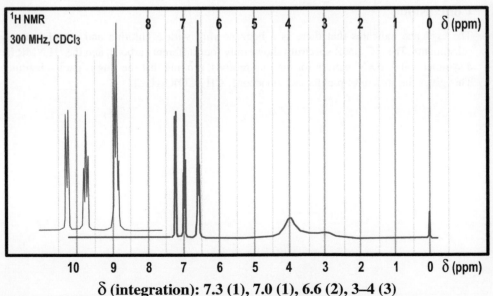

δ (integration): 7.3 (1), 7.0 (1), 6.6 (2), 3–4 (3)

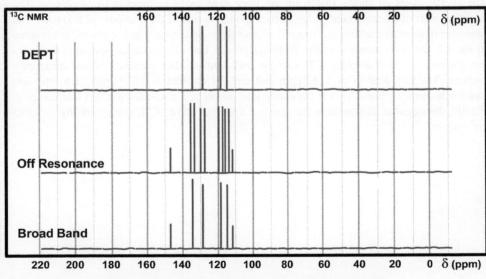

δ: 147.1, 134.7, 128.9, 118.6, 115.1, 111.8

Exercise 039 (087)

Infrared

The bands just above $3000\,\text{cm}^{-1}$ correspond to the $=$C–H stretching vibrations of an aromatic ring. The band near $1600\,\text{cm}^{-1}$ confirms the presence of an aromatic ring (stretching $>$C$=$C$<$). The band at $760\,\text{cm}^{-1}$ represents the deformation (out-of-plane bending) corresponding to the presence of four adjacent aromatic protons, indicating that the compound contains an ortho-disubstituted benzene ring. The two bands at 3350 and $3450\,\text{cm}^{-1}$ result from the symmetric and asymmetric N–H stretching vibrations of an NH_2 group, the band at $2520\,\text{cm}^{-1}$ is the particular S–H stretching characteristic of thiols.

Mass Spectrum

The peak at 125 (100%) is that of the molecular ion. This odd mass shows the presence of an odd number of nitrogen atoms. The molecular ion has isotope peaks at 126 (8.9%) and 127 (4.8%). Only sulfur and silicon could give such a value of M+2 (Cl and Br would give a much stronger M+2 peak). Silicon has isotope ratios for M+1 (5.07) and M+2 (3.36), sulfur has M+1 (0.80) and M+2 (4.44), and nitrogen M+1 (0.36). The size of the observed M+2 peak is larger than expected for a silicon atom, but fits well with the presence of one sulfur atom in the molecule. Also, considering the contribution expected from a nitrogen and six carbons of the benzene ring, silicon would give a M+1 of 12.03 ($0.36 + 5.07 + 6 \times 1.1$). The compound contains sulfur, and in light of the contribution of 4.4% (^{34}S) per atom of S, there is one S in this compound.

^{13}C NMR

The spectrum has signals for six carbon atoms, all in the aromatic region: four CH carbons (doublets in OR) and two quaternary carbons (signals disappear in DEPT). The signal at 147.1 ppm corresponds to a quaternary carbon bearing a nitrogen atom, and the signal at 111.8 ppm corresponds to a quaternary carbon bearing a sulfur atom.

^{1}H NMR

There are signals from four aromatic protons and two protons on heteroatoms proton, with integration ratios of 1:1:2:3 for the peaks at 7.3, 7.0, 6.6 and 3–4 ppm. The signals between 3 and 4 ppm are very broad, corresponding to the hydrogens on the heteroatoms, two on the nitrogen and one on the sulfur. The doublet at 7.3 is the proton ortho to the sulfur, and the triplet at 7.04 is para to the sulfur. The two upfield protons at 6.6 ppm, belong to one that is ortho to, and the one that is para to, the amino group (resonance donor).

Conclusion

The benzene ring has two substituents, a thiol group and a primary amino group in an ortho orientation, giving the structure: *o*-aminothiophenol, $HSC_6H_4NH_2$.

Exercise 040 (070)

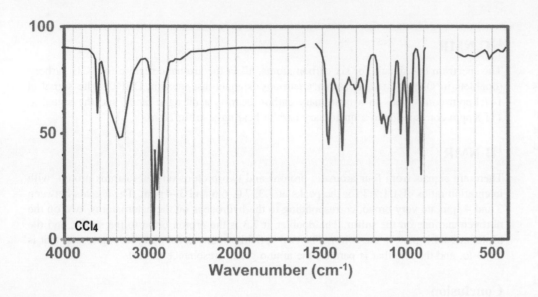

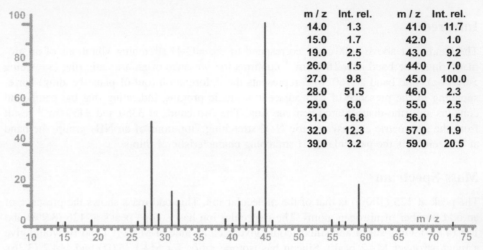

m / z	Int. rel.	m / z	Int. rel.
14.0	1.3	41.0	11.7
15.0	1.7	42.0	1.0
19.0	2.5	43.0	9.2
26.0	1.5	44.0	7.0
27.0	9.8	45.0	100.0
28.0	51.5	46.0	2.3
29.0	6.0	55.0	2.5
31.0	16.8	56.0	1.5
32.0	12.3	57.0	1.9
39.0	3.2	59.0	20.5

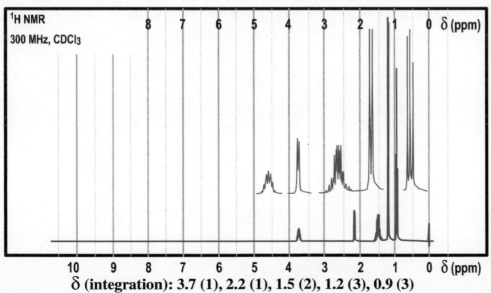

δ (integration): 3.7 (1), 2.2 (1), 1.5 (2), 1.2 (3), 0.9 (3)

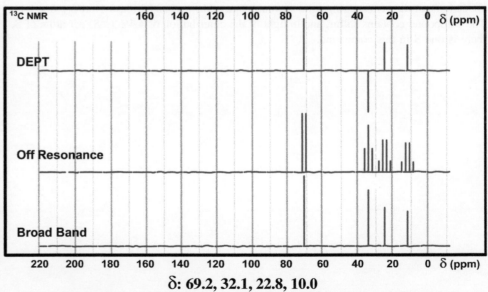

δ: 69.2, 32.1, 22.8, 10.0

Exercise 040 (070)

Infrared

The infrared spectrum gives a clear indication of the functional group in the molecule. The spectrum has bands that are characteristic of an alcohol group in a dilute solution. The sharp band at $3650\,cm^{-1}$ represents the free OH stretching vibration, and the broad band centred near $3350\,cm^{-1}$ represents the hydrogen-bonded OH stretching vibrations. The band near $1100\,cm^{-1}$ corresponds to the C–O stretching of secondary alcohols. The out-of-plane deformation band of the O–H, which is normally seen near $640\,cm^{-1}$, is not observed in this spectrum because the carbon tetrachloride solvent absorbs in this region.

^{13}C NMR

There are signals for four types of carbon in the compound. The DEPT spectrum shows the presence of one CH_2 (inverted signal) at 32 ppm. In the off-resonance spectrum the assignments are straightforward: the doublet at 69 ppm corresponds to a CH bearing the alcohol group, the quartet at 23 ppm corresponds to a CH_3 situated α to the carbon carrying the OH, and the quartet at 10 ppm corresponds to another CH_3.

^{1}H NMR

The ^{1}H NMR spectrum shows the presence of five signals, with integration ratios of 1 (3.7 ppm) : 1 (2.2 ppm) : 2 (1.5 ppm) : 3 (1.2 ppm) : 3 (0.9 ppm). The multiplet of integration 1 at 3.1 ppm corresponds to a CH bearing the hydroxyl group (–OH), to which it is coupled. The OH is seen at 2.2 ppm as a doublet (J = 4 Hz). The CH at 3.1 ppm is also coupled to a CH_3 (signal at 1.2 ppm) and to the CH_2 protons (signal at 1.5 ppm). The multiplet of integration 2 at 1.5 ppm corresponds to the two protons of the CH_2. One might expect the CH_2 to be split into five by the four adjacent protons, but the pattern is much more complex than that. The protons of the CH_2 group are non-equivalent, being next to a chiral carbon. They have different chemical shifts, are mutually coupled and appear as a more complex pattern. Finally, the triplet of integration 3 at 0.9 ppm corresponds to a CH_3 coupled to the two protons of the CH_2 group.

Mass Spectrum

The peak at 73 is not the molecular ion, but a fragment ion due to the loss of a hydrogen atom. The loss of H from the OH is not a characteristic of alcohols. The loss of a hydrogen from the α carbon is possible, but it is usually not the preferred fragmentation. If there are other groups on the α carbon, the larger more stable radicals are preferentially lost. In this example, peaks are found for three α cleavages: M−1 (loss of H), 73 (1%); M−15 (loss of CH_3), 59 (20%); M−29 (loss of CH_3CH_2), 45 (100%). This is compatible with a secondary alcohol having a methyl and an ethyl group on the α carbon. The ion at 28 is probably $CH_2 = CH_2^{+}$, resulting from a rearrangement mechanism where the molecule loses the equivalent of water and ethylene.

Conclusion

The accumulated evidence indicates that the compound is a four–carbon alcohol with a methyl group and ethyl group on the carbon bearing the hydroxyl group. The only alcohol that fits this is 2-butanol, $CH_3CH_2CH(OH)CH_3$.

Exercise 041 (067)

Elemental analysis: C = 84.2% and H = 15.8%

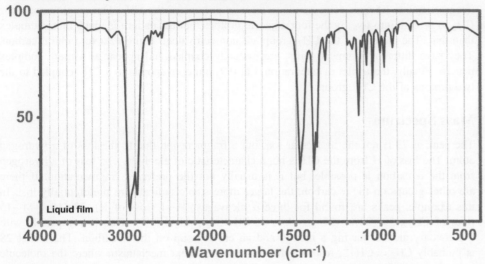

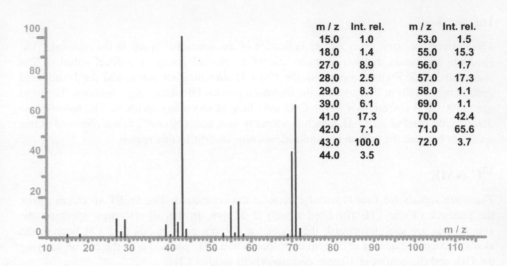

m/z	Int. rel.	m/z	Int. rel.
15.0	1.0	53.0	1.5
18.0	1.4	55.0	15.3
27.0	8.9	56.0	1.7
28.0	2.5	57.0	17.3
29.0	8.3	58.0	1.1
39.0	6.1	69.0	1.1
41.0	17.3	70.0	42.4
42.0	7.1	71.0	65.6
43.0	100.0	72.0	3.7
44.0	3.5		

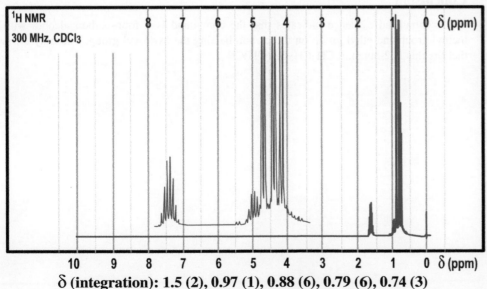

δ (integration): 1.5 (2), 0.97 (1), 0.88 (6), 0.79 (6), 0.74 (3)

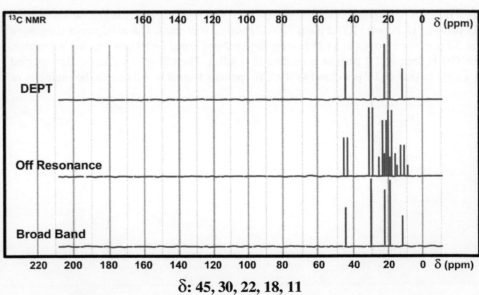

δ: 45, 30, 22, 18, 11

Exercise 041 (067)

Infrared

The infrared spectrum has no bands that identify any specific functional group. The signals between 2990 and 2890 cm^{-1} are sp^3 C–H stretching bands. The characteristic bands of CH$_3$ groups are also present between 1350 and 1450 cm^{-1}. The presence of two bands of similar intensity near 1380 cm^{-1} is characteristic of a pair of geminal methyl groups, such as found in an isopropyl group.

Mass Spectrum

The odd-mass peak at 71 (65.6%) does not appear to be the molecular ion, as there is no evidence of nitrogen in the other spectra. A microanalysis indicates that the compound contains only carbon and hydrogen: (C 84.2% and H 15.8%). The isotope peak at 72 (3.7%) has a intensity relative to the 71 ion of 5.6% ($100 \times 3.7/65.6$) indicates the presence of five carbons in that ion. Five carbons require 11 hydrogens to give the mass of 71 (C$_5$H$_{11}^+$). The microanalysis data give atomic ratios of C:H = 7.01:15.67, (84.2/12.01:15.8/1.008). An integer ratio is obtained when a formula with eight carbons is tried; multiplication by a factor of 8/7.01 yields a ratio of 8:17.9, which rounds to 8:18, for a molecular formula of C$_8$H$_{18}$ (mass 114). The major fragment ions at 43 and 71 represent the C$_3$H$_7^+$ and C$_5$H$_{11}^+$ ions, produced by cleavage at the more highly branched site. The peak at 70 arises by the loss of a propane molecule (C$_3$H$_8$) by a rearrangement reaction.

^{13}C NMR

The broad band spectrum has only five signals, all in the aliphatic region. The DEPT spectrum shows that there are no CH$_2$ groups (no inverted signals) and no quaternary carbons (no missing signals). The OR spectrum shows that there are three for CH$_3$ carbons (quartets) and two CH carbons (doublets). Although the peak intensity is not precisely related to the number of carbons, it is possible to obtain some indication of the number of carbons represented by a signal by comparing relative intensities of peaks of the same type. The two CH signals in this spectrum are clearly of different intensities; the more intense one at 30 ppm represents two carbons, and the one at 45 ppm represents one carbon. Also, the peak for the methyl group at 11 ppm is about half the intensity of the peaks for the other two methyl groups. To summarize, the five signals represent: one CH (45 ppm), two CH (30 ppm), two CH$_3$ (22 ppm), two CH$_3$ (18 ppm) and one CH$_3$ (11 ppm).

^{1}H NMR

There are five signals, with one being far enough from the others to allow an integration ratio of 1:8 to be determined. The lower field signal at 1.5 ppm is split into eight lines (very weak outer lines) indicating that it has seven neighbours. This represents a proton on a tertiary carbon (>CH–) coupled to two methyl groups and one other proton. The methyl groups responsible for this splittiing are found as doublets at 0.88 and 0.79 ppm. Another CH, at 0.97 ppm, is coupled to the single methyl doublet at 0.74 ppm. Coupling between the two methine carbons is seen in each of their signals.

Conclusion

The structure that fits this data is 2,3,4-trimethylpentane, (CH$_3$)$_2$CH(CH$_3$)CHCH(CH$_3$)$_2$. It might appear that the methyl groups of the isopropyl groups should be equivalent, giving the molecule only two types of methyl group instead of three. However, the two methyl groups of each isopropyl group are not equivalent. The carbon to which an isopropyl group is attached has three different substituents (H, CH$_3$ and iPr), so the two methyl groups of the isopropyl experience different environments. There is no symmetry operation that equates the two methyl groups. The central carbon has a plane of symmetry through it, which reflects the separate isopropyl groups, but there is no symmetry operation that equates the methyl groups within the same isopropyl group. The two methyl groups of the isopropyl group are diastereotopic; they have different chemical shifts.

Exercise 042 (063)

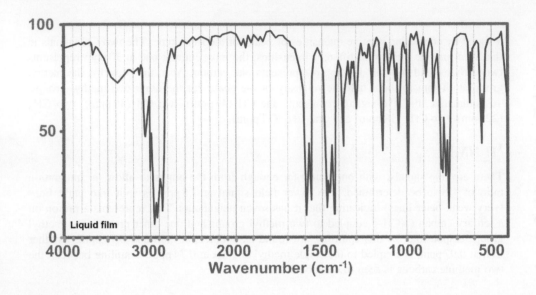

Liquid film

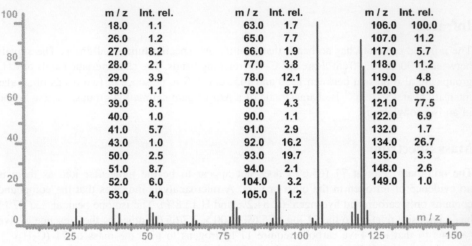

m / z	Int. rel.	m / z	Int. rel.	m / z	Int. rel.
18.0	1.1	63.0	1.7	106.0	100.0
26.0	1.2	65.0	7.7	107.0	11.2
27.0	8.2	66.0	1.9	117.0	5.7
28.0	2.1	77.0	3.8	118.0	11.2
29.0	3.9	78.0	12.1	119.0	4.8
38.0	1.1	79.0	8.7	120.0	90.8
39.0	8.1	80.0	4.3	121.0	77.5
40.0	1.0	90.0	1.1	122.0	6.9
41.0	5.7	91.0	2.9	132.0	1.7
43.0	1.0	92.0	16.2	134.0	26.7
50.0	2.5	93.0	19.7	135.0	3.3
51.0	8.7	94.0	2.1	148.0	2.6
52.0	6.0	104.0	3.2	149.0	1.0
53.0	4.0	105.0	1.2		

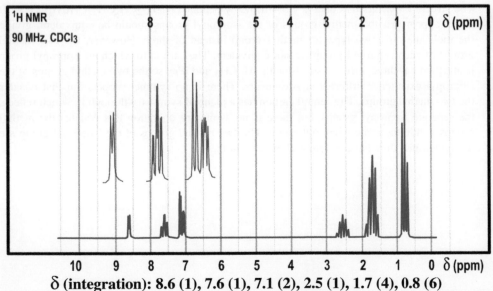

δ (integration): 8.6 (1), 7.6 (1), 7.1 (2), 2.5 (1), 1.7 (4), 0.8 (6)

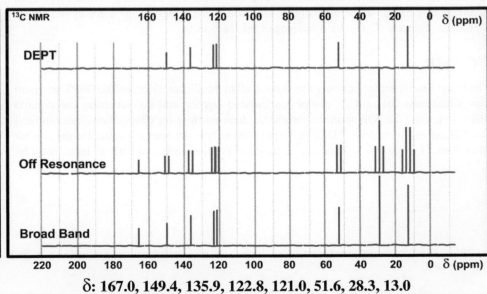

δ: 167.0, 149.4, 135.9, 122.8, 121.0, 51.6, 28.3, 13.0

Exercise 042 (063)

Preliminary Observation

Note that the broad band near 3400 cm^{-1} in the IR spectrum is due to the presence of water in the sample. This sometimes occurs because of moisture in the KBr medium, or from wet cells. A weak broad band in this range should be treated with caution.

Infrared

There are many bands for alkyl groups in this spectrum, but there is no indication of functional groups other than a polar aromatic ring, as indicated by the C–H stretching between 3000 and 3100 cm^{-1}, C=C stretch 1600–1500 cm^{-1} and out-of-plane bending of four adjacent hydrogens near 750 cm^{-1}.

Mass Spectrum

One of the very weak peaks at 148 or 149 may be the molecular ion. These peaks are too weak to provide reliable ratios for isotope analysis. The peak at 134 (149 − 15) arising from the possible loss of a methyl group, suggests that the molecular ion is at 149. The odd mass of the molecular ion would indicate the presence of a nitrogen atom. A mass of 149 gives the formula of $C_{10}H_{15}N$.

^{13}C NMR

There are eight types of carbon present, an aliphatic CH at 51.6 (doublet in OR), a CH_2 at 28.3 (inverted in DEPT, triplet in OR), a CH_3 at 13.0 (quartet in OR) a quaternary carbon at 167 (disappears in DEPT), four CH carbons at 149.4, 135.9, 122.8 and 121.0 (doublets in OR). The presence of five aromatic carbons and a nitrogen suggests the presence of a pyridine ring. There are three positions (α, β, γ) in a pyridine ring relative to the nitrogen, and each has a different chemical shift: the α carbons near 150, the β carbons near 125 and the γ carbon near 135 ppm. This compound has CH carbons at 149.4 (α), 135.9 (γ), 122.8 (β) and 121.0 (β). The other α carbon must be the substituted carbon at 167 ppm.

^{1}H NMR

The six signals have integration ratios of 1:1:2:1:4:6 for a total of 15 protons. The aromatic protons are easily assigned by comparison with the chemical shifts of pyridine: α, 8.59; β, 7.23; γ, 7.62. In this compound, one α proton is seen at 8.55 ppm as a doublet (J $\approx$ 4 Hz, typical α β coupling size in pyridines) with additional weak meta coupling. The γ proton is seen at 7.6 as triplet ($^3J_{ortho}$) with additional meta coupling. The β protons are seen as a poorly resolved triplet and doublet between 7.16 and 6.98 ppm, with a total integration of 2. These chemical shifts and coupling patterns confirm that the pyridine ring has a substituent on the α position. In the aliphatic region, the quintet at 2.5 ppm corresponds to a CH attached to the aromatic ring, and coupled to two isochronous CH_2 groups. The quintet of integration 4 at 1.7 ppm corresponds to two isochronous CH_2 groups coupled to a CH_3 and a CH. The triplet at 0.8 ppm corresponds to two isochronous CH_3 groups each bonded to a CH_2 group.

Summary

The pyridine ring has an alkyl substituent in the α position. The alkyl substituent consists of a CH bearing two ethyl groups. The compound is 2-(1-ethylpropyl)pyridine. In the mass spectrum, the M−1 ion (2.6%) corresponds to the loss of a hydrogen from the benzylic carbon. The alternative cleavage at the same carbon (the preferred loss of an CH_3CH_2 instead of an H) gives the peak at 120 (90%). The peak at 121 (78%) arises from a loss of ethylene (28) by a McLafferty-type rearrangement with the γ hydrogen being taken up by the nitrogen. The base peak (106) most likely arises by the loss of a methyl radical from the 121 ion, which would yield a very stable ion (the 2-vinylpyridinium ion). The ions at 78 and 92 (and their neighbours) represent the pyridine series of fragments similar to the peaks at 77 and 91 observed in the benzene series.

Exercise 043 (092)

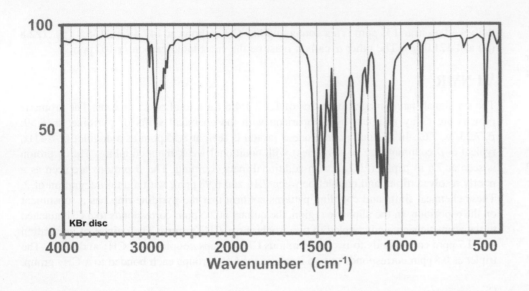

KBr disc

Wavenumber (cm⁻¹)

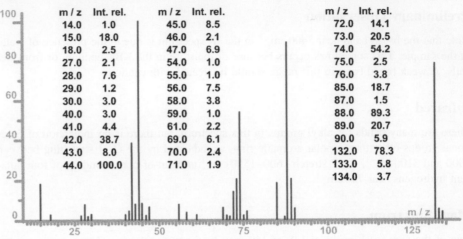

m / z	Int. rel.	m / z	Int. rel.	m / z	Int. rel.
14.0	1.0	45.0	8.5	72.0	14.1
15.0	18.0	46.0	2.1	73.0	20.5
18.0	2.5	47.0	6.9	74.0	54.2
27.0	2.1	54.0	1.0	75.0	2.5
28.0	7.6	55.0	1.0	76.0	3.8
29.0	1.2	56.0	7.5	85.0	18.7
30.0	3.0	58.0	3.8	87.0	1.5
40.0	3.0	59.0	1.0	88.0	89.3
41.0	4.4	61.0	2.2	89.0	20.7
42.0	38.7	69.0	6.1	90.0	5.9
43.0	8.0	70.0	2.4	132.0	78.3
44.0	100.0	71.0	1.9	133.0	5.8
				134.0	3.7

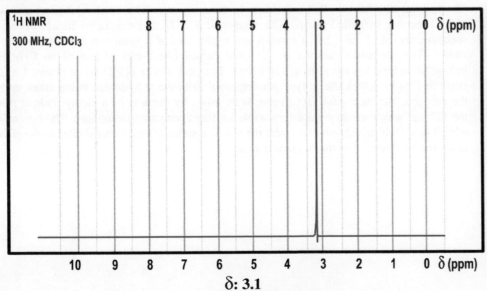

¹H NMR

300 MHz, CDCl₃

δ: 3.1

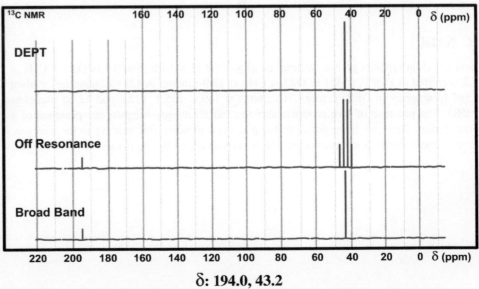

¹³C NMR

DEPT

Off Resonance

Broad Band

δ: 194.0, 43.2

Exercise 043 (092)

Preliminary Observation

The NMR spectra are very simple, indicating that the only hydrogens in the compound are those of methyl groups, and there is only one other kind of carbon. The mass spectrum should provide some critical information.

Mass Spectrum

The molecular ion is at 132 (78.3%) and has isotope peaks at 133 (5.8%) and at 134 (3.7%), for relative intensities of 7.4% and 4.7% of the M+1 and M+2 isotope peaks. Sulfur and silicon could give such a value of M+2 (Cl and Br would give a much stronger M+2 peak). Silicon has isotope ratios for M+1 (5.07%) and M+2 (3.36%), sulfur has M+1 (0.80%) and M+2 (4.44%) and nitrogen has M+1 (0.36%). The size of the observed M+2 peak is larger than expected for a silicon atom, but fits well with the presence of one sulfur atom in the molecule. The major peaks in this mass spectrum are even mass, suggesting that these fragments contain a nitrogen atom. The even molecular mass itself would indicate the presence of an even number of nitrogen atoms. The size of the M+1 and M+2 isotope peaks is compatible with a formula containing one sulfur atom, two nitrogen atoms and five carbons: $M+1 = 0.80 + 2 \times 0.36 + 5 \times 1.1 = 7.02$; and $M+2 = 4.44$. The addition of 12 hydrogens is required to bring the mass to 132, resulting in a formula of $C_5H_{12}N_2S$.

^{13}C NMR

The spectrum has peaks for only two types of carbon, methyl carbons at 43.2 ppm (quartet in OR) and a quaternary carbon at 194.0 ppm (disappears in DEPT). The chemical shift of the quaternary carbon at 194.0 ppm is in the range of an sp^2 carbon, doubly bonded to a sulfur atom in a thioamide or thiourea. The analogous carbon of a urea functional group is more highly shielded (near 150 ppm) than that of a thiourea; this is similar to the case of the strong deshielding of thiones, (260 ppm) compared to ketones (210 ppm on average). The

chemical shift of the methyl groups is in the appropriate range for methyl groups attached to nitrogen.

^1H NMR

The only peak in the spectrum, a singlet at 3.1 ppm, corresponds to four isochronous CH_3 groups on two nitrogen atoms. The chemical shift of the methyl groups is slightly greater than that of the methyl groups on the nitrogen of urea (2.75 ppm). Thus, we have two tertiary nitrogens, $(CH_3)_2N-$, attached to an sp^2 carbon of $>C=S$, giving the complete structure, $(CH_3)_2N(C=S)N(CH_3)_2$, tetramethylthiourea.

Infrared

The C=S bond of the thiocarbonyl group is not very polar and, unlike the analogous carbonyl C=O bond, does not have a distinctive strong absorption band in the infrared spectrum. The C=S stretching is weaker and at lower frequencies, and is thus difficult to identify and use as a diagnostic tool. Thiones such as thiobenzophenone have weak bands between 1224 and 1207 cm^{-1} that are attributed to the C=S stretching vibrations. In thioamides and thioureas, bands between 1500 and 1400, and near 1000 cm^{-1}, may be due to interactions between the C=S and C–N stretching vibrations. The band near 1370 cm^{-1} may be assigned to the methyl group deformations.

Summary

Although there is a distinctive functional group in the molecule, it is one that is difficult to identify in the infrared spectrum. The strong M+2 isotope peak in the mass spectrum gives a good indication of the presence of sulfur, and the base peak at 44 is a significant indicator of nitrogen in the compound ($CH_2 = NHCH_3^+$). The strong peak at 88 (89.3%) results from α cleavage at the $>C=S$ group.

Exercise 044 (086)

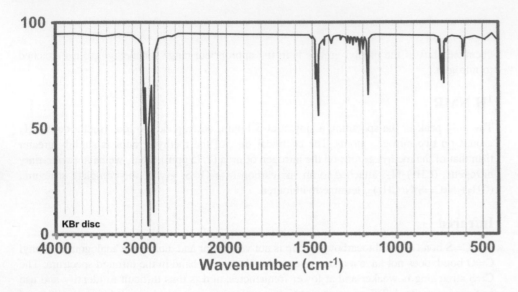

KBr disc

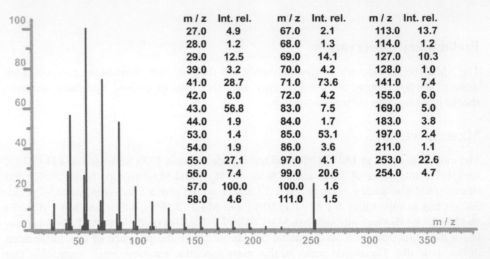

m / z	Int. rel.	m / z	Int. rel.	m / z	Int. rel.
27.0	4.9	67.0	2.1	113.0	13.7
28.0	1.2	68.0	1.3	114.0	1.2
29.0	12.5	69.0	14.1	127.0	10.3
39.0	3.2	70.0	4.2	128.0	1.0
41.0	28.7	71.0	73.6	141.0	7.4
42.0	6.0	72.0	4.2	155.0	6.0
43.0	56.8	83.0	7.5	169.0	5.0
44.0	1.9	84.0	1.7	183.0	3.8
53.0	1.4	85.0	53.1	197.0	2.4
54.0	1.9	86.0	3.6	211.0	1.1
55.0	27.1	97.0	4.1	253.0	22.6
56.0	7.4	99.0	20.6	254.0	4.7
57.0	100.0	100.0	1.6		
58.0	4.6	111.0	1.5		

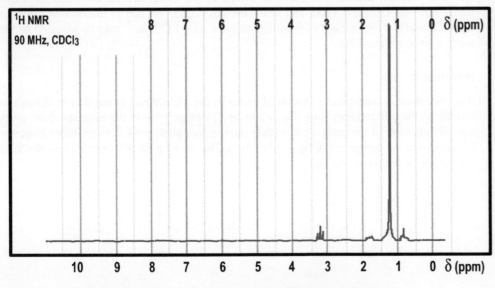

¹H NMR
90 MHz, CDCl₃

δ (integration): 3.2 (2), 1.8 (2), 1.26 (30), 0.9 (3)

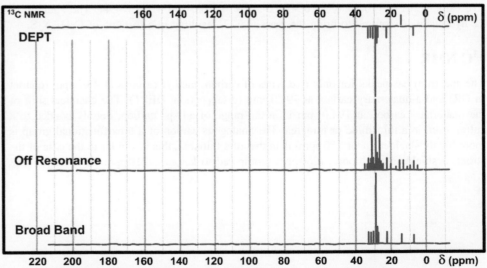

¹³C NMR

DEPT

Off Resonance

Broad Band

δ: 33.6, 32.0, 30.6, 29.7, 29.6, 29.5, 29.4, 28.6, 22.7, 14.1, 7.0

Exercise 044 (086)

Preliminary Observation

The mass spectrum has the classical pattern seen for long straight chain alkyl groups.

Infrared

The spectrum shows no peaks for any group other than a long alkyl chain. The CH stretching bands are at 2800 and $3000\,cm^{-1}$, with intense bands at 2860 and $2910\,cm^{-1}$ due to the CH_2 symmetric and asymmetric stretching respectively. The weaker band at $2960\,cm^{-1}$ is associated with the asymmetric C–H stretching of the CH_3 group. The bands for the CH_2 rocking are found near $720\,cm^{-1}$.

Mass Spectrum

The spectrum shows the very characteristic pattern of linear alkanes, with a base peak at 57 (100%), a fragment at 43 (56.8%) and higher mass ions at intervals of 14 units (CH_2), with regularly decreasing intensity. This is the typical fragmentation pattern of linear alkyl chains. The regularity of the decrease in intensity towards the higher masses indicates the lack of branching in the chain. The mass of 253 is equal to 18 CH_2 units ($18 \times 14 = 252$) $+1$, that is the ion, $C_{18}H_{37}^+$. The ion at 254 is the isotope peak (M+1), with a relative intensity of 20.8% ($100 \times 4.7/22.6$). The contribution from 37 hydrogens is 0.56% (37×0.015), thus the ^{13}C contribution is 20.2 ($20.8 - 0.56$), close to the expected value of 19.8 (18×1.1). The relatively intense peak for the high end ion indicates that the lost substituent is a stable radical. The halides, other than fluorine, give stable radicals. In this case the halide must be iodine, because linear alkyl chlorides and bromides have strong peaks for the cyclic $C_4H_8X^+$ ions, which would be readily detected at 91/93 or 135/137.

^{13}C NMR

Of the 11 different carbon signals, only one is non-inverted in the DEPT spectrum, a CH_3 peak at 14.1 ppm. All the others belong to CH_2 groups in a long aliphatic chain. Many of the carbons near the centre of long chains are nearly isochronous, appearing near 29.7 ppm, resulting in a large peak at this position. A crucial piece of information found in this spectrum is the chemical shift of a particularly highly shielded CH_2. The unusual upfield shift (7.0) indicates that this CH_2 is bonded to an iodine atom (heavy atom effect). The chemical shift of the CH_3 at 14.1 correlates well with the value calculated from single α, β, γ and ε effects (calculated, 14.11 ppm). At the substituted end of the chain, the iodine induces a strong shielding effect on the CH_2 directly bonded to it, and a deshielding effect on the β carbon, the most deshielded carbon in the molecule at 33.6 ppm. The deshielding effect of the iodine diminishes further along the chain.

^{1}H NMR

There are four signals with integration ratios of 2:2:30:3. The signal at 0.9 ppm represents an aliphatic CH_3. The signal of integration 30 at 1.26 ppm corresponds to 15 CH_2 groups, typical of long chains with many nearly isochronous protons. The signal at 1.8 ppm represents a CH_2 β to an iodine atom. The signal at 3.2 ppm corresponds to a CH_2 bonded to an iodine atom. These values correlate well with the values calculated from the tables of α and β effects for methylene groups (CH_2) with an iodine atom as a substituent.

Summary

The mass spectrum has the typical pattern of a linear long-chain alkyl group with peaks 14 units apart, with the intensities increasing towards lower mass, reaching a maximum around four carbon units. Note that the 14-unit difference between the peaks does not mean that they are created by sequential losses of CH_2 units. The fragments are formed by losses of alkene groups such as $CH_2{=}CH_2$. The strong shielding of the terminal carbon (6.95 ppm) by the halogen is a key observation that leads to the identification of the substituent as iodine because of its 'heavy atom effect'.

Exercise 045 (008)

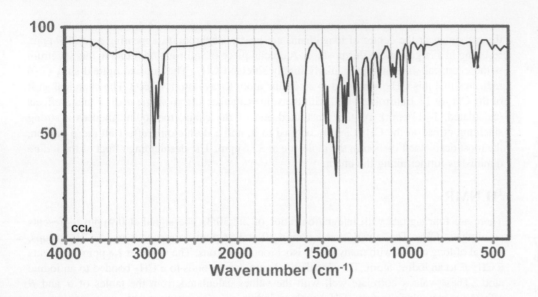

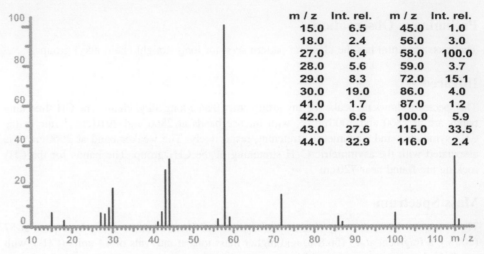

m / z	Int. rel.	m / z	Int. rel.
15.0	6.5	45.0	1.0
18.0	2.4	56.0	3.0
27.0	6.4	58.0	100.0
28.0	5.6	59.0	3.7
29.0	8.3	72.0	15.1
30.0	19.0	86.0	4.0
41.0	1.7	87.0	1.2
42.0	6.6	100.0	5.9
43.0	27.6	115.0	33.5
44.0	32.9	116.0	2.4

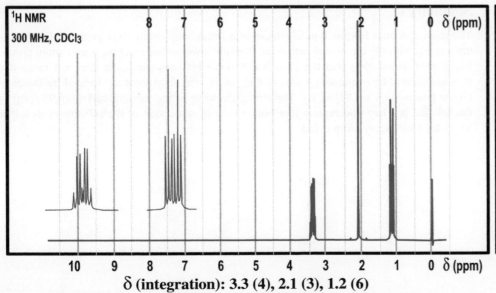

δ (integration): 3.3 (4), 2.1 (3), 1.2 (6)

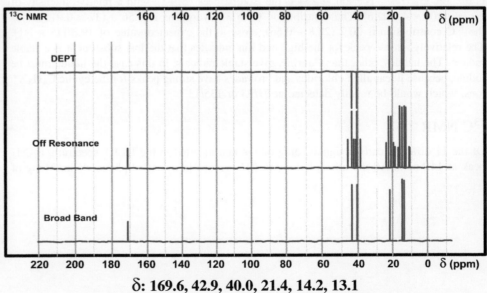

δ: 169.6, 42.9, 40.0, 21.4, 14.2, 13.1

Exercise 045 (008)

Preliminary Observation

Counting the clusters in the mass spectrum indicates that there are eight first-row elements (C, O, N) in the molecule. The strong sharp peak well below $1700\,cm^{-1}$ is significant.

Infrared

The band at $1640\,cm^{-1}$ indicates the presence of a carbonyl group. The relatively low frequency for a carbonyl group indicates that it is an amide, in this case a tertiary amide because no NH bands are seen above $3000\,cm^{-1}$. There is nothing particularly noteworthy that may be discerned from the other bands in the spectrum.

Mass Spectrum

The odd-mass ion at 115 (33.5%) and even-mass fragment ion as a base peak, 58 (100%), suggests that the compound has an odd number of nitrogens. The isotope peak at 116 (2.4%) has a relative intensity 7.2% of the 115 ion ($100 \times 2.4/33.5$) indicating that it contains six or seven carbons.

^{13}C NMR

There are at least six types of carbon in the compound. The DEPT spectrum shows the presence of two CH_2 groups (inverted signals) and one quaternary carbon (signal at 170 ppm disappears). In view of the chemical shift, the quaternary carbon is a carbonyl attached to a heteroatom. The off-resonance spectrum confirms the presence of two CH_2 groups, and also shows three types of CH_3. The methyl group on the carbonyl is more highly deshielded than the other two.

^{1}H NMR

The ^{1}H NMR spectrum shows the presence of five signals. The cluster of peaks at 1.2 consists of two triplets, and the one at 3.3 ppm consists of two quartets. The integration gives a ratio of 4:3:6, for a total of 13 hydrogens. The two quartets at about 3.3 ppm, and the two triplets at about 1.2 ppm, are due to the presence of two ethyl groups, (CH_3CH_2-), that have only slightly different chemical shifts. The singlet at 2.1 ppm has a chemical shift that corresponds to that of a $CH_3C=O$ group.

Summary

From the ^{13}C NMR spectrum, there are six carbons. The infrared spectrum shows the presence of a carbonyl and the ^{1}H NMR integration indicates 13 hydrogens. This gives a partial formula of $C_6H_{13}O$, which adds up to a mass of 101 ($6 \times 12 + 13 + 16$). But the mass spectrum gives a molecular weight of 115, which means that another 14 are needed (one N atom). The molecular formula would then become $C_6H_{13}NO$. The ^{1}H NMR indicates that there are two CH_3CH_2- group and one CH_3CO- group. There is a nitrogen atom, but the IR shows no bands for the N–H stretching, indicating that there is no >N–H or –NH$_2$. This fits with a structure having N attached to two ethyl groups and the acetyl group. The compound is N,N-diethylacetamide. The slight difference that is observed in the chemical shifts of the two ethyl groups reflects hindered rotation about the N–(CO) bond, as is frequently observed in amides. This is due to the resonance donor effect of the nitrogen which increases the double bond character of the C–N bond, restricting rotation, and decreases the double bond character of the C=O bond, lowering the frequency of the C=O vibration in the IR.

Exercise 046 (015)

Elemental analysis: C = 50.85%, H = 8.47% and O = 40.67%

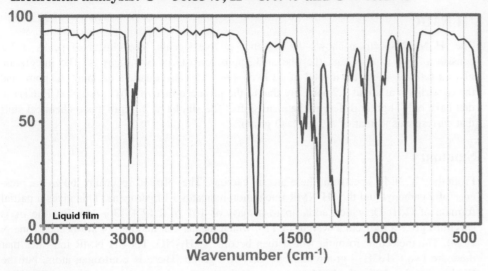

High resolution mass: 91.0395

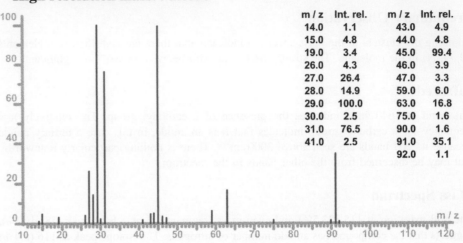

m / z	Int. rel.	m / z	Int. rel.
14.0	1.1	43.0	4.9
15.0	4.8	44.0	4.8
19.0	3.4	45.0	99.4
26.0	4.3	46.0	3.9
27.0	26.4	47.0	3.3
28.0	14.9	59.0	6.0
29.0	100.0	63.0	16.8
30.0	2.5	75.0	1.6
31.0	76.5	90.0	1.6
41.0	2.5	91.0	35.1
		92.0	1.1

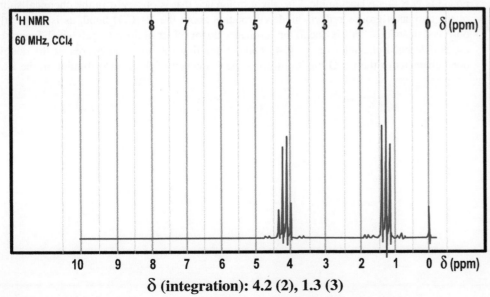

δ (integration): 4.2 (2), 1.3 (3)

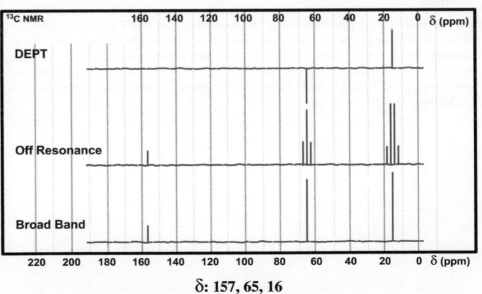

δ: 157, 65, 16

Exercise 046 (015)

Preliminary Observation

The proton NMR has peaks for ethyl groups only, and the IR has evidence of a carbonyl group.

Mass Spectrum

In this mass spectrum there is an accurate mass measurement of 91.0395 reported for the peak at 91. A check of the values in a table of accurate masses reveals that the best fit is for a formula of $C_3H_7O_3$. This cannot be the molecular ion as it has an odd number of hydrogens, and no nitrogen. The fact that the principal peaks (29, 31 and 45) are odd suggests that the peak at 91 is also a fragment ion. The elemental analysis of C = 50.85%, H = 8.47% and O = 40.67% gives atomic ratios of C:H:O = 4.23:8.40:2.54, which can be converted to integer ratios of C:H:O = 5:10:3, for an empirical formula of $C_5H_{10}O_3$ (mass 118). Thus the peak at 91 is due to the loss of 27 from the molecular ion. This arises from a characteristic fragmentation of esters, where the protonated acid is produced by a rearrangement in which two oxygen atoms are protonated. The protonated acid formed in this case is $CH_3CH_2OC(OH)_2{}^+$.

^{13}C NMR

There are signals for only three types of carbon, a CH_3 at 16 ppm (quartet in OR), a CH_2 at 65 ppm (triplet in OR, inverted in DEPT) and a quaternary carbon at 157 ppm (disappears in DEPT). The chemical shift (157 ppm) of the quaternary carbon is close to the value expected for the carbonyl of an ester, more precisely to the carbonyl of a carbonate ester. The chemical shift (65 ppm) of the CH_2 carbon is in the proper range for a CH_2 attached to the oxygen of an ester.

^1H NMR

The spectrum has two sets of coupled protons at 4.2 ppm (two H) and 1.3 ppm (three H). The multiplicities, a quartet and triplet, indicate that we have an ethyl group, CH_3CH_2-, and the chemical shift of the quartet suggest that it is part of an ester group, $CH_3CH_2O-C=O$.

Infrared

The band at 1745 cm^{-1} suggests the presence of an ester function. This is supported by the bands between 1000 and 1250 cm^{-1} which represent the =C–O and C–O stretching vibrations.

Summary

As the peak for the molecular ion is absent, the elemental analysis is used to obtain the molecular formula, $C_5H_{10}O_3$. The ^{13}C NMR and the ^1H spectra give convincing evidence for the presence of an ethyl group bonded to the oxygen, CH_3CH_2O-. The ^{13}C and infrared spectra show that there is a carbonyl of an ester present. If there are two ethoxy groups present in the molecule, then we need only add the C=O to obtain the requisite number of atoms of the molecular formula. This gives $(CH_3CH_2O)_2C=O$ diethyl carbonate.

Exercise 047 (019)

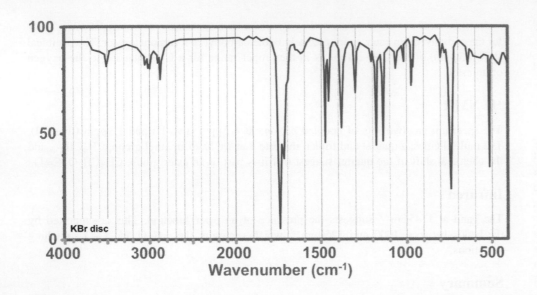

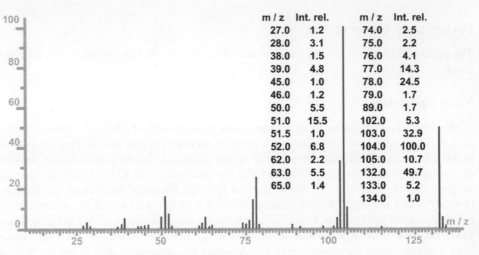

m/z	Int. rel.	m/z	Int. rel.
27.0	1.2	74.0	2.5
28.0	3.1	75.0	2.2
38.0	1.5	76.0	4.1
39.0	4.8	77.0	14.3
45.0	1.0	78.0	24.5
46.0	1.2	79.0	1.7
50.0	5.5	89.0	1.7
51.0	15.5	102.0	5.3
51.5	1.0	103.0	32.9
52.0	6.8	104.0	100.0
62.0	2.2	105.0	10.7
63.0	5.5	132.0	49.7
65.0	1.4	133.0	5.2
		134.0	1.0

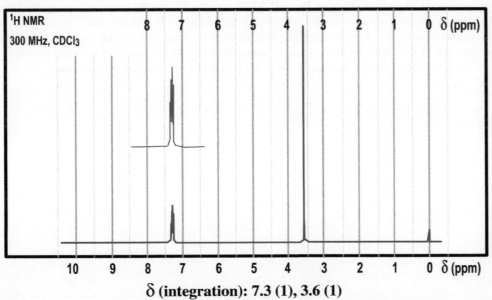

δ (integration): 7.3 (1), 3.6 (1)

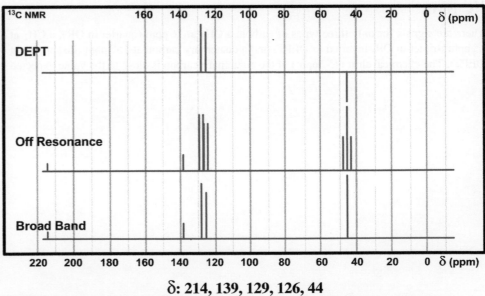

δ: 214, 139, 129, 126, 44

Exercise 047 (019)

Preliminary Observation

The small sharp peak at 3500 cm^{-1} in the IR spectrum is due to the overtone of the carbonyl peak at 1750 cm^{-1}.

Mass Spectrum

The mass spectrum has a peak for the molecular ion at 132 (49.7%) and an M+1 isotope peak at 133 (5.2%). The relative percentage of the isotope peak is 10.5% ($100 \times 5.2/49.7$) indicating that the compound has 9 or 10 carbons. The base peak at 104 (100%) arises from the loss of 28 (C=O). Nine carbons and one oxygen add up to a mass of 124, which requires eight hydrogens to complete the molecular formula of C_9H_8O, for a mass of 132.

Infrared

The frequency of the carbonyl peak at 1750 cm^{-1} is very significant. The mass spectrum reveals that there is only one oxygen atom in the molecule, so the carbonyl group cannot be part of an ester, yet it is at a higher frequency than that of a normal ketone. While an ester might be expected to be found near 1740 cm^{-1}, a normal acyclic ketone or unstrained cyclic ketone absorb near 1715 cm^{-1}. The two factors that most commonly affect the frequencies of ketones, are ring strain and conjugation. Conjugation lowers the frequency of a ketone band by about 30 cm^{-1}, whereas ring strain raises the frequency. The typical frequencies of simple ketones with different ring sizes are: six-membered ring (cyclohexanone), 1715 cm^{-1}; five-membered ring (cyclopentanone), 1745 cm^{-1}; four-membered ring (cyclobutanone), 1780 cm^{-1}. It appears that this compound has a five-membered cyclic ketone. The spectrum has bands typical of an aromatic ring, with a strong band at 740 cm^{-1}, typical of an ortho disubstituted benzene ring. Note the peaks on both sides of 3000 cm^{-1} for the aromatic and aliphatic CH stretching.

^{13}C NMR

The five carbon signals are found in three regions, the carbonyl region, the aromatic region and the aliphatic region. The peak at 214 ppm (disappears in DEPT) belongs to the ketone carbon. The benzene ring is obviously symmetrically substituted, as it has only three carbon peaks, two large CH signals (doublets in OR) at 126 and 129 ppm and a quaternary signal (disappears in DEPT) at 139 ppm. The only peak in the aliphatic region is from the carbon and is a methylene group (inverted in DEPT, doublet in OR).

^{1}H NMR

The signal for the aromatic protons is at the normal position of benzene protons (7.3), indicating that the ring is not substituted with any polar groups. The singlet at 3.6 ppm for the aliphatic protons is at the appropriate shift for being deshielded by a carbonyl group and a benzene ring. The estimate from Shoolery's rule gives a value of 3.78 ppm ($0.23 + 1.70 + 1.85$).

Summary

The spectral evidence indicates the presence of a benzene ring, a cyclic ketone and methylene groups. The ketone is in a five-membered ring which must share two of its carbons with the benzene ring. This gives 2-indanone as the structure. The benzene ring has two ortho methylene groups as substituents, which would have little effect on the chemical shifts of the ring protons, and would give the correct peak for out-of-plane bending in the IR spectrum. The carbonyl peak is at the correct position for an unconjugated ketone in a five-membered ring.

Exercise 048 (032)

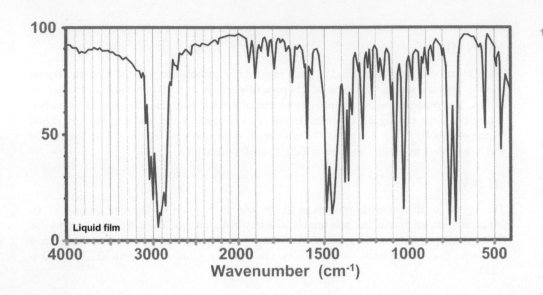

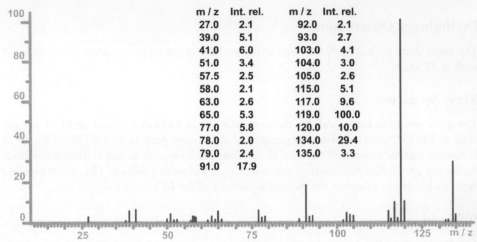

m / z	Int. rel.	m / z	Int. rel.
27.0	2.1	92.0	2.1
39.0	5.1	93.0	2.7
41.0	6.0	103.0	4.1
51.0	3.4	104.0	3.0
57.5	2.5	105.0	2.6
58.0	2.1	115.0	5.1
63.0	2.6	117.0	9.6
65.0	5.3	119.0	100.0
77.0	5.8	120.0	10.0
78.0	2.0	134.0	29.4
79.0	2.4	135.0	3.3
91.0	17.9		

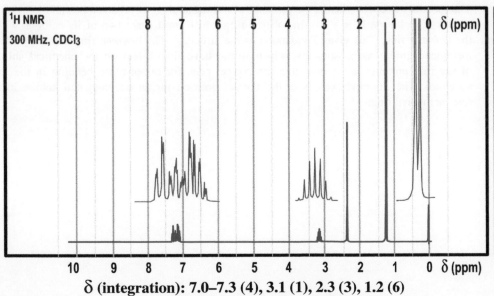

δ (integration): 7.0–7.3 (4), 3.1 (1), 2.3 (3), 1.2 (6)

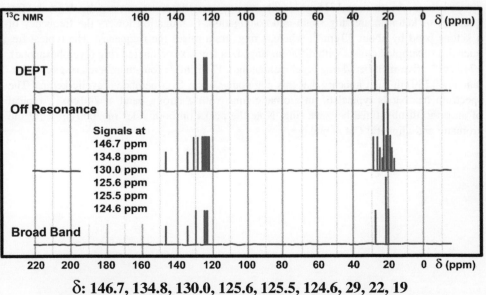

Signals at
146.7 ppm
134.8 ppm
130.0 ppm
125.6 ppm
125.5 ppm
124.6 ppm

δ: 146.7, 134.8, 130.0, 125.6, 125.5, 124.6, 29, 22, 19

Exercise 048 (032)

Preliminary Observation

There is an indication of alkyl and aryl hydrogens from the peaks straddling $3000\,cm^{-1}$. There is no evidence of any other functional group.

Mass Spectrum

The molecular ion at 134 (29.4%) has an M+1 isotope peak, 135 (3.3%), that has a relative intensity of 11% indicating the presence of 10 carbons. The base peak at 119 (100%) is due to the loss of a methyl group from the molecular ion. One of the peaks has a m/z value of 57.5, which obviously belongs to a doubly charged ion (115). These doubly charged ions are sometimes observed in molecules that have a large system of non-bonding or π electrons.

^{13}C NMR

In counting the carbons in the aromatic region, note that the three peaks near 126 ppm are all doublets in the OR spectrum. There are six carbons in the aromatic region, two of them quaternary, suggesting the presence of a non-symmetrically disubstituted benzene ring. There are peaks for three types of aliphatic carbon, one methine carbon and two kinds of methyl group.

^{1}H NMR

The integration ratios are 4:1:3:6 respectively for the signals at 7.0 to 7.3, 3.1, 2.3 and 1.2 ppm. The doublet at 1.22 ppm represents two CH_3 groups attached to a CH carbon. The singlet at 2.33 ppm represents a CH_3 on an aromatic ring. The multiplet (septet) at 3.12 ppm represents the ring substituent CH, which bears the two CH_3 groups. The signals between 7.0 and 7.3 ppm represent the ring CH protons. The $^{3}J_{ortho}$ coupling can be seen to split the signals into a triplet (7.06), a doublet (7.11), a triplet (7.16) and a doublet (7.23). The meta couplings broaden the peaks, but are not clearly resolved.

Infrared

The aromatic ring is indicated by the bands between 3000 and $3100\,cm^{-1}$, at 1500 and $1600\,cm^{-1}$, and at 720 and $760\,cm^{-1}$ (out-of-plane bending of four adjacent hydrogens). The doublet of the same intensity near $1380\,cm^{-1}$ is due to the presence of the *gem*-dimethyl group $(CH_3)_2CH-$.

Summary

The evidence indicates that the benzene ring has two substituents, a methyl group and an isopropyl group. They cannot be in a para position because there are six different aromatic carbon signals in the ^{13}C spectrum, and the ^{1}H NMR data are not compatible with symmetrical substitution. The ^{1}H NMR splitting pattern indicates that the ring is ortho substituted. The appearance of two triplets and two doublets of $^{3}J_{ortho}$ couplings in the ^{1}H NMR spectrum is compatible with ortho substitution, but not meta substitution, which would have one proton with no ortho coupling. The presence of out-of-plane bending bands between 740 and $770\,cm^{-1}$ in the IR spectrum support this assignment. The compound is *o*-cymene, $CH_3C_6H_4CH(CH_3)_2$.

Exercise 049 (034)

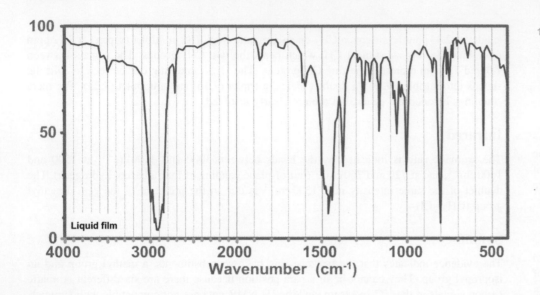

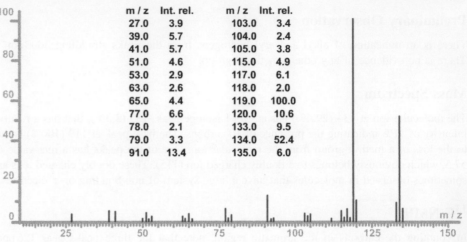

m / z	Int. rel.	m / z	Int. rel.
27.0	3.9	103.0	3.4
39.0	5.7	104.0	2.4
41.0	5.7	105.0	3.8
51.0	4.6	115.0	4.9
53.0	2.9	117.0	6.1
63.0	2.6	118.0	2.0
65.0	4.4	119.0	100.0
77.0	6.6	120.0	10.6
78.0	2.1	133.0	9.5
79.0	3.3	134.0	52.4
91.0	13.4	135.0	6.0

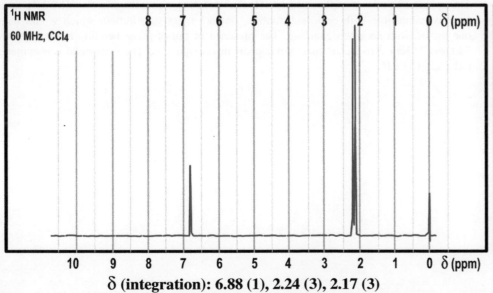

δ (integration): 6.88 (1), 2.24 (3), 2.17 (3)

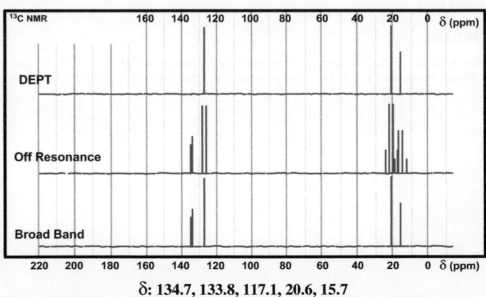

δ: 134.7, 133.8, 117.1, 20.6, 15.7

Exercise 049 (034)

Preliminary Observation

There is evidence in all the regions where bands for an aromatic ring are expected. There is no indication of any other functional group.

Mass Spectrum

The peak at 134 (52.4%) is a candidate for the molecular ion. It is of even mass, and the base peak at 119 (100%) would be the fragment ion from the loss of a methyl group (M−15). The isotope peak at 135 (6%) has a relative intensity of 11.5%, indicating the presence of 10 carbons. Counting along the clusters on the plot leads to the same conclusion; there are 10 carbons in the molecule. The peaks at 77, at 91 (77 + 14) and at 105 (91 + 14) suggest the presence of alkyl fragments on a benzene ring.

^{13}C NMR

There are five signals in the BB spectrum, two in the aliphatic region and three in the aromatic region. Both of the aliphatic carbons are methyl groups (unchanged in DEPT, and quartets in OR). The aromatic carbons consist of two quaternary and one CH carbon. The two types of CH_3 group correlate with the presence of two types of aromatic quaternary carbon (the ring carbons bearing methyl groups). The quartets at 20.6 and 15.7 ppm represent the two different types of CH_3 group attached to the aromatic ring. Each quartet in reality represents two equivalent CH_3 groups. The doublet at 117.1 ppm in the off-resonance spectrum represents two aromatic CH carbons. Finally, the signals at 134.7 and 133.8 ppm represent the two pairs of aromatic carbons bearing CH_3 groups.

^{1}H NMR

The proton NMR has peaks in the same regions as seen in the carbon NMR, with one peak in the aromatic region (6.88 ppm) and two in the aliphatic region (2.24 and 2.17 ppm). The integration gives a ratio of 1:6 for the aromatic to aliphatic protons. The molecule must have an element of symmetry and these values have to be doubled to obtain an even number of hydrogens. We therefore have two equivalent aromatic hydrogens, two pairs of equivalent CH_3 groups, for a total of four methyl groups on the aromatic ring.

Infrared

The bands between 3000 and 3100 cm^{-1} for an aromatic ring (=C–H stretching) are difficult to see because of the large overlapping alkyl CH stretching band. The other typical aromatic bands are visible, the small peaks in the aromatic fingerprint region, ring >C=C< stretching at near 1600 cm^{-1} and 1500 cm^{-1}, and the band at 800 cm^{-1} for the out-of-plane bending of two adjacent hydrogens on the tetrasubstituted benzene ring.

Summary

The mass spectrum gives a peak for the molecular ion at 134 for a molecular formula of $C_{10}H_{14}$. Evidence for an aromatic ring is found in the IR, ^{13}C NMR and ^{1}H NMR spectra. The appearance of only two peaks for four methyl groups in both the ^{13}C and ^{1}H NMR spectra, and only two peaks for the four quaternary carbons indicates that there is a symmetry element in the molecule. There is only one substitution pattern that has four methyl groups on a benzene ring leading to two pairs of equivalent methyl groups, that is 1,2,3,4-tetramethylbenzene. There are two other isomers; the 1,2,3,5- isomer would have three different kinds of methyl group, and the 1,2,4,5-isomer would have only one type of methyl group.

Exercise 050 (033)

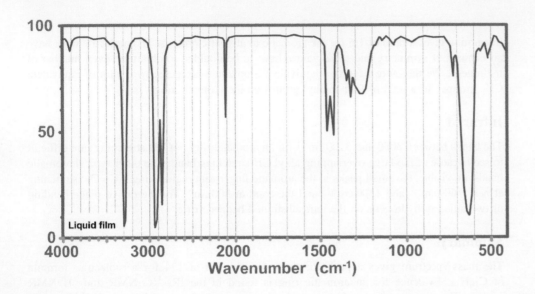

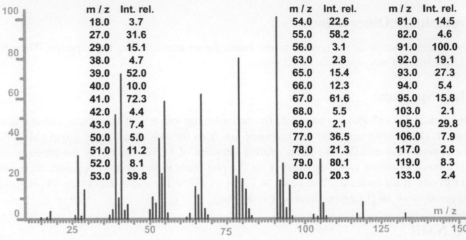

m/z	Int. rel.	m/z	Int. rel.	m/z	Int. rel.
18.0	3.7	54.0	22.6	81.0	14.5
27.0	31.6	55.0	58.2	82.0	4.6
29.0	15.1	56.0	3.1	91.0	100.0
38.0	4.7	63.0	2.8	92.0	19.1
39.0	52.0	65.0	15.4	93.0	27.3
40.0	10.0	66.0	12.3	94.0	5.4
41.0	72.3	67.0	61.6	95.0	15.8
42.0	4.4	68.0	5.5	103.0	2.1
43.0	7.4	69.0	2.1	105.0	29.8
50.0	5.0	77.0	36.5	106.0	7.9
51.0	11.2	78.0	21.3	117.0	2.6
52.0	8.1	79.0	80.1	119.0	8.3
53.0	39.8	80.0	20.3	133.0	2.4

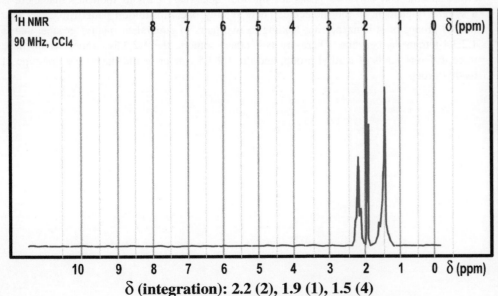

δ (integration): 2.2 (2), 1.9 (1), 1.5 (4)

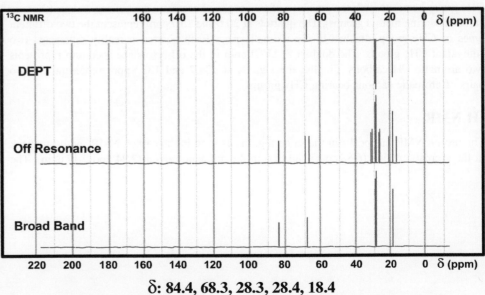

δ: 84.4, 68.3, 28.3, 28.4, 18.4

Exercise 050 (033)

Preliminary Observation

Note that there are two peaks in the signal near 30 ppm on the ^{13}C NMR chart. Remember that hydrogen migrations are common in the mass spectra of compounds with unsaturated alkyl chains. The functional group is most clearly evident in the IR spectrum.

Infrared

The band at 3300 cm^{-1} is that of the H–C stretching of a terminal alkyne (H–C≡C–), and the band at 620 cm^{-1} is that of the bending vibration. The triple bond is confirmed by the sharp peak at 2100 cm^{-1} due to the C≡C stretching vibration. Note the lack of a peak near 1380 cm^{-1} for a methyl group.

Mass Spectrum

In considering the possibility that the ion at 133 is the molecular ion, one sees that the principal fragments are odd (27, 39, 41, 53, 55, 67, 79 and 91), suggesting that the 133 itself is also a fragment ion. The peak at 133 could be M−1, resulting from a loss of hydrogen, which is typical for terminal alkyne groups.

^{13}C NMR

There are peaks for five different carbons, one quaternary carbon at 84.4 ppm (disappears in DEPT), one CH carbon at 68.3 ppm (doublet in OR) and three CH$_2$ carbons at 28.3, 28.4 and 18.4 ppm (triplets in OR, inverted in DEPT). The peaks at 68.3 and 84.4 ppm have the correct chemical shifts for the carbons of a terminal alkyne group (–C≡CH). The triplet at 18.4 ppm corresponds to a CH$_2$ group α to an alkyne. As there are only five carbon signals, and the mass spectrum indicates a molecular mass greater than 133, the molecule must be symmetrical, and in fact have 10 carbons. The observed carbon splittings (three CH$_2$ and one CH) requires seven protons for the five carbons, or 14 for the molecule, C$_{10}$H$_{14}$.

^1H NMR

There are three signals with integration ratios of 2:1:4 at 2.2, 1.9 and 1.5 ppm, respectively. The signal at 1.93 ppm (a finely split triplet by ^4J coupling to the α CH$_2$) corresponds to the hydrogen of a terminal alkyne. The α CH$_2$ coupling to this alkyne ≡CH is seen at 2.18 ppm as a triplet of fine doublets. The broad peak at 1.5 ppm corresponds to the other alkyl CH$_2$ groups. The coupling is not well resolved in the CH$_2$ signals.

Summary

The highest mass ion observed in the spectrum is a 133, but it is not the molecular ion. There is no evidence for the presence of nitrogen, but there is clear evidence in the IR spectrum for the presence of a terminal alkyne group. This group readily loses a hydrogen atom in the mass spectrum, thus explaining the odd mass ion at 133. The actual molecular mass is 134. The ^{13}C NMR shows peaks for three CH$_2$ groups, one CH and one quaternary carbon, for a total of C$_5$H$_7$ and a mass of 67, indicating that the molecule is symmetrical ($2 \times 67 = 134$).

In the proton NMR spectrum, the alkyne hydrogen is seen to be split into a fine triplet as split by a CH$_2$ attached to the triple bond. There are two other methylene groups to be added, giving the substructure, HC≡CCH$_2$CH$_2$CH$_2$–. This leads to one possible structure, HC≡CCH$_2$CH$_2$CH$_2$CH$_2$CH$_2$C≡CH, 1,9-decadiyne.

The mass spectrum of this compound might be somewhat misleading in that the fragments at 119, 105 and 91 (base peak) arising from loss of CH$_3$, CH$_3$CH$_2$ and CH$_3$CH$_2$CH$_2$ radicals might suggest that these alkyl groups are present in the molecule. However, it is well known that hydrogen migration occurs readily in unsaturated alkyl chains. The significant peaks at 39 (C$_3$H$_3$)$^+$ and 53 (C$_4$H$_5$)$^+$ are equivalent to unsaturated three- and four-carbon alkyl chains (different from the saturated chains by −4H because of the triple bond), HC≡CCH$_2$$^+$, 39; (HC≡CCH$_2CH_2$)$^+$, 53.

Exercise 051 (039)

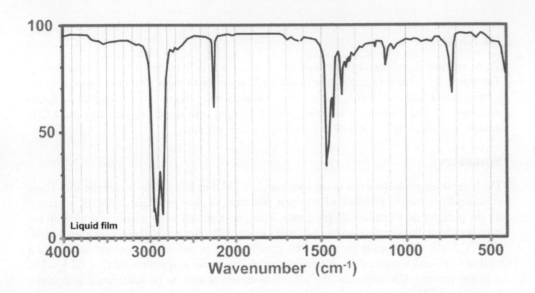

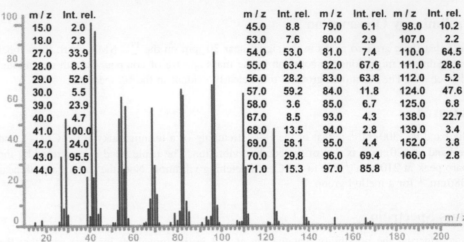

m/z	Int. rel.		m/z	Int. rel.	m/z	Int. rel.	m/z	Int. rel.
15.0	2.0		45.0	8.8	79.0	6.1	98.0	10.2
18.0	2.8		53.0	7.6	80.0	2.9	107.0	2.2
27.0	33.9		54.0	53.0	81.0	7.4	110.0	64.5
28.0	8.3		55.0	63.4	82.0	67.6	111.0	28.6
29.0	52.6		56.0	28.2	83.0	63.8	112.0	5.2
30.0	5.5		57.0	59.2	84.0	11.8	124.0	47.6
39.0	23.9		58.0	3.6	85.0	6.7	125.0	6.8
40.0	4.7		67.0	8.5	93.0	4.3	138.0	22.7
41.0	100.0		68.0	13.5	94.0	2.8	139.0	3.4
42.0	24.0		69.0	58.1	95.0	4.4	152.0	3.8
43.0	95.5		70.0	29.8	96.0	69.4	166.0	2.8
44.0	6.0		71.0	15.3	97.0	85.8		

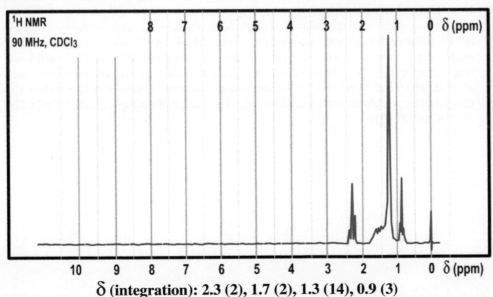

δ (integration): 2.3 (2), 1.7 (2), 1.3 (14), 0.9 (3)

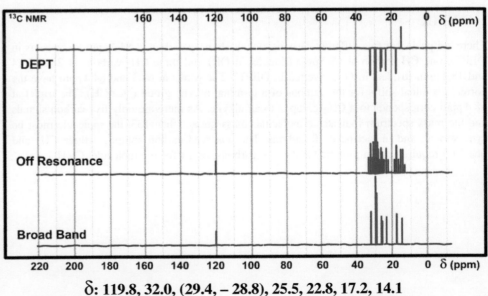

δ: 119.8, 32.0, (29.4, – 28.8), 25.5, 22.8, 17.2, 14.1

Exercise 051 (039)

Preliminary Observations

The mass spectral pattern of a continuum of peaks 14 units apart is typical of a long-chain alkyl group. The peak at $2260 \, cm^{-1}$ is very distinctive.

Infrared

The sharp peak at $2260 \, cm^{-1}$ is in the triple bond stretching region, and might belong to either a nitrile or an alkyne. However, if it were an alkyne, such a strong band could only come from a terminal alkyne (near $2100 \, cm^{-1}$), but there is no peak for $\equiv$C–H stretching near $3300 \, cm^{-1}$. This peak must belong to a nitrile group. The rest of the spectrum is characteristic of a long-chain alkyl group. The CH_2 'rocking' band at $720 \, cm^{-1}$ indicates the presence of a chain of at least four CH_2 groups.

Mass Spectrum

If the peak at 166 is the molecular ion it is difficult to explain the next peak at 152. The loss of a fragment of 14 mass units from a molecular ion would be extremely unusual. This suggests that the 166 ion is a fragment ion. There are several fragment ions of even mass, which indicates that they contain a nitrogen atom. There are also several ions of odd mass which would have lost the nitrogen. The presence of nitrogen is supported by the nitrile band in the IR spectrum. It is likely that the molecular mass is 167, and that the ion at 166 comes from the loss of a hydrogen (M−1). The 152 ion would come from the loss of a methyl group (M−15), and the 138 fragment comes from the loss of HCN (M−29). The molecular formula is $C_{11}H_{21}N$.

^{13}C NMR

The ^{13}C NMR indicates the presence of at least eight carbons, but the signals between 29.42 and 28.81 ppm have intensities clearly stronger than the others, suggesting the presence more than one carbon with identical or very similar chemical shifts. The DEPT indicates the presence of one methyl carbon at 14 ppm, and one quaternary carbon at 120 ppm. The chemical shift of this signal suggests the presence of a C$\equiv$N bond, but C$\equiv$N groups are often missed by ^{13}C NMR as they can be confused with the signals of sp^2 carbons. Fortunately, its characteristic peak is visible in the IR spectrum. The absence of peaks between 70 and 90 ppm confirms the absence of C$\equiv$C. All of the other carbons are methylene (CH_2) carbons.

^{1}H NMR

There are four signals, with integration ratios of 2:2:14:3 for the peaks at 2.3, 1.7, 1.3 and 0.9 ppm respectively. The distorted triplet at 0.89 ppm has the typical shape of a CH_3 in a linear aliphatic chain of at least four carbons. The signal at 2.32 ppm is assigned to a CH_2 group α to a nitrile (C$\equiv$N). The signal at 1.65 ppm is that of the CH_2 situated β to the nitrile group, and the signal between 1.65 and 1.27 ppm represents all the other CH_2 groups of the aliphatic chain.

Summary

The only functional group present appears to be a nitrile. The mass spectrum has a high end peak at 166, but the presence of nitrogen in the molecule indicates that the mass should be odd. Alkyl nitriles are susceptible to proton loss from the α carbon, which suggests that the molecular ion is actually 167 ($C_{11}H_{21}N$). The structure of the alkyl chain is easily established from the NMR spectra. There is only one methyl group, and only one quaternary carbon (C$\equiv$N) in the compound, indicating that it has an unbranched alkyl chain. The compound is $C_{11}H_{21}N$, undecanenitrile.

As well as the general weakness of the molecular ion of long-chain hydrocarbons, in this case there is also the propensity of alkyl nitriles to undergo α cleavage (M−1 peak) and β cleavage by McLafferty type rearrangement (base peak at 41). With this 'false' even-mass molecular ion and the regular decrease of fragments, the spectrum seems at first sight to be that of a linear hydrocarbon. The sequence of even-mass ions at 110, 124 and 138 indicates the presence of nitrogen-containing fragments. The mass spectrum could be quite misleading, but with the C$\equiv$N peak in IR, the 21 protons count in the ^{1}H NMR spectrum, and the C$\equiv$N peak in the ^{13}C spectrum, the presence of nitrogen is definitely established.

Exercise 052 (028)

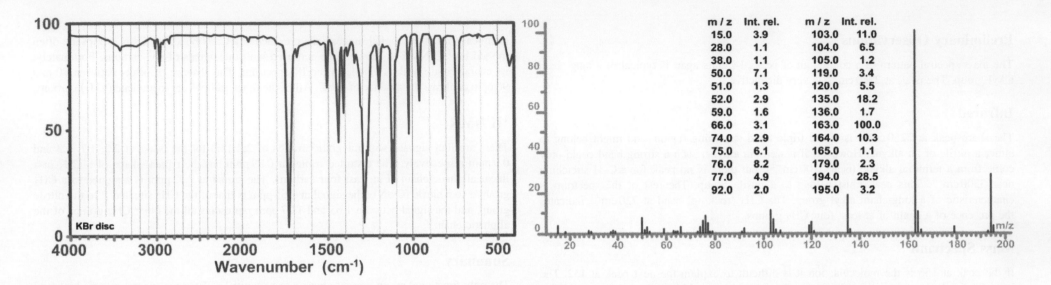

m / z	Int. rel.	m / z	Int. rel.
15.0	3.9	103.0	11.0
28.0	1.1	104.0	6.5
38.0	1.1	105.0	1.2
50.0	7.1	119.0	3.4
51.0	1.3	120.0	5.5
52.0	2.9	135.0	18.2
59.0	1.6	136.0	1.7
66.0	3.1	163.0	100.0
74.0	2.9	164.0	10.3
75.0	6.1	165.0	1.1
76.0	8.2	179.0	2.3
77.0	4.9	194.0	28.5
92.0	2.0	195.0	3.2

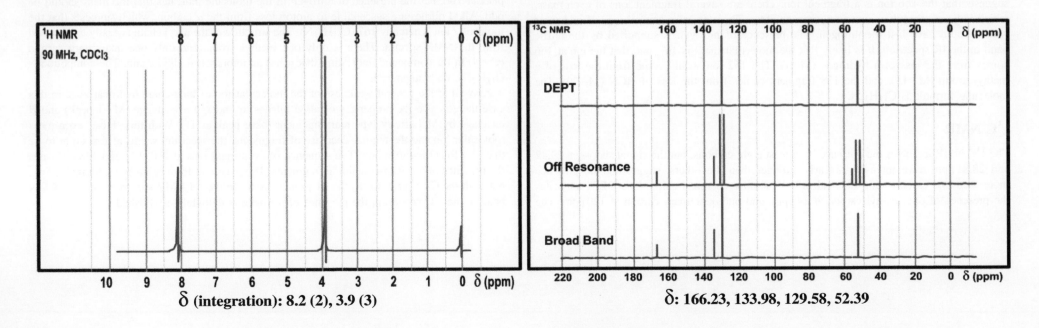

δ (integration): 8.2 (2), 3.9 (3)

δ: 166.23, 133.98, 129.58, 52.39

Exercise 052 (028)

Infrared

The strong sharp peak at $1725\,cm^{-1}$ indicates the presence of a carbonyl group. The position of the carbonyl peak and the strong peak at $1280\,cm^{-1}$ (C–O stretch) indicate that it is the carbonyl of an ester, more precisely a conjugated ester. A simple aliphatic ester absorbs around $1740\,cm^{-1}$, but conjugation at the carbonyl lower the frequency by about $20\,cm^{-1}$. The bands near $1500\,cm^{-1}$ show the presence of an aromatic ring (>C=C< stretching). The aromatic functionality is barely discernible in this spectrum, the characteristic band at $1600\,cm^{-1}$ is absent. This is generally true in the case of para-disubstituted benzene rings having the same substituents. A relatively weak band is found at $810\,cm^{-1}$ for the out-of-plane bending band expected for a para-substituted benzene ring.

^{13}C NMR

There are four peaks in the decoupled spectrum, one in the ester carbonyl region (166), two in the aromatic region (134, 130) and one in the aliphatic region (52). The chemical shift of the methyl carbon (quartet in OR) at 52 ppm indicates that it is the methoxy carbon of an ester. The aromatic carbons consist of two types, a quaternary carbon (disappears in DEPT) at 135 ppm, and a methine carbon (doublet in OR) at 130 ppm. The chemical shift of the methyl group is significant in that it is at a very low field for a methyl group.

Mass Spectrum

The molecular ion is observed at 194 (28.5%). The M+1 isotope peak at 195 (3.2%) has a relative intensity of 11.2% ($100 \times 3.2/28.5$), which indicates that there are 10 carbons present. The base peak at 163 (100%) arises from the loss of a fragment of 31 mass units. This corresponds to the loss of OCH_3, clearly from the ester group (α cleavage at the carbonyl). The peak at 135 corresponds to the loss of $COOCH_3$.

^{1}H NMR

The aromatic protons at 8.2 ppm are downfield from the normal position of benzene protons. This limits the possibilities for the ring substituent to an electron-withdrawing group, in this case a carbonyl group. A directly bonded oxygen atom is a resonance donor to the π system, thus would shield the ring protons moving the signal upfield. The singlet at 3.9 ppm is that of the methyl group. The chemical shift of these protons is very far downfield, suggesting that the carbon is bonded to the oxygen of an ester. The integration ratio of aromatic protons : methyl protons of 2:3 indicates that there are two methyl groups per four protons of a disubstituted benzene ring.

Conclusion

The spectra show that the compound has a symmetrically substituted benzene ring and two methyl ester groups. The fact that all of the unsubstituted ring positions are equivalent dictates that the ring is para substituted. The compound is dimethyl terephthalate, $CH_3OOCC_6H_4COOCH_3$.

Exercise 053 (042)

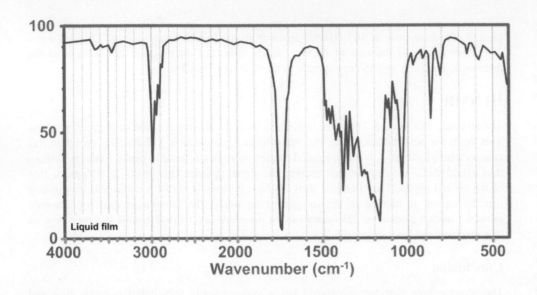

Liquid film

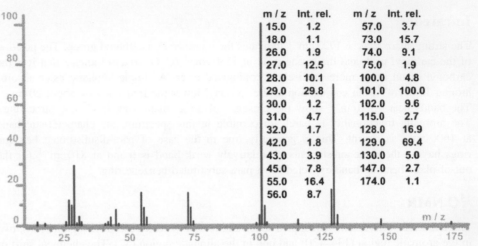

m/z	Int. rel.	m/z	Int. rel.
15.0	1.2	57.0	3.7
18.0	1.1	73.0	15.7
26.0	1.9	74.0	9.1
27.0	12.5	75.0	1.9
28.0	10.1	100.0	4.8
29.0	29.8	101.0	100.0
30.0	1.2	102.0	9.6
31.0	4.7	115.0	2.7
32.0	1.7	128.0	16.9
42.0	1.8	129.0	69.4
43.0	3.9	130.0	5.0
45.0	7.8	147.0	2.7
55.0	16.4	174.0	1.1
56.0	8.7		

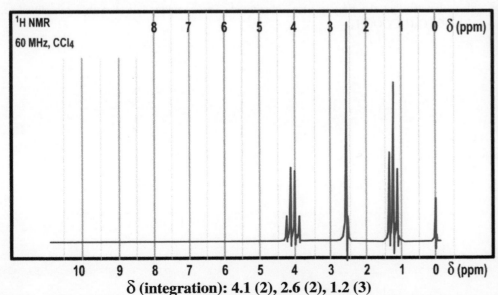

¹H NMR
60 MHz, CCl₄

δ (integration): 4.1 (2), 2.6 (2), 1.2 (3)

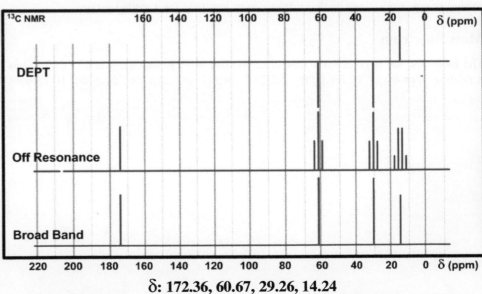

¹³C NMR

DEPT

Off Resonance

Broad Band

δ: 172.36, 60.67, 29.26, 14.24

Exercise 053 (042)

Infrared

The spectrum has a very strong band at $1736\,cm^{-1}$, typical of a carboxylic acid ester. It appears to be a non-conjugated ester. Conjugation on the carbonyl side lowers the frequency, while conjugation with the alkyl oxygen raises the frequency. An ester also has bands for the C–O single bond vibrations which are found in the region of 1000 to $1250\,cm^{-1}$.

^{13}C NMR

There are peaks for four different carbons, a quaternary carbon at 172 ppm (C=O of ester), a methylene carbon at 60 ppm ($-CH_2-O$), a methylene carbon at 29 ppm ($-CH_2-C=O$) and a methyl carbon at 14 ppm (CH_3-C).

^{1}H NMR

There are only three signals, a triplet at 1.2 ppm that is coupled to a quartet at 4.1 ppm, and finally, a singlet at 2.6 ppm. The integration yields ratios of 2:2:3 for a total of seven hydrogens, an odd number which suggests the presence of a symmetry element that creates a molecule with an even number of protons. The mutually coupled quartet and triplet are the protons of an ethyl group bonded to the oxygen of the ester: $CH_3CH_2O(C=O)–R$. The singlet at 2.6 ppm belongs to the CH_2 bonded to the carbonyl of the ester, giving a substructure of $-CH_2(C=O)OCH_2CH_3$. The complete molecule can be made by joining two of these substructures. The two CH_2 groups situated on both sides of the centre of symmetry are isochronous and show no coupling.

Mass Spectrum

There is a peak for the molecular ion ($C_8H_{14}O_4$) at 174. Esters generally have very weak peaks for their molecular ions. The fragment at 147 is due to a rearrangement, with the loss of 27 to give the $CH_3CH_2O(CO)CH_2CH_2C(OH)_2^+$ ion. This type of ion, a protonated acid, is a typical product of the fragmentation of esters. The large peak at 129 corresponds to M−45, from the loss of the OCH_2CH_3 radical by the classical α cleavage at the carbonyl. A further loss of 28 yields the base peak at 101. The peak at 73 is likely due to the $CH_3CH_2O(C=O)$ ion, resulting from α cleavage.

Summary

The very strong sharp peak at $1736\,cm^{-1}$ in the infrared spectrum, and the peaks for the strongly deshielded ethyl group in the ^{1}H and ^{13}C NMR spectra reveal the presence of an ethyl ester. The only other peaks in the NMR spectra are those of a CH_2 group. Two methylene groups are needed between the esters groups because of the molecular mass and the chemical shift of the CH_2. A single CH_2 would be deshielded by two carbonyl groups (about 3.4 ppm), whereas a two-carbon bridge would have CH_2 groups bonded to only one carbonyl (about 2.6 ppm). The ^{1}H NMR integration and the mass spectrum provide more direct evidence for the presence of two mehylene groups between the esters. The compound is diethyl succinate, $CH_3CH_2O(CO)CH_2CH_2(CO)OCH_2CH_3$.

Exercise 054 (036)

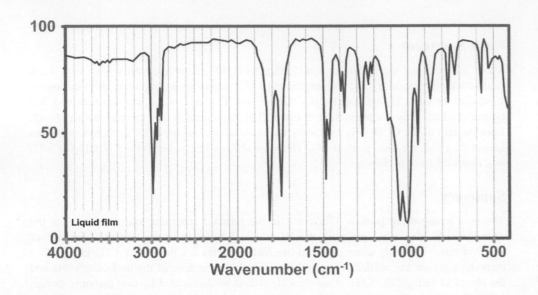

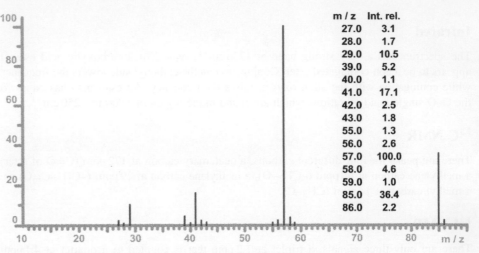

m / z	Int. rel.
27.0	3.1
28.0	1.7
29.0	10.5
39.0	5.2
40.0	1.1
41.0	17.1
42.0	2.5
43.0	1.8
55.0	1.3
56.0	2.6
57.0	100.0
58.0	4.6
59.0	1.0
85.0	36.4
86.0	2.2

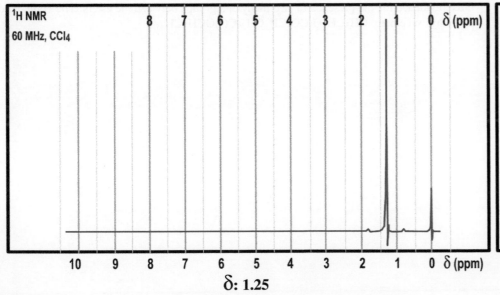

δ: 1.25

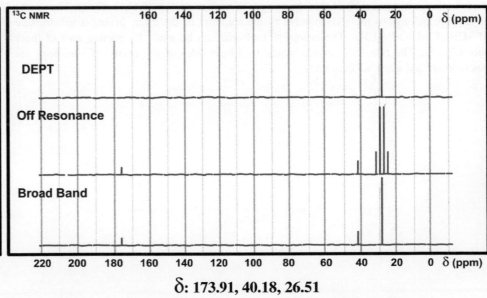

δ: 173.91, 40.18, 26.51

Exercise 054 (036)

Preliminary Observation

This looks like a very simple compound. The ^{1}H NMR spectrum has one single peak and the ^{13}C NMR spectrum shows only three types of carbon. The IR spectrum has a good indication of the functional group.

Infrared

The bands at 1745 and 1810 cm^{-1} are characteristic of the carbonyls (>C=O) of the anhydride group RCO–O–CO–R$'$. The >C=O band at higher frequency being more intense than the lower one, indicates that it is a non-cyclic anhydride. The bands between 1000 and 1200 cm^{-1} represent C–O–C stretching, and confirms the presence of an anhydride group. The doublet at 1380 cm^{-1} of different intensities, with the lower frequency band more intense, is typical of a *tert*-butyl group –C(CH$_3$)$_3$.

Mass Spectrum

The odd-mass ion at 85 is probably not the molecular ion. The isotope peak at 86 has a relative intensity of 6%, which indicates that it contains five carbon atoms. The fragment below this one occurs at 57, resulting from the loss of 28, which is usually carbon monoxide (CO) or ethylene (CH$_2$ = CH$_2$). The ion at 57 is probably due to the alkyl cation C$_4$H$_9{}^+$, in this case the *tert*-butyl ion, considering the ^{1}H NMR spectrum, which shows a single peak for the methyl groups.

^{13}C NMR

The spectrum has signals for three types of carbon, a quaternary carbon at 174 ppm (peak disappears in DEPT), a quaternary carbon at 40 ppm (peak disappears in DEPT) and a CH$_3$ carbon at 27 ppm (quartet in OR, unchanged in DEPT). The quaternary carbon at 174 ppm is in the range of carbonyl groups of esters and anhydrides, and the quaternary carbon at 40 ppm is in the range for a carbon α to a carbonyl group.

^{1}H NMR

The spectrum has only one peak, a singlet at 1.25 ppm. The lack of coupling tells us that the CH$_3$ groups are on a quaternary carbon. It must be a *tert*-butyl group: –C(CH$_3$)$_3$.

Summary

The infrared spectrum gives us the functional group, an anhydride. That assertion is supported by the carbon NMR spectrum with a peak at 174 ppm corresponding to the quaternary carbon of a carbonyl group. The ^{13}C NMR has two other signals, one for another quaternary carbon (α to the carbonyl group at 40.18 ppm) and one from the carbon of a CH$_3$ group (26.51 ppm). These three carbons make a part of the molecule (CH$_3$)$_3$C–C=O which is seen as a peak in the mass spectrum at 85. In view of the functional group and the symmetry of the molecule, the structure must be ((CH$_3$)$_3$CC=O)$_2$O, pivalic anhydride.

Exercise 055 (048)

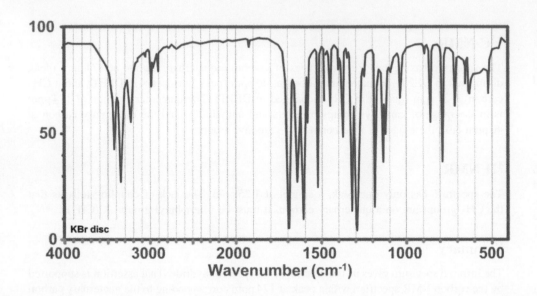

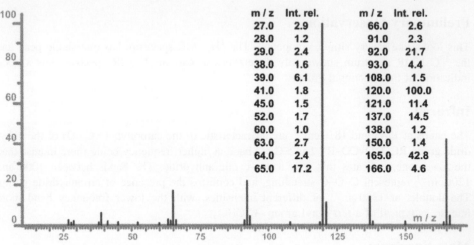

m/z	Int. rel.	m/z	Int. rel.
27.0	2.9	66.0	2.6
28.0	1.2	91.0	2.3
29.0	2.4	92.0	21.7
38.0	1.6	93.0	4.4
39.0	6.1	108.0	1.5
41.0	1.8	120.0	100.0
45.0	1.5	121.0	11.4
52.0	1.7	137.0	14.5
60.0	1.0	138.0	1.2
63.0	2.7	150.0	1.4
64.0	2.4	165.0	42.8
65.0	17.2	166.0	4.6

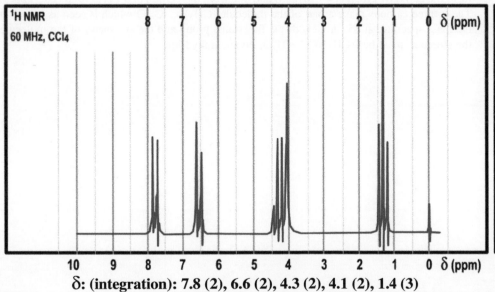

δ: (integration): 7.8 (2), 6.6 (2), 4.3 (2), 4.1 (2), 1.4 (3)

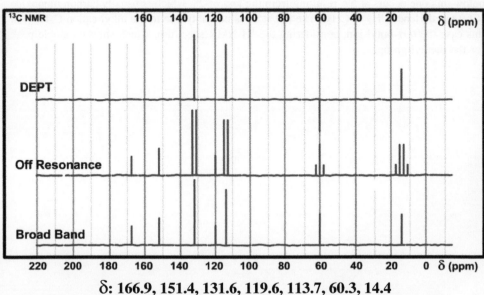

δ: 166.9, 151.4, 131.6, 119.6, 113.7, 60.3, 14.4

Exercise 055 (048)

Preliminary Observation

The strong sharp at $1690\,cm^{-1}$ in the IR spectrum indicates there is a carbonyl in the molecule, probably conjugated. The two peaks near 3400 and $3300\,cm^{-1}$ suggest the presence of an NH_2 group. The nitrogen is also indicated by what appears to be an odd-mass molecular ion. The IR and NMR spectra indicate that there is an aromatic ring in the compound.

Mass Spectrum

The molecule has a molecular mass of 165. The odd mass number agrees with the presence of a nitrogen atom. The peaks at 150 (M−15), 137 (M−28, loss of $CH_2=CH_2$, perhaps by McLafferty rearrangement) and the base peak at 120 (M−45, loss of OCH_2CH_3 by α cleavage) indicate the presence of an ethyl ester. The loss of carbon monoxide (C≡O) from the base peak (120 − 28) gives the fragment at 92, ($C_6H_6N^+$), which is similar to the tropylium ion ($C_7H_7^+$, 91). Loss of HC≡N (27) from the $C_6H_6N^+$ ion gives the peak at 65, and the loss of HC≡CH (26) gives the weak peak at 66.

^{1}H NMR

The five signals have integration ratios of 2:2:2:2:3 (7.8, 6.6, 4.3, 4.1, 1.4 ppm) for a total count of 11 protons. The doublets at 7.8 and 6.6 ppm represent four protons of a para-disubstituted aromatic ring (AA′XX′ system). The two protons at 7.8 ppm are ortho to a deshielding group (carbonyl group), while the two protons at 6.6 ppm are ortho to a shielding group (NH_2). The mutually coupled triplet (1.4 ppm) and quartet (4.1 ppm) are the protons of an ethyl group. The chemical shifts are compatible with the ethoxy chain $CH_3CH_2O–(C=O)–$ of an ester. The final signal is a singlet at 4.3 ppm, corresponding to an NH_2 group.

^{13}C NMR

Three peaks disappear in the DEPT spectrum, one from a carbonyl carbon (167) and two from quaternary aromatic carbons (119, 151). The carbon at 151 ppm is shifted downfield by the inductive withdrawal effect of its substituent NH_2 (ipso $\Delta\delta = +18.2$) and the resonance withdrawal from the ester carbonyl (para $\Delta\delta = +3.9$). The other quaternary ring carbon (119) is slightly deshielded by its substituent ester carbonyl (ipso $\Delta\delta = +2.1$) and shielded by the resonance donation from the NH_2 (para $\Delta\delta = -10.0$). The carbons (114) ortho to the NH_2 group are shielded (ortho $\Delta\delta = -13.4$) by the NH_2 and slightly shielded by the ester (meta $\Delta\delta = -0.5$). The carbons (131) ortho to the ester are slightly deshielded (ortho $\Delta\delta = +1.0$) by the carbonyl and by the ester ($\Delta\delta = +0.8$). The signal at 60 ppm represents the CH_2 of the ethyl group, and at 14 ppm represents the CH_3 group.

Infrared

The band at $1690\,cm^{-1}$ corresponds to the carbonyl of a conjugated ester. The other peaks of the ester are found between 1000 and $1300\,cm^{-1}$. Peaks for the aromatic ring are found at 3000–3100, between 1500 and 1650, and at $780\,cm^{-1}$. The strong bands at 3430 and $3350\,cm^{-1}$ are due to N–H, the asymmetric and symmetric stretching vibrations characteristic of an NH_2 group. The third band at $3230\,cm^{-1}$ is probably due to an overtone of the N–H bending vibration. The presence of the NH_2 group is supported by the very broad band below $750\,cm^{-1}$ representing the >N–H deformations.

Summary

The compound is ethyl p-aminobenzoate, $CH_3CH_2O(CO)C_6H_4NH_2$. The structure is easily solved by recognizing its components: the ethyl ester, the primary amine and the para-substituted benzene ring. The ester carbonyl is at a very low wavenumber ($1690\,cm^{-1}$) because of conjugation to the aromatic ring involving the resonance donation to the carbonyl group by the para NH_2. Note that the location of the peak for an NH_2 in the ^{1}H NMR spectrum is variable, depending upon sample conditions, such as solvent and temperature, and may appear at different frequencies and with different peak widths.

Exercise 056 (030)

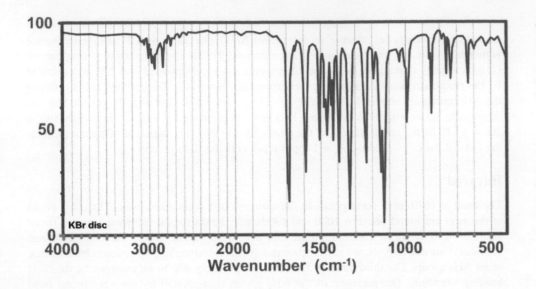

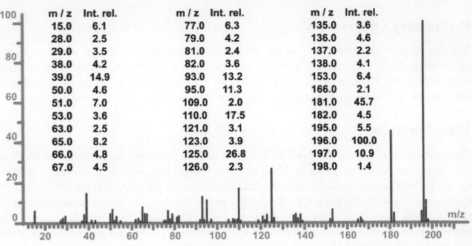

m / z	Int. rel.	m / z	Int. rel.	m / z	Int. rel.
15.0	6.1	77.0	6.3	135.0	3.6
28.0	2.5	79.0	4.2	136.0	4.6
29.0	3.5	81.0	2.4	137.0	2.2
38.0	4.2	82.0	3.6	138.0	4.1
39.0	14.9	93.0	13.2	153.0	6.4
50.0	4.6	95.0	11.3	166.0	2.1
51.0	7.0	109.0	2.0	181.0	45.7
53.0	3.6	110.0	17.5	182.0	4.5
63.0	2.5	121.0	3.1	195.0	5.5
65.0	8.2	123.0	3.9	196.0	100.0
66.0	4.8	125.0	26.8	197.0	10.9
67.0	4.5	126.0	2.3	198.0	1.4

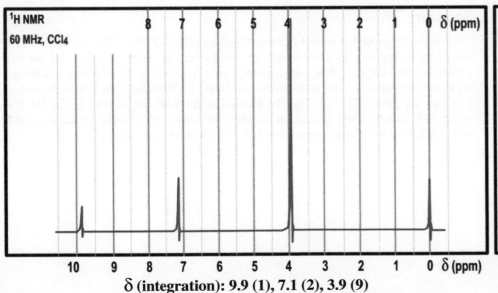

δ (integration): 9.9 (1), 7.1 (2), 3.9 (9)

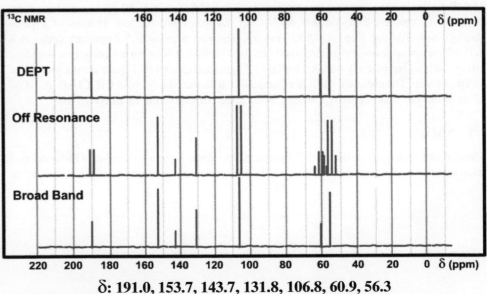

δ: 191.0, 153.7, 143.7, 131.8, 106.8, 60.9, 56.3

Exercise 056 (030)

Preliminary Observations

There are several indications of the presence of an aldehyde group. The IR spectrum has a carbonyl band, the ^{1}H NMR has a peak at 9.8 ppm for the aldehyde proton, and the ^{13}C NMR OR spectrum has a doublet for the CH in the carbonyl region.

Mass Spectrum

The molecular ion at 196 (100%) has a M+1 of 10.9%, suggesting the presence of 10 carbons (10.9/1.1 = 9.9). The isotope peak at M+2 (1.4%) consists of ions containing two ^{13}C atoms or one ^{18}O atom. Ions with two ^{13}C atoms contribute 0.5% and those with ^{18}O contribute 0.9%. Each oxygen atom contributes 0.2% to P+2, suggesting the presence of about four oxygen atoms. However, determining the number of oxygens from the P+2 is unreliable because the small contribution of an oxygen atom is close to the margin of error of the peak intensity. The ion at 181 (45.7%) corresponds to M−15, for the loss of a CH$_3$ group. There is a significant M−1 peak at 195 (5.5%) for the loss of a hydrogen atom, characteristic of aldehydes. The ions at 77 (6.3%) and 65 (8.2%) are typical fragments from a benzene ring.

^{13}C NMR

The spectrum shows seven types of carbon: four aromatic carbons, three of which (153.7, 143.7, 131.8) are quaternary (signals disappear in DEPT), and one (106.8) a CH (doublet in OR); one carbonyl (191.0) of an aldehyde (doublet in OR); and, finally, two (60.9, 56.3) methyl carbons (quartets in OR). The two lower field quaternary carbons (153.7 and 143.7) are strongly deshielded, being bonded to oxygen. The higher relative intensity of the signal at 153.7 ppm suggests that this signal belongs to two equivalent carbons. Of the two peaks responsible for the methoxy carbons (56.3, 60.9), the more intense peak at 56.3 is most likely the one that belongs to the two isochronous methoxy groups. To summarize, there are six aromatic carbon atoms, three of which bear methoxy groups (CH$_3$O), and one which bears an aldehyde carbon.

^{1}H NMR

There are only three peaks in the spectrum, all apparent singlets. The integration ratios of 1:2:9 for the peaks at 9.9, 7.1 and 3.9, give a total of 12 protons. The peak at 7.13 ppm corresponds to two aromatic protons shielded by an ortho-methoxy group (OCH$_3$), and deshielded by an ortho-aldehyde group. The peak at 9.87 ppm is at the characteristic shift of an aldehyde proton. The peak at 3.93 ppm actually consists of two peaks (3.943 and 3.934 ppm) that are not separated by the resolution of this spectrum. This is the typical chemical shift for methoxy groups on an aromatic ring.

Infrared

The typical bands for aromatic rings are seen between 3000 and 3100 cm^{-1} and at 1600 and 1500 cm^{-1}. The bands at 2850 and 1250 cm^{-1} are assigned to the methoxy groups on the aromatic ring. The band at 1690 cm^{-1} corresponds to a carbonyl group conjugated to the aromatic ring. Between 2900 and 2700 cm^{-1} one might expect to see the O=C–H stretching bands characteristic of aldehydes and a band (2820–2850 cm^{-1}) corresponding to CH$_3$O groups attached to an aromatic ring, but it is difficult to distinguish the separate bands.

Summary

The evidence for the aldehyde group is unequivocal. The IR spectrum has a carbonyl peak, the ^{1}H NMR has a peak for the aldehyde proton, and the ^{13}C NMR has a doublet for the aldehyde carbon in the OR spectrum. Evidence for the aromatic ring is present in the IR bands, the ^{1}H NMR, the ^{13}C NMR chemical shifts, and in the mass spectrum. The methoxyl groups are clearly evident from the chemical shifts in the proton and carbon NMR spectra. Having one aldehyde and three methoxy groups, a symmetrical benzene ring can have only two possible substitution patterns, with the two remaining protons either ortho or meta to the aldehyde. If the protons were ortho to the aldehyde, they would be subjected to the electron-withdrawing effect of the aldehyde, and the electron-donating effect of the methoxy group; the chemical shift would be close to the normal benzene position of 7.3 ppm. If the protons were meta to the aldehyde, they would be shielded by the resonance donor effect of three methoxy groups, two ortho and one para, resulting in a strong upfield shift of at least 1 ppm. The observed shift of the protons is 7.13 ppm, thus they are ortho to the aldehyde, and the compound is 3,4,5-trimethoxybenzaldehyde.

Exercise 057 (099)

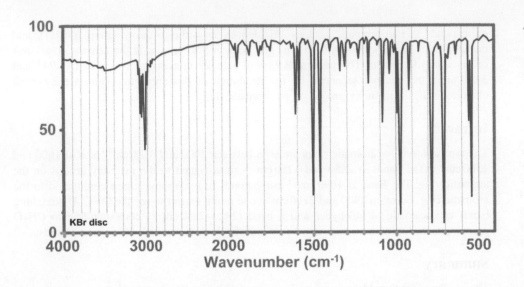

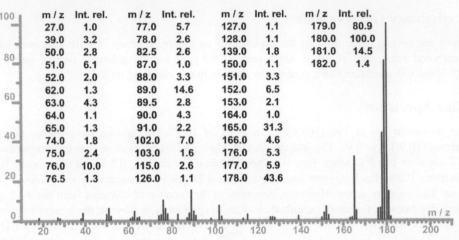

m/z	Int. rel.	m/z	Int. rel.	m/z	Int. rel.	m/z	Int. rel.
27.0	1.0	77.0	5.7	127.0	1.1	179.0	80.9
39.0	3.2	78.0	2.6	128.0	1.1	180.0	100.0
50.0	2.8	82.5	2.6	139.0	1.8	181.0	14.5
51.0	6.1	87.0	1.0	150.0	1.1	182.0	1.4
52.0	2.0	88.0	3.3	151.0	3.3		
62.0	1.3	89.0	14.6	152.0	6.5		
63.0	4.3	89.5	2.8	153.0	2.1		
64.0	1.1	90.0	4.3	164.0	1.0		
65.0	1.3	91.0	2.2	165.0	31.3		
74.0	1.8	102.0	7.0	166.0	4.6		
75.0	2.4	103.0	1.6	176.0	5.3		
76.0	10.0	115.0	2.6	177.0	5.9		
76.5	1.3	126.0	1.1	178.0	43.6		

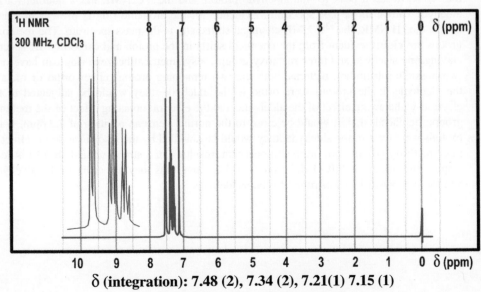

δ (integration): 7.48 (2), 7.34 (2), 7.21(1) 7.15 (1)

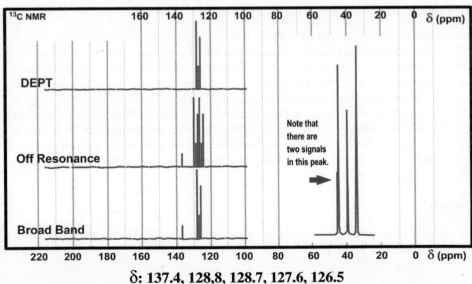

δ: 137.4, 128,8, 128.7, 127.6, 126.5

Exercise 057 (099)

Preliminary Observation

The IR and NMR spectra have peaks for an aromatic ring, but no aliphatic groups or other functional group are evident.

Mass Spectrum

The strong peak for the molecular ion at 180 (100%) suggests a π system with extensive conjugation. The M+1 peak (14.5%) appears to indicate the presence of 13 carbons. A more careful analysis shows that the number of carbons is 14. The peak at 180 contains the isotope peak of the M−1 ion at 179 (80.9%) as well as the molecular ion. Subtracting the contribution of the M−1 isotope peak gives a molecular ion of 87.5%, $(100 − 14 \times 80.9 \times 1.1/100)$. The M+1 peak at 181 also contains a contribution of 0.88% $(1.09 \times 80.9/100)$ from the M+2 isotope peak (two ^{13}C) of the M−1 ion. Recalculation of the relative percentage of the M+1/M peak gives 15.6% $(100 \times (14.5 − 0.88)/87.5)$. This is close to the value expected for 14 carbons, 15.4% (14×1.1). The number of carbons can also be obtained by counting along the ion clusters, which confirms a count of 14 carbons. The molecular formula is therefore, $C_{14}H_{12}$.

^{13}C NMR

Five signals are seen in the spectrum, but the compound contains 14 carbons; obviously the compound has some elements of symmetry. The peak at 137.4 ppm represents a quaternary carbon (disappears in DEPT), and the other four peaks at 128.8, 128.7, 127.6 and 126.5 ppm are all CH carbons (doublets in OR).

^1H NMR

There are peaks in the region for protons on sp^2 carbons only. The splitting pattern of the peaks between 7.2 and 7.5 ppm is that of a monosubstituted benzene ring. The five protons are seen in three signals: a doublet at 7.48 ppm for the ortho protons, a triplet at 7.34 ppm for the meta protons, and a triplet at 7.21 ppm for the para proton. The doublets and triplet splitting refers to the $^3J_{ortho}$ couplings. Each of these patterns would also show finer $^4J_{meta}$ coupling. The singlet at 7.2 ppm represents two protons isolated from those of the benzene ring.

Infrared

The aromatic ring is evident from the bands above 3000, the overtone-combination between 2000 and 1600 cm^{-1}, the ring stretching at 1600 and 1500 cm^{-1}, and the out-of-plane bending bands at 695 and 770 cm^{-1} for the monosubstituted benzene ring.

Conclusion

There is clear evidence for a monosubstituted benzene ring in the IR and NMR spectra. The compound has a formula of $C_{14}H_{12}$, which requires two phenyl groups and C_2H_2. The two additional protons are seen in the ^1H NMR as a singlet at 7.1 ppm. This chemical shift is in the correct region for the protons on a double bond bearing a pair of phenyl groups. The phenyl groups must be on different carbons because the molecule has no CH_2 groups. The structure is $C_6H_5CH=CHC_6H_5$, stilbene, which has two possible configurations, *cis* (Z) and *trans* (E). In most cases the double bond configuration can be determined from the size of the coupling between the two protons $(^3J_{trans} \approx 17\,Hz, ^3J_{cis} \approx 11\,Hz)$. In this case, no coupling is observed because the two protons are isochronous. Estimation of the chemical shifts of the isomers from a table of substituent factors gives 7.09 ppm for the *trans* isomer, and 6.66 ppm for the *cis* isomer. Shift $= 5.25 + \Delta\delta_{gem} + \Delta\delta_{cis} + \Delta\delta_{trans}$, with the factors for a phenyl group being $\Delta\delta_{gem} = +1.38$, $\Delta\delta_{cis} = +0.36$ and $\Delta\delta_{trans} = -0.07$. The observed chemical shift of 7.1 ppm indicates that the compound is the *trans* isomer. The position of the out-of-plane bending band in the IR spectrum provides another way to determine the configuration of olefins. A *trans*-disubstituted alkene has a band near 970 cm^{-1}, whereas a *cis* isomer has a band near 700 cm^{-1}. The IR spectrum has a strong sharp band at 970 cm^{-1}, which indicates that the double bond is *trans* substituted. The compound is *trans*-stilbene.

In the mass spectrum there is a significant peak at 165 for M−15. This unexpected loss of a methyl group is a dramatic illustration of the need for caution in interpreting the loss of 15 in a mass spectrum. When there is a stable molecular ion, a deep-seated rearrangement may occur, resulting in the loss of a methyl group although none is present in the original molecule. However, the presence of a CH_3 is easily detected in the ^{13}C NMR. The loss of 15 is a characteristic of stilbenes, substituted or not. The loss is associated with the rearrangement involving the ortho protons, resulting in a stable ion (fluorene ring less one H).

Exercise 058 (096)

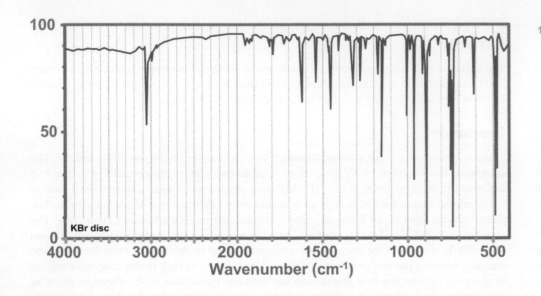

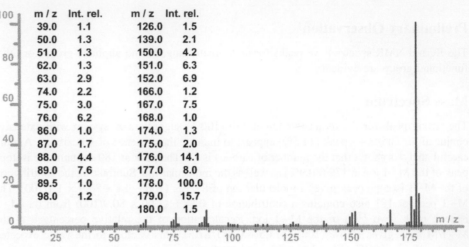

m / z	Int. rel.	m / z	Int. rel.
39.0	1.1	126.0	1.5
50.0	1.3	139.0	2.1
51.0	1.3	150.0	4.2
62.0	1.3	151.0	6.3
63.0	2.9	152.0	6.9
74.0	2.2	166.0	1.2
75.0	3.0	167.0	7.5
76.0	6.2	168.0	1.0
86.0	1.0	174.0	1.3
87.0	1.7	175.0	2.0
88.0	4.4	176.0	14.1
89.0	7.6	177.0	8.0
89.5	1.0	178.0	100.0
98.0	1.2	179.0	15.7
		180.0	1.5

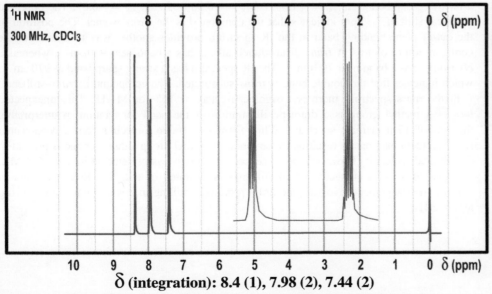

δ (integration): 8.4 (1), 7.98 (2), 7.44 (2)

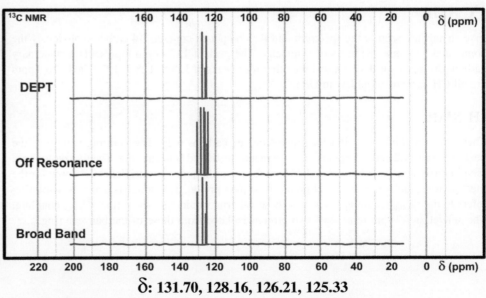

δ: 131.70, 128.16, 126.21, 125.33

Exercise 058 (096)

Preliminary Observations

The infrared spectrum shows clear evidence of aromatic rings, and no evidence of other functional groups or alkyl groups.

Mass Spectrum

The base peak in the spectrum is that of the molecular ion at 178 (100%). The M+1 ion at 179 has a relative intensity of 15.7%, which is close to that expected for 14 carbons (14 × 1.1 = 15.4%). The strongly conjugated aromatic system is so stable that it is able to support a double charge. The peaks at 89 and 89.5 represent the doubly charged molecular ion and isotope ion. The molecular ion has very few fragmentations. The small peaks at 177 and 176 arise from the loss of a hydrogen radical and a hydrogen molecule respectively. The molecular formula is $C_{14}H_{10}$, (14 × 12 + 10).

^{13}C NMR

There are peaks for only four types of carbon in the spectrum, all aromatic, one quaternary at 131.7 ppm and three CH carbons at 128.2, 126.2 and 125.3 ppm. Since the compound has 14 carbons, it must have some elements of symmetry.

^1H NMR

The spectrum has a singlet at 8.4 ppm, and a set of symmetrical multiplets typical of an AA′XX′ system. The singlet implies that it is on a ring that has no ortho or meta protons that would split the signal. The AA′XX′ system has two isochronous protons (AA′) that are coupled to two other isochronous protons (XX′), but not equally coupled to the each member of the pair. In such a system, $J_{AX} = J_{A'X'}$ and $J_{AX'} = J_{A'X}$, but $J_{AA'} \neq J_{XX}$. This is the type of coupling system that one finds in a symmetrical ortho-substituted ring such as o-dichlorobenzene. Although the two signals of the AA′XX′ system have the same integration, the lower field signal appears less intense because it is broadened by very fine coupling to the proton at 8.4 ppm.

Infrared

The spectrum has bands in all of the regions where an aromatic ring is expected to absorb. The CH stretching absorption is found at 3080 cm^{-1}, the weak overtone-combination between 2000 and 1600 cm^{-1}, the ring stretching at 1620 and 1540 cm^{-1} and the out-of-plane bending between 740 and 890 cm^{-1}. The band at 890 cm^{-1} represents an isolated aromatic H between two substituents, and the band at 740 cm^{-1} represents four adjacent aromatic hydrogens. There is no indication of any other functional group in the molecule.

Summary

The aromatic system has a symmetrically substituted benzene ring with four adjacent protons (ortho AA′XX′ coupling pattern in the ^1H NMR spectrum, out-of-plane band in IR). The two identical ortho substituents can only be the carbons of another ring. Fourteen carbons allow for two benzene rings and two carbons to connect them. This gives an aromatic condensed ring system such as anthracene or phenanthrene. Anthracene has the requisite pair of benzene rings with four adjacent protons, and a central ring with two isolated para protons. Phenanthrene also has a pair of benzene rings with four adjacent protons, but they are non-equivalent, and not an AA′XX′ system. Also, the phenanthrene ring has one plane of symmetry passing through the central ring which divides the molecule into two halves, giving seven different carbons. The anthracene ring has two planes of symmetry, giving four different carbons. The compound in this exercise is anthracene. The signal at 131.7 ppm corresponds to four isochronous quaternary carbons. The signal at 128.2 ppm corresponds to four isochronous CH carbons. The signal at 126.2 ppm corresponds to two isochronous CH carbons. Finally, the signal at 125.3 ppm corresponds to four isochronous CH carbons.

There are two-fold rotational axes in the plane of the aromatic system, one passing through the two CH carbons of the central ring, (the vertical axis), and one passing through the centres of the three rings (the horizontal axis). The two sets of four equivalent CH groups exchange positions by rotation on both axes. The quaternary carbons are also equated by rotation on both axes. One may do the same analysis by using mirror planes perpendicular to the aromatic plane.

Exercise 059 (097)

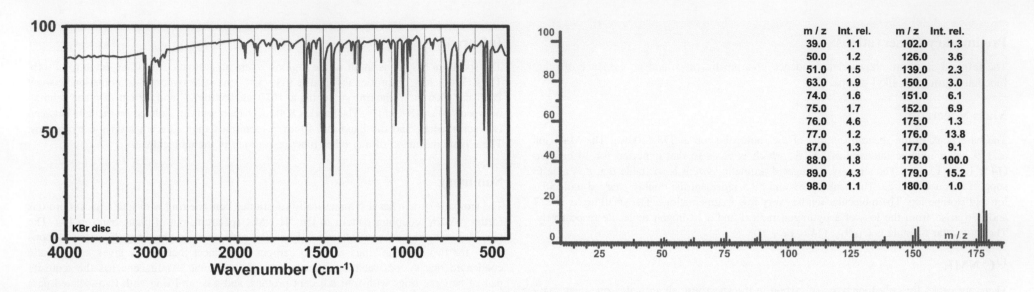

m / z	Int. rel.	m / z	Int. rel.
39.0	1.1	102.0	1.3
50.0	1.2	126.0	3.6
51.0	1.5	139.0	2.3
63.0	1.9	150.0	3.0
74.0	1.6	151.0	6.1
75.0	1.7	152.0	6.9
76.0	4.6	175.0	1.3
77.0	1.2	176.0	13.8
87.0	1.3	177.0	9.1
88.0	1.8	178.0	100.0
89.0	4.3	179.0	15.2
98.0	1.1	180.0	1.0

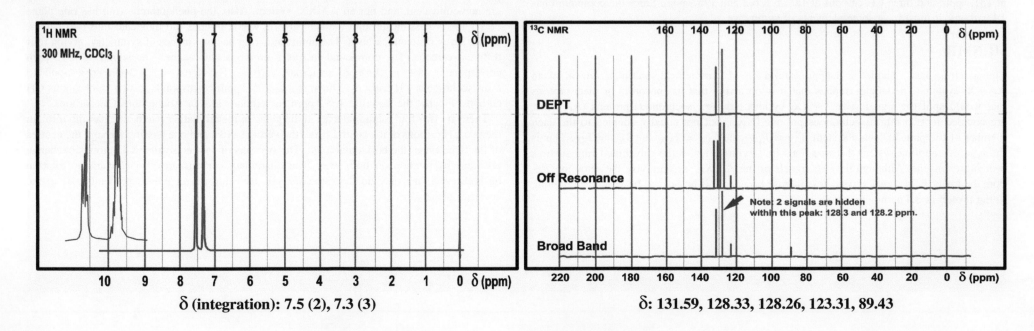

δ (integration): 7.5 (2), 7.3 (3)

Note: 2 signals are hidden within this peak: 128.3 and 128.2 ppm.

δ: 131.59, 128.33, 128.26, 123.31, 89.43

Exercise 059 (097)

Preliminary Observation

The infrared spectrum has evidence of an aromatic ring in all of the typical frequency ranges. The ^{1}H NMR spectrum has peaks for the aromatic protons, but no others.

Mass Spectrum

The compound has a very stable molecular ion at 178 (100%). The M+1 isotope peak has a relative intensity of 15.2%, which corresponds to 14 carbons, thus requiring 10 H to complete the formula ($178 - 14 \times 12 = 10$ H), for a molecular formula, $C_{14}H_{10}$. The molecule has 10 centres of unsaturation.

Infrared

The infrared spectrum has all the typical bands of an aromatic ring. Note the position of the strong out-of-plane bending bands. There is no indication of any other functional group, and no band between 3000 and 2900 cm^{-1} that would indicate the presence of aliphatic groups.

^{1}H NMR

The two sets of peaks in the aromatic region have an integration ratio of 2 (7.5 ppm) : 3 (7.3 ppm). This pattern is typical of a monosubstituted benzene ring in which the two ortho protons are slightly more deshielded than the other three protons.

^{13}C NMR

There are four signals in the aromatic region between 123 and 132 ppm. The signal at 123.3 ppm corresponds to the substituted carbon of a phenyl ring (quaternary, signal disappears in DEPT spectrum). The other three signals in the aromatic region (128.2, 128.3, 131.6) correspond to the para carbon and the pairs of ortho and meta carbons (all doublets in the OR spectrum). The signal at 89.4 ppm is in the range for the carbons of an alkyne group. Note that a triple-bonded carbon is a shielding substituent at the *ipso* position, which explains why the chemical shift of this carbon is at a slightly higher field (123.3 ppm) than the others.

Conclusion

The only structural features evident from the spectra are the phenyl rings and alkyne carbons. An alkyne bond with a phenyl group on each end gives the required molecular formula, $C_{14}H_{10}$. The compound is diphenylacetylene, $C_6H_5C{\equiv}CC_6H_5$. The compound has a highly conjugated stable molecular ion with small peaks for M−1 (9%) and M−2 (14%) ions. The ions at 76 and 77 that are typically found for phenyl compounds are present but very weak. Note the ion at 152 (P−26) which results from the loss of acetylene from a retro Diels–Alder at a benzene ring.

The IR bands at 695 and 760 cm^{-1} are due to the out-of-plane bending of the five adjacent hydrogens. No alkyne C≡C stretching band near 2100 cm^{-1} is observed because the alkyne is symmetrically substituted. The stretching does not result in a dipole moment change, therefore no absorption occurs.

Exercise 060 (098)

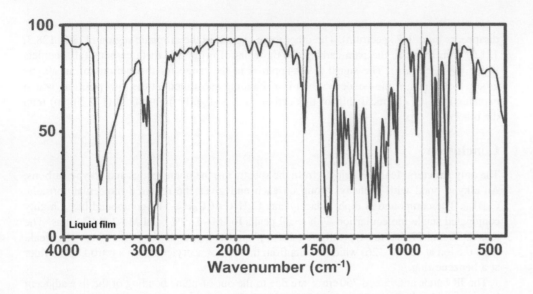

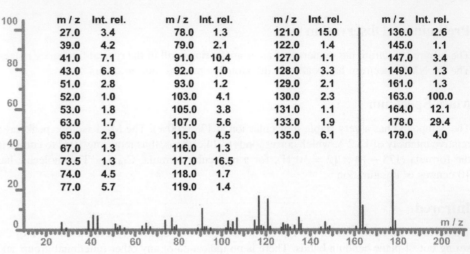

m/z	Int. rel.	m/z	Int. rel.	m/z	Int. rel.	m/z	Int. rel.
27.0	3.4	78.0	1.3	121.0	15.0	136.0	2.6
39.0	4.2	79.0	2.1	122.0	1.4	145.0	1.1
41.0	7.1	91.0	10.4	127.0	1.1	147.0	3.4
43.0	6.8	92.0	1.0	128.0	3.3	149.0	1.3
51.0	2.8	93.0	1.2	129.0	2.1	161.0	1.3
52.0	1.0	103.0	4.1	130.0	2.3	163.0	100.0
53.0	1.8	105.0	3.8	131.0	1.1	164.0	12.1
63.0	1.7	107.0	5.6	133.0	1.9	178.0	29.4
65.0	2.9	115.0	4.5	135.0	6.1	179.0	4.0
67.0	1.3	116.0	1.3				
73.5	1.3	117.0	16.5				
74.0	4.5	118.0	1.7				
77.0	5.7	119.0	1.4				

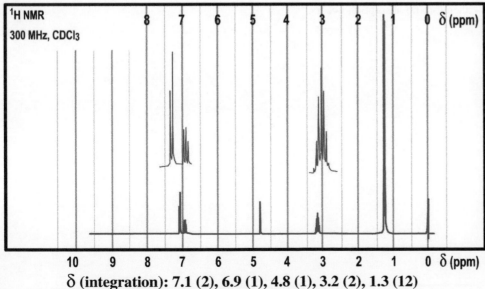

δ (integration): 7.1 (2), 6.9 (1), 4.8 (1), 3.2 (2), 1.3 (12)

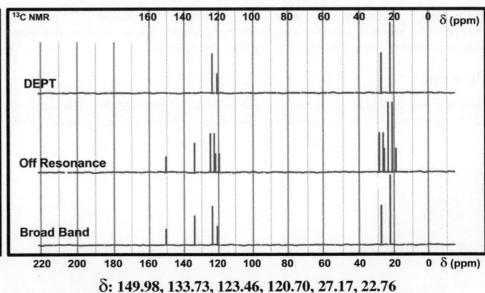

δ: 149.98, 133.73, 123.46, 120.70, 27.17, 22.76

Exercise 060 (098)

Preliminary Observation

A preliminary scan of the IR spectrum indicates that the compound has aliphatic groups, an aromatic ring and a hydroxy group.

^{13}C NMR

The spectrum shows six types of carbon, two aliphatic and four aromatic. Two of the aromatic carbons are quaternary (signals disappear in DEPT), representing the substituted ring carbons. The chemical shift of one (150) indicates that it is bonded to an oxygen, while the chemical shift of the other (134) indicates that it has a carbon as its substituent. The other peaks in the aromatic region (124, 121) represent the unsubstituted carbons (doublets in OR). The peak at 27 ppm represents aliphatic CH carbons (OR doublet), and the peak at 23 ppm represents CH_3 groups (OR quartet). Although there is no integration in the ^{13}C NMR spectrum, the comparison of peak size from atoms of the same type may give a tentative indication of peaks that represent more than one carbon.

Mass Spectrum

The spectrum has a relatively stable molecular ion with a peak at 178 (29.4%). The base peak, 163(100%), results from the loss of a methyl group. Its isotope peak at 164 (12.1%) indicates that the ion contains 11 carbons, which gives a total of 12 carbons when the methyl group is added. One should be cautious in using the P+1 of a fragment ion as an isotope peak. It may be an ion from another fragmentation path. In this case the M:M+1 ratio gives the same result; the relative intensity of 179 (4.0%) is 13.6%, for 12 carbons. The saturated molecule with 12 carbons and one oxygen would have 26 hydrogens, and a mass of 186 ($12 C + 26 H + 1 O$). This molecular mass (178) is eight less, indicating that the molecule has four sites of unsaturation – exactly that required for one benzene ring. The molecular formula is $C_{12}H_{18}O$.

^{1}H NMR

The doublet at 7.1 ppm represents two protons split by one ortho proton. The triplet at 6.9 ppm represents the ortho proton split by these two neighbours. The signal at 4.8 ppm is that of OH proton of the phenol group. The septet at 3.2 ppm and doublet at 1.3 ppm together represent an isopropyl group.

Infrared

The very broad band between 3700 and 3300 cm^{-1} is due to the O–H stretching of a phenol. The aromatic C–H stretching bands are found at 3050 cm^{-1} and the large bands just below 3000 cm^{-1} represent aliphatic CH stretching. The ring-stretching vibrations are seen at 1600 and 1480 cm^{-1}. The doublet of equal intensities at 1380 cm^{-1} is associated with a *gem*-dimethyl group, $>C(CH_3)_2$, possibly an isopropyl group $–CH(CH_3)_2$.

Summary

The molecule has molecular formula, $C_{12}H_{18}O$ consisting of a benzene ring substituted with two isopropyl groups and a hydroxy group. The isochronous signals of the isopropyl groups show that the molecule is symmetrical. This requires the phenol to be symmetrically substituted, either 2,6-disubstituted (ortho) or 3,5-disubstituted (meta) to the phenol. In the ^{1}H NMR spectrum, the para proton will be half the intensity of the other two equivalent ring protons. The triplet at 6.9 ppm (integration 1) is the peak of the para proton, and the doublet at 7.1 ppm (integration 2) is the peak of the other two ring protons. The coupling of the para proton shows that it has two ortho neighbours, therefore the compound is 2,6-diisopropylphenol. The para proton (6.9 ppm) is at a higher field than the meta protons (7.1 ppm) because of the greater effect of the OH on the para position. In the ^{13}C NMR spectrum, the less intense peak of the para CH is also at a higher field than the peak of the other unsubstituted carbons. This is expected for carbons that are meta to the OH ($\Delta\delta+1.4$), otherwise if they were ortho to the OH ($\Delta\delta - 12.8$) they would be at a higher field than the para carbon ($\Delta\delta - 7.4$). This substitution assignment is also supported by the infrared spectrum which has out-of-plane deformation bands at 750 and 780 cm^{-1}, in agreement with 1,2,3 substitution.

Exercise 061 (077)

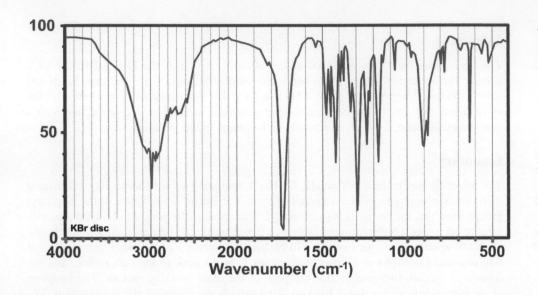

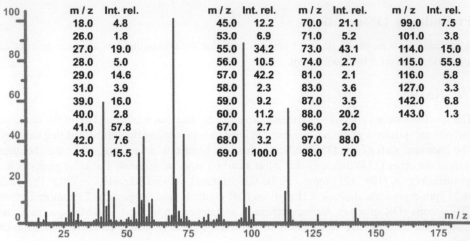

m / z	Int. rel.	m / z	Int. rel.	m / z	Int. rel.	m / z	Int. rel.
18.0	4.8	45.0	12.2	70.0	21.1	99.0	7.5
26.0	1.8	53.0	6.9	71.0	5.2	101.0	3.8
27.0	19.0	55.0	34.2	73.0	43.1	114.0	15.0
28.0	5.0	56.0	10.5	74.0	2.7	115.0	55.9
29.0	14.6	57.0	42.2	81.0	2.1	116.0	5.8
31.0	3.9	58.0	2.3	83.0	3.6	127.0	3.3
39.0	16.0	59.0	9.2	87.0	3.5	142.0	6.8
40.0	2.8	60.0	11.2	88.0	20.2	143.0	1.3
41.0	57.8	67.0	2.7	96.0	2.0		
42.0	7.6	68.0	3.2	97.0	88.0		
43.0	15.5	69.0	100.0	98.0	7.0		

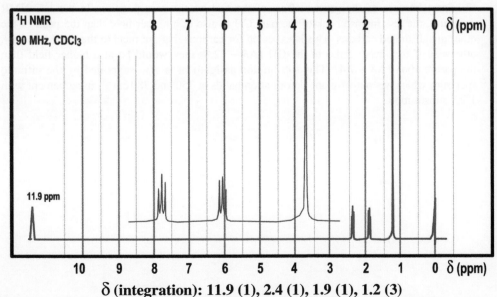

δ (integration): 11.9 (1), 2.4 (1), 1.9 (1), 1.2 (3)

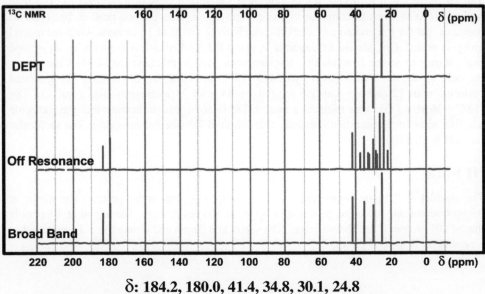

δ: 184.2, 180.0, 41.4, 34.8, 30.1, 24.8

Exercise 061 (077)

Infrared

The very broad band between 2300 and 3500 cm^{-1} has a pattern that gives a strong indication of the functional group. Other bands that characterize this group are found at 1730, 1290, 1420 and 900 cm^{-1}. The two bands of the same intensity at 1369 and 1388 cm^{-1} are significant.

^{13}C NMR

There are peaks for six types of carbon in the compound, two of which are clearly within functional groups. The off-resonance and DEPT spectra show that the carbons of the two functional groups are quaternary, and that there is a quaternary carbon in the aliphatic chain. There is also evidence for a methyl carbon and two methylene carbons.

^{1}H NMR

The ^{1}H NMR spectrum shows peaks for the carboxylic acid protons (11.9 ppm), as well as the two methylene groups and the methyl groups. The integration ratios indicate that there are two acid protons and two methyl groups per methylene group. The mutual splitting of the methylene signals show that the CH_2 groups are bonded to one another.

Mass Spectrum

The peak at 142 may be the molecular ion, but the IR spectrum indicates the presence of a functional group that readily loses a neutral molecule. The evidence from the other spectra indicates the presence of two carboxylic acids (2 COOH), one aliphatic quaternary carbon (C), two methyl groups (2 CH_3) and two methylene groups (2 CH_2), which gives a molecular formula of $C_7H_{12}O_4$ and a mass of 160. The ion at 142 is not the molecular ion, but the product arising from the loss of water, a common fragmentation reaction of carboxylic acids.

Conclusion

The parts of the structure can be put together by joining the two methylene carbons, adding the quaternary carbon containing the two methyl groups, and attaching carboxyl groups to the two ends of the chain: $HOOCCH_2CH_2C(CH_3)_2COOH$, 2,2-dimethylglutaric acid.

Notes

The extremely broad band of 2300 and 3500 cm^{-1} (asymmetric, centred around 3000 cm^{-1}), and the associated carbonyl and C–O stretching bands, are characteristic of a carboxylic acid. The two bands of the same intensity at 1369 and 1388 cm^{-1} are indicative of a pair of methyl groups on the same carbon.

The protons of the pair of non-equivalent CH_2 groups form an AA′XX′ system which appears as a symmetrical pattern of two four-line signals. In this example, the central lines are overlapping because of the similar sizes of J_{AX} and $J_{AX'}$.

In the mass spectrum, the ion at 115 (M−45) is the result of cleavage at the quaternary carbon. The ion at 97 (88%) comes from the additional loss of a molecule of water (18). The base peak at 69 corresponds to the stable allylic ion, $(CH_3)_2C{=}CHCH_2^+$, resulting from the loss of carbon monoxide (CO) from the 97 ion. A McLafferty rearrangement yields the peak at 88, $(CH_3)_2CHCOOH$, which then loses water to give the ion at 70.

Exercise 062 (037)

Elemental analysis : C% : 50.21, N% : 5.86, H% : 3.79 , O% : 40.14

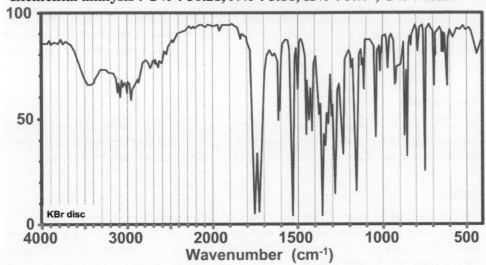

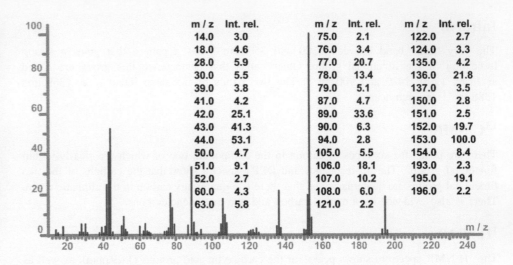

m / z	Int. rel.	m / z	Int. rel.	m / z	Int. rel.
14.0	3.0	75.0	2.1	122.0	2.7
18.0	4.6	76.0	3.4	124.0	3.3
28.0	5.9	77.0	20.7	135.0	4.2
30.0	5.5	78.0	13.4	136.0	21.8
39.0	3.8	79.0	5.1	137.0	3.5
41.0	4.2	87.0	4.7	150.0	2.8
42.0	25.1	89.0	33.6	151.0	2.5
43.0	41.3	90.0	6.3	152.0	19.7
44.0	53.1	94.0	2.8	153.0	100.0
50.0	4.7	105.0	5.5	154.0	8.4
51.0	9.1	106.0	18.1	193.0	2.3
52.0	2.7	107.0	10.2	195.0	19.1
60.0	4.3	108.0	6.0	196.0	2.2
63.0	5.8	121.0	2.2		

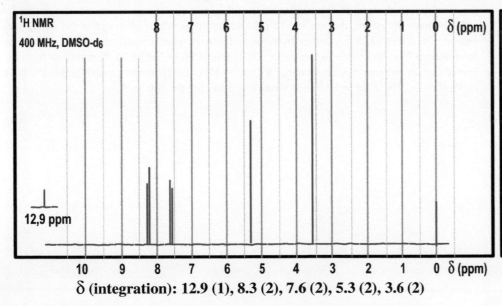

δ (integration): 12.9 (1), 8.3 (2), 7.6 (2), 5.3 (2), 3.6 (2)

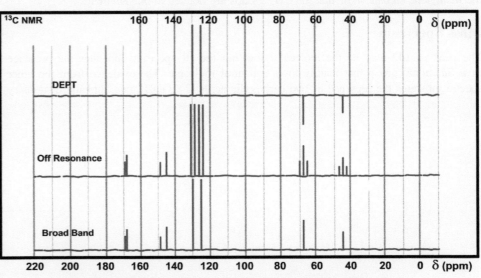

δ: 167.9, 166.7, 147.1, 143.6, 128.4, 123.5, 64.8, 41.4

Exercise 062 (037)

Preliminary Observations

The infrared spectrum has two very distinctive carbonyl bands. Note the odd-mass ion at the high end of the mass spectrum. The proton NMR has a very distinct splitting pattern for the aromatic protons, and a very low-field chemical shift (12.9 ppm) for one of the other protons.

Infrared

The very broad OH stretching band between 3700 and 2400 cm^{-1} and the carbonyl group at 1720 cm^{-1} can be used to identify one of the functional groups. Another functional group is associated with the carbonyl band at 1750 cm^{-1}. The presence of a nitro group is indicated by the bands near 1530 and 1350 cm^{-1} in the IR spectrum.

^{1}H NMR

The peak at 12.9 ppm is indicative of a very acidic proton. The peaks at 8.3 and 7.6 ppm have a splitting pattern that is typical of the protons of a para-disubstituted benzene ring, with one of the substituents being strongly electron-withdrawing. Two isolated methylene groups are found at 5.3 and 3.6 ppm. Note the very strong deshielding of the CH_2 at 5.3 ppm.

^{13}C NMR

The strong peaks at 123 and 128 ppm represent the isochronous CH carbons of the para-substituted benzene ring. The DEPT and OR spectra allow the assignment of the other signals to four quaternary and two methylene carbons. The quaternary carbons consist of two aromatic and two carbonyl carbons. The CH_2 at 65 ppm is clearly bonded to an oxygen, and the CH_2 at 41 ppm is deshielded by two carbonyl groups.

Mass Spectrum

The ion at 195 could be the molecular ion, but with the presence of two carbonyl groups, it is likely that this is a fragment ion. The microanalysis results of C% : 50.21, N% : 5.86, H% : 3.79, O% : 40.14 may help to resolve this question. Conversion to atomic ratios gives: C:H:N:O = 4.18:3.76:0.418:2.51, which simplifies to C:H:N:O = 10:9:1:6, for a formula of $C_{10}H_9NO_6$, and a mass of 239.

Summary

The substructure units derived from the spectral analysis: para-disubstituted benzene ring (X–C_6H_4–Y), acid (–COOH), ester (–COOR), two methylene groups (CH_2) and a nitro group (NO_2) add up to a molecular formula of $C_{10}H_9NO_6$, which is in agreement with the elemental analysis. This mass (239) is equal to the highest observed mass peak of 195 plus 44 (CO_2). Carbon dioxide is readily lost from a carboxylic acids that has a β carbonyl group, which in this case would be the carbonyl group of the ester, HO(O=C)CH_2(C=O)OR. The peak at 3.5 ppm represents the CH_2 group between the two carbonyl groups, one of an ester (–COOR) and one of an acid (–COOH). The para-substituted benzene ring has the nitro group as one of the substituent, and a methylene group as the other (–CH_2–C_6H_4–NO_2). The CH_2 at 5.3 ppm is bonded to the oxygen of the ester oxygen and is also deshielded by the strongly electron-withdrawing para nitro group of the benzene ring. The complete structure is thus, $O_2NC_6H_4CH_2O(CO)CH_2COOH$, 4-nitrophenyl hydrogen malonate.

The aromatic protons have an AA′XX′ coupling pattern with a J_{AX} of about 8 Hz (ortho coupling). The protons at 7.6 ppm are only slightly shifted downfield from the benzene chemical shift of 7.3 ppm, but the protons at 8.2 ppm are shifted by 0.9 ppm. The table of substituent effects reveals that the nitro group deshields ortho protons by 0.95 ppm.

The ion at 195 arises from the loss of 44 (CO_2), a common reaction of carboxylic acids having a carbonyl group in the β position. The base peak at 153 comes from the loss of ketene (CH_2=C=O) from the 195 ion.

Exercise 063 (080)

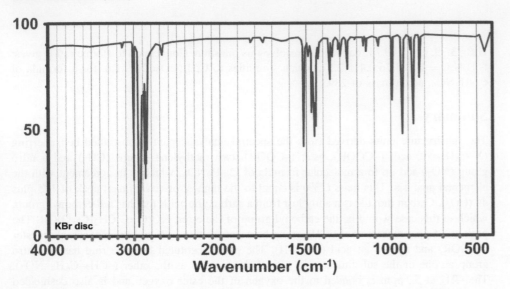

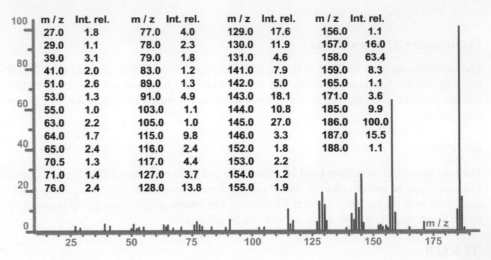

m/z	Int. rel.	m/z	Int. rel.	m/z	Int. rel.	m/z	Int. rel.
27.0	1.8	77.0	4.0	129.0	17.6	156.0	1.1
29.0	1.1	78.0	2.3	130.0	11.9	157.0	16.0
39.0	3.1	79.0	1.8	131.0	4.6	158.0	63.4
41.0	2.0	83.0	1.2	141.0	7.9	159.0	8.3
51.0	2.6	89.0	1.3	142.0	5.0	165.0	1.1
53.0	1.3	91.0	4.9	143.0	18.1	171.0	3.6
55.0	1.0	103.0	1.1	144.0	10.8	185.0	9.9
63.0	2.2	105.0	1.0	145.0	27.0	186.0	100.0
64.0	1.7	115.0	9.8	146.0	3.3	187.0	15.5
65.0	2.4	116.0	2.4	152.0	1.8	188.0	1.1
70.5	1.3	117.0	4.4	153.0	2.2		
71.0	1.4	127.0	3.7	154.0	1.2		
76.0	2.4	128.0	13.8	155.0	1.9		

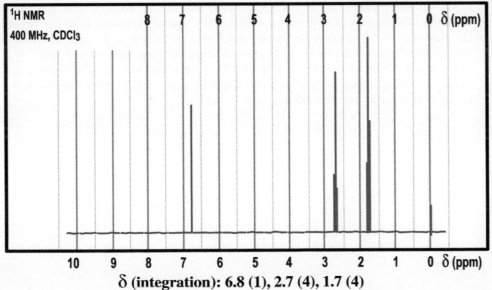

δ (integration): 6.8 (1), 2.7 (4), 1.7 (4)

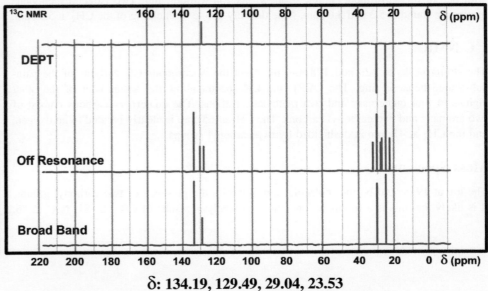

δ: 134.19, 129.49, 29.04, 23.53

Exercise 063 (080)

Preliminary Observation

The IR and NMR spectra are uncomplicated, yet the mass spectrum indicates a fairly large molecular weight. This suggests that the molecule is quite symmetrical.

Mass Spectrum

The spectrum has a very strong peak for the molecular ion at 186 (100%), with an M+1 isotope peak of 15.5, indicating that the molecule contains 14 carbons. The 14 carbons contribute 168 u (14×12) to the molecular mass, requiring 18 to make up the total of 186, and a formula of $C_{14}H_{18}$, which indicates that the molecule has six unsaturations. The very stable molecular ion is associated with the presence of an electron-rich π system. The largest peak after the base peak is the M−28 ion at 158 (63.4%) from the loss of a molecule of ethylene from a rearrangement reaction.

Infrared

The presence of an aromatic ring is indicated by the bands at 3000 and 1500 cm^{-1}. The 1600 cm^{-1} band is absent because of the symmetry of the molecule, this stretching being very weak in the absence of polarizing groups. The band at 861 cm^{-1} is in the range expected for the out-of-plane bending of an isolated aromatic proton. Bands at 2850 and 2930 cm^{-1} are quite specific for CH_2 groups, and there is no band for CH_3 groups between 1350 and 1400 cm^{-1}.

^{1}H NMR

The three signals have integration ratios of 1 (6.8) : 4 (2.7) : 4 (1.7) for a total of nine protons. Doubling this number gives the 18 hydrogens needed to complete the formula derived from the mass spectral data ($C_{14}H_{18}$). The triplet of integration at 1.7 ppm corresponds to four isochronous CH_2 groups, and the signal at 2.7 corresponds to another four isochronous CH_2 groups. These pairs of methylene groups are coupled to each other, creating triplets. The most highly deshielded of the two signals (2.7 ppm) corresponds to the CH_2 groups α to the aromatic ring. The singlet at 6.8 ppm corresponds to two isochronous aromatic CH protons. They are slightly upfield compared to benzene (7.3 ppm) due to the shielding effect of the methylene groups (ortho, −0.17; meta, −0.09).

^{13}C NMR

There are signals for four different carbons in the compound, two sp^3 carbons (2 types of CH_2) and two sp^2 carbons (C and CH). The two different methylene carbon signals represent four carbons each, and the methine carbon signal represents two carbons, for a total of 10 hydrogen-bearing carbons. The other four carbons of the molecule are represented by signals for the sp^2 quaternary carbons. In summary, the triplet at 23.5 ppm in the off-resonance spectrum corresponds to four CH_2 groups, the triplet at 29.0 ppm to four other CH_2 groups, the doublet at 129.5 ppm to two aromatic CH carbons and the singlet at 134.2 ppm to four quaternary aromatic carbons.

Summary

The IR spectrum shows bands for both sp^2 and sp^3 CH stretching as well as bands for an aromatic ring. The molecular formula of $C_{14}H_{18}$ indicates the presence of six unsaturations in the molecule, four of which are accounted for by one benzene ring. The remaining two unsaturations are accounted for by the two rings needed to accommodate the remaining eight carbons (eight CH_2 groups). Adding the two rings to a benzene ring to create a molecule with two sets of four equivalent methylene units gives 1,2,3,4,5,6,7,8-octahydroanthracene. In this molecule there are four equivalent α CH_2 groups and four equivalent β CH_2 groups, as well as the two equivalent CH carbons.

Exercise 064 (31)

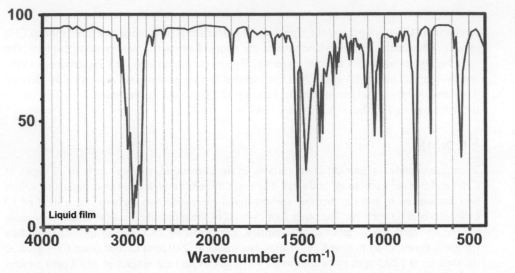

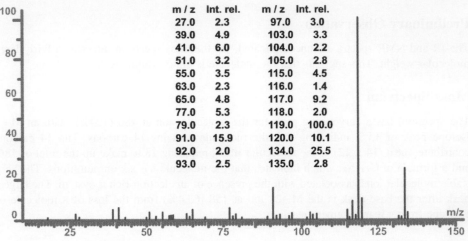

m/z	Int. rel.	m/z	Int. rel.
27.0	2.3	97.0	3.0
39.0	4.9	103.0	3.3
41.0	6.0	104.0	2.2
51.0	3.2	105.0	2.8
55.0	3.5	115.0	4.5
63.0	2.3	116.0	1.4
65.0	4.8	117.0	9.2
77.0	5.3	118.0	2.0
79.0	2.3	119.0	100.0
91.0	15.9	120.0	10.1
92.0	2.0	134.0	25.5
93.0	2.5	135.0	2.8

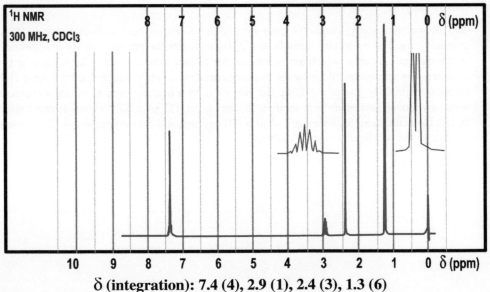

δ (integration): 7.4 (4), 2.9 (1), 2.4 (3), 1.3 (6)

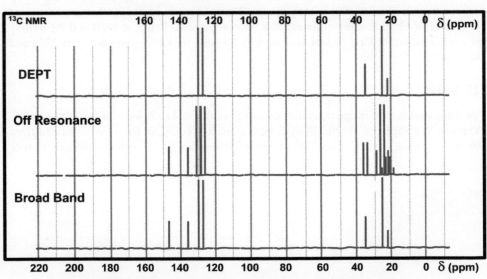

δ: 145.89, 135.11, 129.05, 126.31, 33.78, 24.12, 20.94

Exercise 064 (031)

Preliminary Observation

Evidence for the presence of an aromatic ring can be seen in each spectrum. There is no evidence for the presence of a carbonyl, NH, OH or any other functional group.

Mass Spectrum

The peak for the molecular ion is found at 134 (25.5%), and has an ^{13}C isotope peak at 135 (2.8%). Calculation of the relative intensity of M+1/M, gives 11% ($100 \times 2.8/25.5$), which is in good agreement with the presence of 10 carbons. This gives a molecular formula of $C_{10}H_{14}$. The peaks at 77, at 91 (77+14) and at 105 (91+14) indicate the presence of alkyl fragments on a benzene ring. The base peak at M−15, 119 (100%), suggests that there is a methyl group which is readily lost, possibly from the α position of the aromatic ring.

^{13}C NMR

There are peaks for four sp^2 carbons (aromatic) and three sp^3 carbons. Two of the aromatic carbons are methine (CH) carbons (doublets in OR), and two are quaternary carbons (peaks disappear in DEPT). The different chemical shifts of the two quaternary carbons indicate that there are two different substituents on the aromatic ring. The fact that there are only two types of CH carbon, indicates that the benzene ring is para substituted. The carbons in the sp^3 region consist of one CH (doublet in OR) and two types of CH_3 (two quartets of different intensities in OR spectrum). The smaller CH_3 signal belongs to a single methyl carbon, and the larger to a pair of equivalent methyl carbons.

^1H NMR

There are four signals for the protons: a singlet (7.4), a septet (2.9), a singlet (2.4) and a doublet (1.3); with integration ratios of 4:1:3:6. The signal at 7.4 ppm corresponds to the aromatic protons which are fortuitously isochronous, a common occurrence in benzene rings that have only alkyl substituents. The environments of the protons are so similar that the signals appear to be isochronous. The doublet at 1.3 ppm corresponds to the two equivalent CH_3 groups of an isopropyl group, $-CH(CH_3)_2$. These CH_3 protons are split by the CH proton, which itself appears as a septet (2.9), split by six methyl protons (the weak outer lines may be difficult to see). The last CH_3 group is found as a singlet at 2.1 ppm, as expected for a methyl group directly bonded to the aromatic ring.

Infrared

The bands between 3000 and 3100 cm^{-1} for the aromatic C–H stretching are difficult to see because of the intense alkyl CH stretching band. The band for the aromatic ring $>C=C<$ stretching is visible at 1500 cm^{-1}, and the band for the out-of-plane bending is seen at 816 cm^{-1}, which is indicative of a para-substituted ring. Note that in para-substituted compounds with non-polar substituents the band at 1600 cm^{-1} is weak or absent. The doublet of the same intensity near 1380 cm^{-1} indicates the presence of a *gem*-dimethyl group, $(CH_3)_2CH-$.

Structure

The molecular formula of $C_{10}H_{14}$ indicates that the molecule has four unsaturations, which is exactly that required for one benzene ring. The other four carbons provide the alkyl groups that are the ring substituents. These alkyl groups consist of a methyl group and an isopropyl group. The ^{13}C spectrum shows that the ring is para substituted (only two types of aromatic CH carbon). The out-of-plane bending band at 816 cm^{-1} supports this substitution pattern. The compound is *p*-cymene.

Exercise 065 (074)

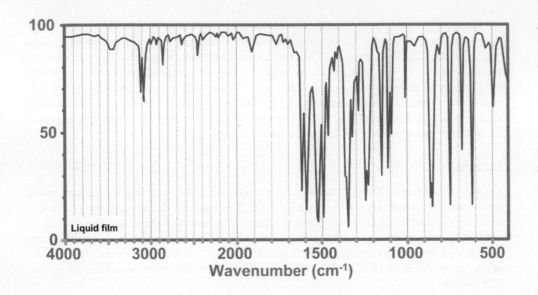

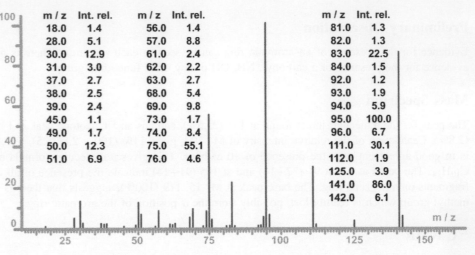

m / z	Int. rel.	m / z	Int. rel.	m / z	Int. rel.
18.0	1.4	56.0	1.4	81.0	1.3
28.0	5.1	57.0	8.8	82.0	1.3
30.0	12.9	61.0	2.4	83.0	22.5
31.0	3.0	62.0	2.2	84.0	1.5
37.0	2.7	63.0	2.7	92.0	1.2
38.0	2.5	68.0	5.4	93.0	1.9
39.0	2.4	69.0	9.8	94.0	5.9
45.0	1.1	73.0	1.7	95.0	100.0
49.0	1.7	74.0	8.4	96.0	6.7
50.0	12.3	75.0	55.1	111.0	30.1
51.0	6.9	76.0	4.6	112.0	1.9
				125.0	3.9
				141.0	86.0
				142.0	6.1

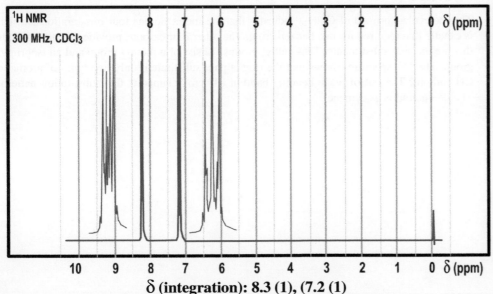

δ (integration): 8.3 (1), (7.2 (1)

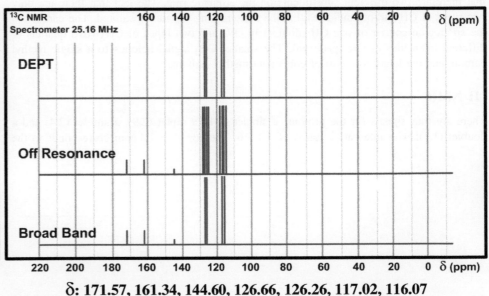

δ: 171.57, 161.34, 144.60, 126.66, 126.26, 117.02, 116.07

Exercise 065 (074)

Preliminary Observation

A quick scan of the IR and NMR spectra indicates that the compound has an aromatic ring. The IR spectrum is noteworthy in its lack of absorption bands just below $3000\,cm^{-1}$ in the region of sp^3 CH stretching. There is also no evidence for the presence of carbonyl groups or functions with an NH or an OH.

Mass Spectrum

The strong odd-mass peak at 141 (86%) indicates the presence of nitrogen in the molecule. The M+1 isotope peak 142 (6.1%) has a relative intensity of 7.1%, indicating that the molecule has six carbon atoms (6.6 for six C and 0.4 for one N). The series of peaks for the loss of 16, 30 and 46 are typical of aromatic nitro compounds. The small peak at 125 (3.9%) is due to the loss of an oxygen atom, the peak at 111 (30.1%) from the loss of nitrogen monoxide (NO) and the peak at 95 (100%) from the loss of nitrogen dioxide. The peak at 83(22.5%) is due to the loss of carbon monoxide from the 111 ion. Benzene has a mass of 78, and replacing one hydrogen with an NO_2 group increases the mass to 123. This is 18 mass units less than the required molecular mass of 141. If a hydrogen of the benzene ring is replaced by a substituent, that substituent would need to have a mass of 19, which is just that of a fluorine atom. The peak at 75 (55.1%) is due to the loss of HF from the base peak.

^{1}H NMR

One pair of the ring protons is found near the normal position for benzene, and the other pair at a strongly deshielded position as expected of protons to a nitro group. The splitting pattern of the protons is interesting. The upper peak has four lines with splittings of 9.2 and 8.2 Hz, and the lower signal has four lines with splittings of 9.2 and 4.9 Hz. The mutual splitting of 9.2 Hz between the two protons is that of $^3J_{ortho}$ proton–proton coupling. The other splittings are caused by coupling to the fluorine atom, $^3J_{HF} = 8.2\,Hz$ and $^4J_{HF} = 4.9\,Hz$.

^{13}C NMR

There are six carbons in a benzene ring; with two different substituents one would expect four carbons signals for para substitution, six for meta substitution, and six for ortho substitution. In this case there are seven lines in the broad band spectrum. However, in counting the number of carbons from the lines, one must take into account the coupling to the fluorine atom. One can see a doublet (171.57 and 161.34) centred at 166.5 ppm with a splitting of 257 Hz, a doublet (126.66 and 126.26) centred at 126.5 ppm with a splitting of 10 Hz, and a doublet (117.02 and 116.07) centred at 116.5 with a splitting of 24 Hz. The doublet centred at 166.5 ppm and the singlet at 144.6 ppm are quaternary carbons (disappear in DEPT). The one at 166.5 is clearly the one bearing the fluorine because of the large splitting ($^1J = 257\,Hz$). The other quaternary carbon, which has such a small coupling it is not observed, bears the NO_2 group. The doublet at 117 ppm is ortho ($^3J = 24\,Hz$) to the fluorine, and the one at 127 is meta ($^4J = 10\,Hz$) to the fluorine.

Infrared

The absence of bands between 3000 and $2900\,cm^{-1}$ indicates that there are no alkyl groups in the molecule. The typical peaks for an aromatic ring are present: bands between 3000 and $3100\,cm^{-1}$ for C–H stretching, weak overtone-combination bands between 2000 and $1600\,cm^{-1}$, bands at 1500 and $1600\,cm^{-1}$ for the >C=C< stretching, and bands in the out-of-plane bending region. The bands typical of an aromatic nitro group are found at 1350 and $1520\,cm^{-1}$. The C–F stretching vibration is found at $1230\,cm^{-1}$.

Conclusion

The compound is *p*-fluoronitrobenzene.

Exercise 066 (051)

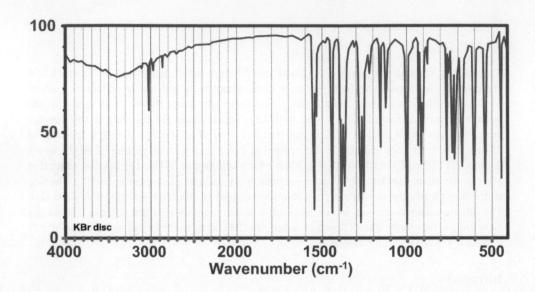

KBr disc

Wavenumber (cm⁻¹)

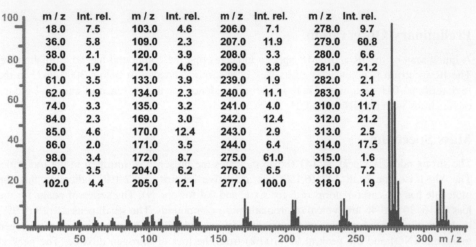

m / z	Int. rel.	m / z	Int. rel.	m / z	Int. rel.	m / z	Int. rel.
18.0	7.5	103.0	4.6	206.0	7.1	278.0	9.7
36.0	5.8	109.0	2.3	207.0	11.9	279.0	60.8
38.0	2.1	120.0	3.9	208.0	3.3	280.0	6.6
50.0	1.9	121.0	4.6	209.0	3.8	281.0	21.2
61.0	3.5	133.0	3.9	239.0	1.9	282.0	2.1
62.0	1.9	134.0	2.3	240.0	11.1	283.0	3.4
74.0	3.3	135.0	3.2	241.0	4.0	310.0	11.7
84.0	2.3	169.0	3.0	242.0	12.4	312.0	21.2
85.0	4.6	170.0	12.4	243.0	2.9	313.0	2.5
86.0	2.0	171.0	3.5	244.0	6.4	314.0	17.5
98.0	3.4	172.0	8.7	275.0	61.0	315.0	1.6
99.0	3.5	204.0	6.2	276.0	6.5	316.0	7.2
102.0	4.4	205.0	12.1	277.0	100.0	318.0	1.9

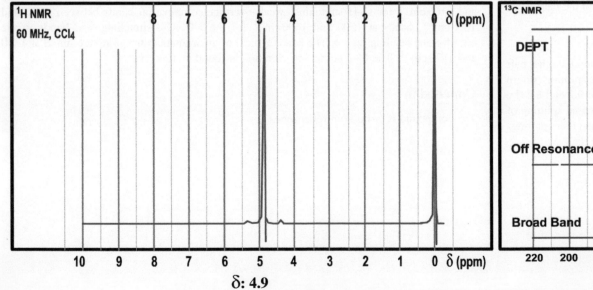

¹H NMR

60 MHz, CCl₄

δ: 4.9

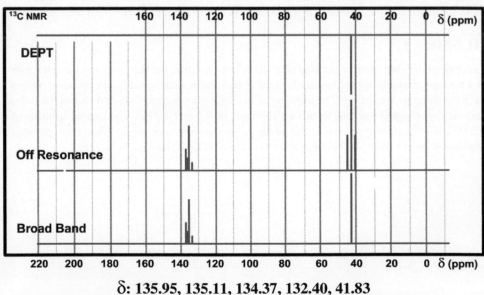

¹³C NMR

DEPT

Off Resonance

Broad Band

δ: 135.95, 135.11, 134.37, 132.40, 41.83

Exercise 066 (051)

Preliminary Observation

The extremely broad sloping IR band centred near $3400\,\text{cm}^{-1}$ is an artefact of sample preparation. The bands between 1600 and $1500\,\text{cm}^{-1}$ indicate the presence of an aromatic ring, but there is no evidence of functional groups containing oxygen or nitrogen.

Mass Spectrum

The peak for the molecular ion at 310 has isotope peaks at M+2, M+4, M+6 and M+8, indicating the presence of chlorine or bromine atoms. A table of isotope patterns could be used to determine the composition, or one could simply count the number of isotope peaks to determine the number of bromine and chlorine atoms. However, the highest isotope peak may be missed if it is very weak, as in the case of multiple chlorines. The base peak (277) is one of the isotope peaks of the M−Cl fragment. The lowest isotope peak (all ^{35}Cl) of this fragment has a mass of 275. The peaks at 240 and 205 (M−2Cl, and M−3Cl) still show the presence of halogens. Peaks are found for ions for the loss of chlorine atoms at: 275 (61%), 240 (11%), 205 (12%), 170 (12%), 135 (3%). The table of isotope compositions for multiple chlorine atoms, shows that the isotope pattern for the molecular ion (310) corresponds to that of six chlorines. The fragment ions also have the appropriate patterns for five Cl at 275, four Cl 240, three Cl at 205 and two Cl at 170. For the ions having four and fewer chlorine atoms, there is a series of peaks one mass unit lighter, corresponding to fragmentation by the loss of HCl. Subtracting the mass of the six chlorines from the molecular mass (310 − 210) gives a mass of 100 for the rest of the molecule. An eight-carbon unit ($C_8H_4 = 100$) gives a molecular formula of $C_8H_4Cl_6$. Calculation of the number of unsaturations gives four centres, equivalent to one benzene ring.

^{1}H NMR

The single peak at 4.9 ppm represents the four equivalent protons of two CH_2 groups. The chemical shift is close to that estimated ($0.23 + 1.85 + 2.53 = 4.61$) for CH_2 groups bonded to an aromatic ring and to a chlorine atom.

^{13}C NMR

There are five types of carbons in the compound, four quaternary carbons in the aromatic region (disappear in DEPT), and one methylene carbon (42 ppm, inverted in DEPT, triplet in OR). In the aromatic region, the most intense peak (134.37 ppm) is associated with the two ring carbons bearing the $-CH_2Cl$ groups (the protons increase the intensity through the nuclear Overhauser effect, and the decrease in the longitudinal relaxation time). The other pair of isochronous carbons is found at 135.95 ppm. The last two carbons are found at 135.11 and 132.40 ppm, the more intense of these (135.11) is associated with the carbon between the two substituted carbons.

Infrared

The extremely broad band near $3400\,\text{cm}^{-1}$ is caused by dispersion from the sample and the presence of water in the KBr disc. There are many bands from the highly polarized bonds, but not many that are easily interpreted. The band at $3020\,\text{cm}^{-1}$ might be mistaken for a hydrogen on an sp^2 carbon, but there is no CH showing in the ^{13}C NMR spectra. It is in fact due to the asymmetric C–H stretching vibration of the CH_2Cl group on the aromatic ring. The only peaks that can be assigned to the aromatic ring are those at 1550 and $1450\,\text{cm}^{-1}$, as there are no ring hydrogens. The C–Cl stretching bands are found between 1000 and $1200\,\text{cm}^{-1}$ for the chlorines on the sp^2 carbons, but they are lower for the Cl–CH$_2$ (near $700\,\text{cm}^{-1}$).

Conclusion

The compound has a benzene ring with two chloromethyl groups ($-CH_2Cl$) and four chlorine atoms as substituents. As there are four different ring carbons in the ^{13}C NMR spectrum the ring must have the chloromethyl groups meta to one another. Para substitution would give only two carbon types, and ortho would give three types. The structure is $\alpha,\alpha',2,4,5,6$-hexachloro-m-xylene.

Exercise 067 (049)

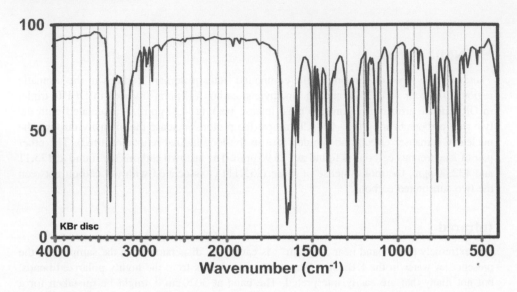

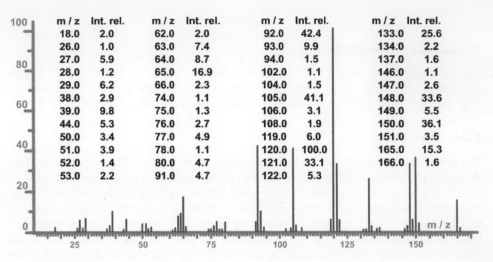

m/z	Int. rel.	m/z	Int. rel.	m/z	Int. rel.	m/z	Int. rel.
18.0	2.0	62.0	2.0	92.0	42.4	133.0	25.6
26.0	1.0	63.0	7.4	93.0	9.9	134.0	2.2
27.0	5.9	64.0	8.7	94.0	1.5	137.0	1.6
28.0	1.2	65.0	16.9	102.0	1.1	146.0	1.1
29.0	6.2	66.0	2.3	104.0	1.5	147.0	2.6
38.0	2.9	74.0	1.1	105.0	41.1	148.0	33.6
39.0	9.8	75.0	1.3	106.0	3.1	149.0	5.5
44.0	5.3	76.0	2.7	108.0	1.9	150.0	36.1
50.0	3.4	77.0	4.9	119.0	6.0	151.0	3.5
51.0	3.9	78.0	1.1	120.0	100.0	165.0	15.3
52.0	1.4	80.0	4.7	121.0	33.1	166.0	1.6
53.0	2.2	91.0	4.7	122.0	5.3		

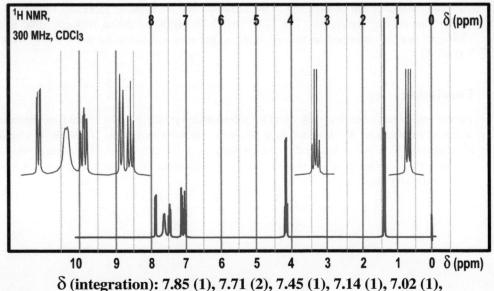

δ (integration): 7.85 (1), 7.71 (2), 7.45 (1), 7.14 (1), 7.02 (1), 4.2 (2), 1.45 (3)

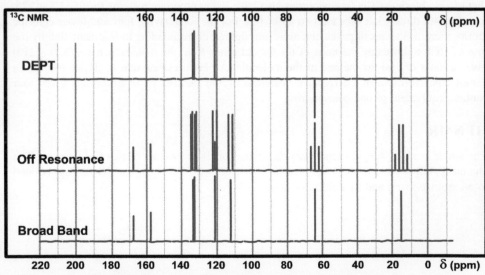

δ: 167.4, 157.4, 133.3, 132.6, 121.1, 121.0, 112.4, 64.7, 14.9

Exercise 067 (049)

Preliminary Observation

The IR spectrum has a carbonyl band below $1700\,cm^{-1}$, which is in the amide region. The pair of peaks near 3200 and $3400\,cm^{-1}$ suggests that the compound is a primary amide.

^{1}H NMR

The signal at 7.7 ppm corresponds to two hydrogens bonded to nitrogen (broad shape). The chemical shift of primary amide protons is variable, depending on sample conditions. The mutually coupled triplet–quartet system represents an ethoxy group ($CH_3CH_2O–$). The chemical shift of the CH_2 quartet (4.2) is at the appropriate position for an ethoxy group attached to a phenyl ring. The coupling patterns of the four aromatic protons show that the benzene ring is ortho disubstituted. There are two doublets with a splitting of about 9 Hz from the ortho coupling (also showing weak meta coupling). These doublets (7.85, 7.14) represent aromatic protons having only one ortho neighbour. The one at 7.85 represents the proton ortho to the electron-withdrawing amide carbonyl, and the one at 7.14 ortho to the resonance-donating oxygen. There are also two triplets (showing additional weak meta coupling) for the protons having two ortho neighbours. The one at 7.45 ppm represents the proton para to the amide, and the one at 7.02 the proton para to the oxygen.

Mass Spectrum

The molecule has a molecular mass of 165. The odd mass is expected for a compound containing nitrogen. For an isolated ethoxy group on an aromatic ring, the principal fragmentation is the loss of ethylene (CH_2CH_2), and for an isolated primary amide group, the principal fragmentation is the loss of NH_2, followed by the loss of carbon dioxide to yield an M−44 ion. However, when two substituents are ortho to one another on the ring, they often interact to give different fragmentation patterns. In this case, the ether group fragments by cleavage α to the oxygen to lose a methyl group (M−15, 150), and the amide group loses ammonia (M−17) to form the ion at 148, which then loses CO to form the base

peak at 120. The other major peaks may be explained as: 133 (150 –NH_3), 121 (149 –CO), 105 (133 –CO), 92 (120 –CO). The series of ions at 77, 65 and 51 is indicative of a benzene ring.

^{13}C NMR

Three peaks disappear in the DEPT spectrum: two quaternary aromatic carbons (121.0, 157.4), and one carbonyl carbon of an amide group (167.4). Signals at 112.4, 121.1, 132.6 and 133.3 ppm each represent an aromatic CH carbon. The signal at 157.4 ppm belongs to a carbon bearing an inductively electron-withdrawing oxygen atom; and the one at 121.1 belongs to the other aromatic carbon, carrying the amide function. The CH carbons at 133.3 and 132.6 ppm, are para and ortho to the amide group (deshielded by the electron-withdrawing resonance effect), and those at 121.1 and 112.4 ppm are para and ortho to the $–OCH_2CH_3$ (shielded by the resonance-donor effect). The peak at 64.7 ppm corresponds to a CH_2 bonded to oxygen, and the peak at 14.9 ppm to the CH_3 group.

Infrared

The intense bands at 3200 and $3400\,cm^{-1}$ are the symmetric and asymmetric stretching bands characteristic of an NH_2 group. The band at $1650\,cm^{-1}$ is the amide C=O stretching band, and that at $1630\,cm^{-1}$ is an NH_2 deformation band, called the 'amide II' band. In secondary amides, this 'amide II' band is found near $1550\,cm^{-1}$. The band at $1250\,cm^{-1}$ represents the C–O–C band of the aromatic ether. The aromatic ring is seen in the bands at 1600 and $1500\,cm^{-1}$, and in the out-of-plane bending band at $760\,cm^{-1}$ (the position characteristic of four adjacent hydrogens, ortho substitution).

Summary

The compound has a benzene ring carrying a primary amide group and an ethoxy group ortho to one another. The compound is o-ethoxybenzamide, $CH_3CH_2OC_6H_4(CO)NH_2$.

Exercise 068 (047)

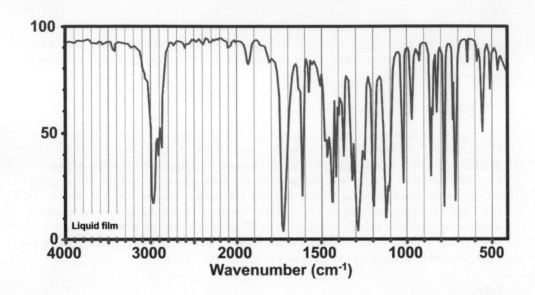

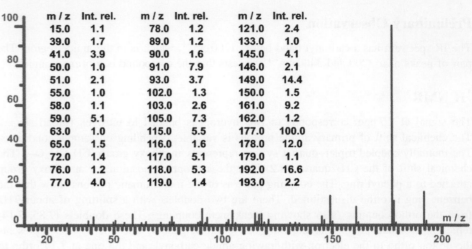

m / z	Int. rel.	m / z	Int. rel.	m / z	Int. rel.
15.0	1.1	78.0	1.2	121.0	2.4
39.0	1.7	89.0	1.2	133.0	1.0
41.0	3.9	90.0	1.6	145.0	4.1
50.0	1.0	91.0	7.9	146.0	2.1
51.0	2.1	93.0	3.7	149.0	14.4
55.0	1.0	102.0	1.3	150.0	1.5
58.0	1.1	103.0	2.6	161.0	9.2
59.0	5.1	105.0	7.3	162.0	1.2
63.0	1.0	115.0	5.5	177.0	100.0
65.0	1.5	116.0	5.1	178.0	12.0
72.0	1.4	117.0	5.1	179.0	1.1
76.0	1.2	118.0	5.3	192.0	16.6
77.0	4.0	119.0	1.4	193.0	2.2

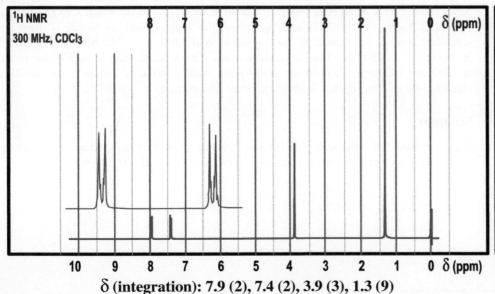

δ (integration): 7.9 (2), 7.4 (2), 3.9 (3), 1.3 (9)

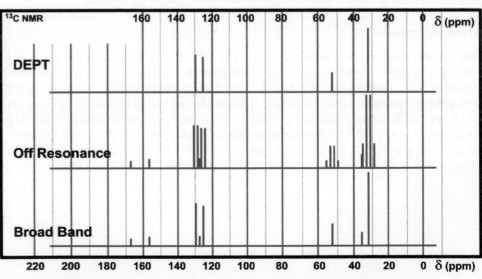

δ: 167.14, 156.54, 129.45, 127.42, 125.32, 51.92, 35.08, 31.13

Exercise 068 (047)

Preliminary Observation

There is evidence for an aromatic ring in the NMR and IR spectra, and the strong peak for a carbonyl below 1725 cm^{-1} indicates the presence of a carbonyl group.

^{13}C NMR

There are four quaternary carbons, two in the aromatic region (156, 129), one in the ester carbonyl region (167) and one in the alkyl region (35). The peak at 156 ppm corresponds to an aromatic carbon bearing a carbonyl. There are two CH carbons in the aromatic region (125, 129) and two methyl peaks (52, 35) of unequal intensity.

Mass Spectrum

The mass spectrum shows a peak for the molecular ion at 192 (16.6%). The ion at 177 (100%) corresponds to M−15, which is equivalent to the loss of a methyl group. The ion at 161 arises from the loss of 31 (OCH$_3$). The loss of carbon monoxide (C≡O) from the 177 ion leads to the ion at 149.

^{1}H NMR

The peaks at 7.96 and 7.45 ppm represent four protons on a para-disubstituted aromatic ring. The two protons at 7.96 ppm are deshielded by an ortho carbonyl group. The other two protons, at 7.45 ppm, are only slightly deshielded. The four protons form an AA′XX′ coupling system with the main coupling of about 9 Hz. The singlet at 3.9 ppm is at the chemical shift characteristic of the methoxy group of an aromatic ester (Ar–COOCH$_3$). Finally, the signal at 1.3 ppm (integration 9) corresponds to the three methyl groups of a *tert*-butyl group (–C(CH$_3$)$_3$).

Infrared

The shoulder above 3000 cm^{-1} represents the aromatic C–H stretching vibrations. The other typical bands of the aromatic ring are found at 1600 and 1500 cm^{-1}, representing stretching of the >C=C< bonds, and in the out-of-plane bending region. The bands at 1100 and 1280 cm^{-1} corresponds to the C–O stretching vibrations of an ester. Finally, the peak at 1725 cm^{-1} corresponds to the C=O stretching of a conjugated ester. The observed frequency is lower than normal (1740 cm^{-1}) because of conjugation to the aromatic ring.

Summary

There is a para-substituted benzene ring with a *tert*-butyl group and a carbonyl group as substituents. The only other component of the structure is a methoxy group. The compound is methyl *p-tert*-butylbenzoate.

Exercise 069 (050)

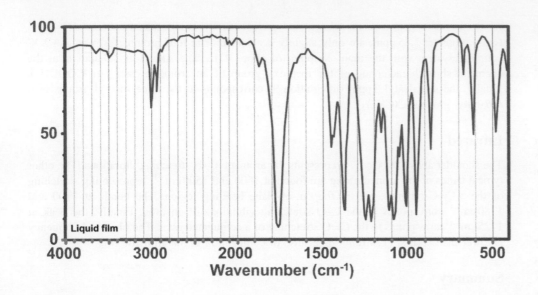

Liquid film

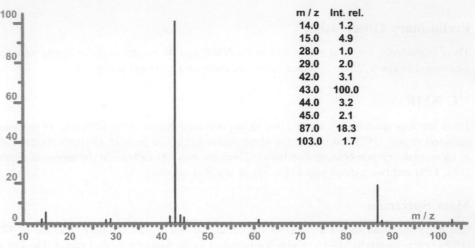

m / z	Int. rel.
14.0	1.2
15.0	4.9
28.0	1.0
29.0	2.0
42.0	3.1
43.0	100.0
44.0	3.2
45.0	2.1
87.0	18.3
103.0	1.7

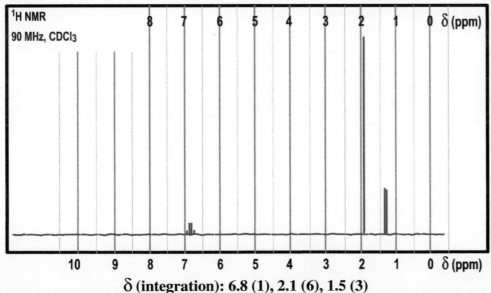

^{1}H NMR

90 MHz, CDCl$_3$

δ (integration): 6.8 (1), 2.1 (6), 1.5 (3)

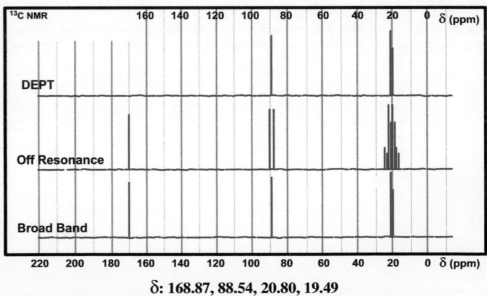

^{13}C NMR

DEPT

Off Resonance

Broad Band

δ: 168.87, 88.54, 20.80, 19.49

Exercise 069 (050)

Preliminary Observation

The strong peak at $1760\,\text{cm}^{-1}$ in the IR spectrum is probably that of an ester, and the base peak of 43 might be that of the acylium ion, $(CH_3C{\equiv}O^+)$.

^{13}C NMR

Of the four signals showing in the spectrum, one is of a quaternary carbon (signal disappears in DEPT) with a chemical shift (169 ppm) in the region expected for an ester carbonyl. The CH (doublet in OR) at 89 ppm is more highly deshielded than expected for a CH bonded to the oxygen of an ester (expected near 65 ppm). The other two carbons are identified as methyl groups by the off-resonance spectrum. The different intensities suggest that one of the signals is due to two isochronous CH_3 groups.

^{1}H NMR

The quartet at 6.8 ppm and the doublet at 1.5 ppm are mutually coupled, thus the molecule contains a $>CHCH_3$ substructural unit. The chemical shift of the CH (6.8 ppm) is at remarkably low field. The singlet of integration 6 at 2.1 ppm is in the correct location of the methyl of an acetyl ester ($CH_3COO–$, acetate).

Infrared

The band at $1760\,\text{cm}^{-1}$ is at the appropriate wavenumber for an ester carbonyl. The other bands expected for the ester are found between 1250 and $1000\,\text{cm}^{-1}$ for the =C–O and O–C stretching bands respectively.

Mass Spectrum

The base peak in the spectrum is at 43, $[CH_3C{\equiv}O]^+$, due to the expected α cleavage at the carbonyl. There does not appear to be a peak for the molecular ion, which is not unusual in a compound with an easily fragmented ester group. The peak at 103 is a fragment ion.

Conclusion

The proton NMR shows that the CH is coupled to a CH_3 group, thus the molecule has the substructure, $>CHCH_3$. The only other peak in the spectrum is from six equivalent protons of two acetyl groups. Adding the two acetyl groups to the unfilled valences of the $>CHCH_3$ group gives the complete structure, $(CH_3COO)_2CHCH_3$, ethylidene diacetate.

This compound has a molecular mass of 146, but the highest mass ion seen in the mass spectrum is the barely perceptible peak at 103 (1.7%), which corresponds to M−43. The ion at 87 corresponds to the loss of an acetyl radical by α cleavage at the methine carbon (CH) yielding the $[CH_3COOCH(CH_3)]^+$.

Exercise 070 (029)

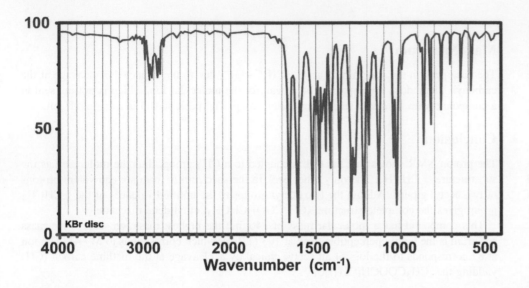

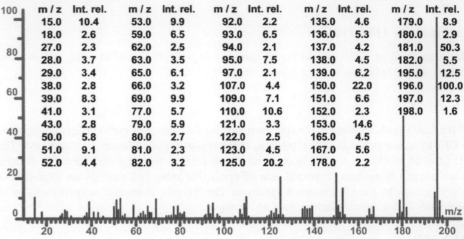

m/z	Int. rel.	m/z	Int. rel.	m/z	Int. rel.	m/z	Int. rel.	m/z	Int. rel.
15.0	10.4	53.0	9.9	92.0	2.2	135.0	4.6	179.0	8.9
18.0	2.6	59.0	6.5	93.0	6.5	136.0	5.3	180.0	2.9
27.0	2.3	62.0	2.5	94.0	2.1	137.0	4.2	181.0	50.3
28.0	3.7	63.0	3.5	95.0	7.5	138.0	4.5	182.0	5.5
29.0	3.4	65.0	6.1	97.0	2.1	139.0	6.2	195.0	12.5
38.0	2.8	66.0	3.2	107.0	4.4	150.0	22.0	196.0	100.0
39.0	8.3	69.0	9.9	109.0	7.1	151.0	6.6	197.0	12.3
41.0	3.1	77.0	5.7	110.0	10.6	152.0	2.3	198.0	1.6
43.0	2.8	79.0	5.9	121.0	3.3	153.0	14.6		
50.0	5.8	80.0	2.7	122.0	2.5	165.0	4.5		
51.0	9.1	81.0	2.3	123.0	4.5	167.0	5.6		
52.0	4.4	82.0	3.2	125.0	20.2	178.0	2.2		

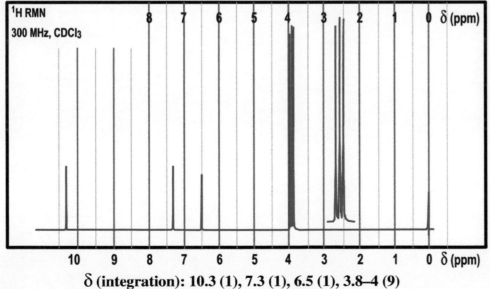

δ (integration): 10.3 (1), 7.3 (1), 6.5 (1), 3.8–4 (9)

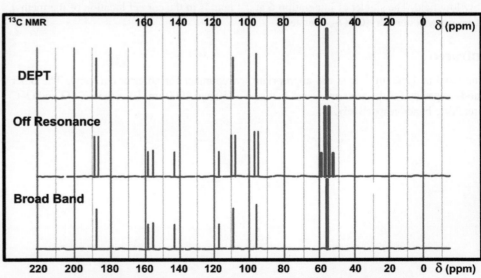

δ: 187.9, 158.7, 155.9, 143.7, 117.5, 109.2, 96.2, 56.34, 56.26, 56.21

Exercise 070 (029)

Preliminary Observation

Note that in the ^{13}C NMR spectrum the peak at 56 ppm on the chart contains three peaks, 56.21, 56.26 and 56.34 ppm. The IR has peaks for a carbonyl and an aromatic ring.

^{13}C NMR

There are 10 types of carbon in the compound, three in the aliphatic region (56.34, 56.26, 56.21), six in the aromatic region (158.7, 155.9, 143.7, 117.5, 109.2, 96.2) and one in the carbonyl region (187.9). The three peaks near 56 ppm are all CH_3 groups (quartets in OR) and have the chemical shift of methoxy groups. The DEPT spectrum shows that four of the aromatic carbons are quaternary. Three of these are quite strongly deshielded, having oxygen as substituents (OCH_3). The fourth (117.5) corresponds to an aromatic carbon bearing a carbonyl group. The carbon of the carbonyl group appears as a doublet at 187.9 ppm in the off-resonance spectrum, showing that it is an aldehyde group. The OR spectrum also has doublets for two aromatic CH carbons (96.2 and 109.2 ppm). To summarize, there are six aromatic carbon atoms, three of them having methoxy groups (CH_3O) and one bearing an aldehyde group (–CHO).

^{1}H NMR

The downfield singlet at 10.3 ppm represents the proton of the aldehyde group. The lack of splitting indicates that the carbon is attached to a quaternary carbon (aromatic ring). The absence of any coupling in the two aromatic protons (7.3 and 6.5. ppm) suggests that they are para to one another. There is only one way to place the four substituents on a benzene ring so that the two protons are para, that gives 2,4,5-trimethoxybenzaldehyde. The proton at 7.3 ppm has a chemical shift near that of benzene, being influenced primarily by opposing ortho groups, a shielding methoxy and a deshielding aldehyde (calculated shift, 7.24). The proton at 6.5 ppm is shielded by two ortho methoxy groups (calculated value, 6.53).

Infrared

Peaks indicative of the aromatic ring are found in the expected regions; a weak C–H stretching band just above 3000 cm^{-1}, ring bond stretching bands near 1600 and 1500 cm^{-1} and out-of-plane bending bands between 850 and 870 cm^{-1}. The peak at 2840 cm^{-1} is typical of a methoxy group on an aromatic ring. The presence of this group is supported by the bands between 1250 and 1200 cm^{-1}, which represent C–O–C= stretching. The band at 1680 cm^{-1} corresponds to a carbonyl group conjugated to the aromatic ring. The two peaks for the aldehyde C–H stretching vibrations are found at the expected positions near 2820 and 2730 cm^{-1}.

Mass Spectrum

The base peak at 196 is the molecular ion. Its isotope peak at M+1 of 12.3% suggests that there are 11 carbons, but the evidence from the other spectra shows unequivocally that there are 10 carbons in the molecule. The precision of the intensity measurements in mass spectra may be susceptible to instrument settings, and conclusions drawn from them should be verified by other spectral evidence. The stability of the molecular ion is in accord with a strongly conjugated aromatic system. The ion at 181 corresponds to an M−15, (M−CH_3). There is a significant M−1 peak at 195 (12.5%) for the loss of a hydrogen atom, which is expected of an aromatic aldehyde.

Structure and Notes

The combined evidence indicates that the compound has a benzene ring with three methoxy groups and an aldehyde as substituents. The lack of symmetry of the ring dictates that only two of the five possible isomers are allowed: the 2,3,5- and the 2,4,5-rimethoxy derivatives. The 2,3,5 isomer would show meta coupling, and the proton chemical shifts would be near 6.39 and 6.96 ppm. The 2,4,5 isomer would not show meta coupling and would have chemical shifts of about 7.24 and 6.53 ppm in agreement with the observed values. The compound is 2,4,5-trimethoxybenzaldehyde.

Exercise 071 (035)

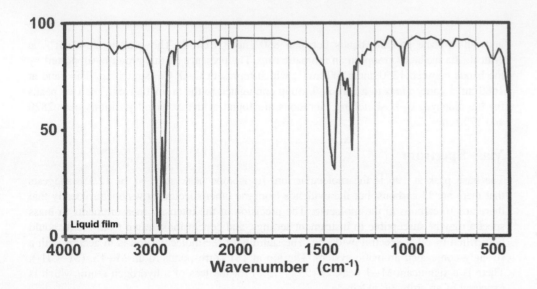

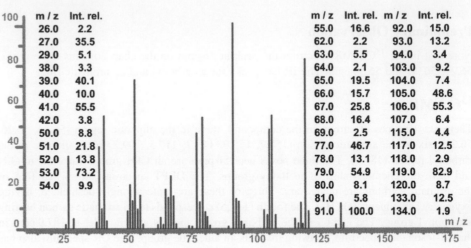

m / z	Int. rel.		m / z	Int. rel.	m / z	Int. rel.
26.0	2.2		55.0	16.6	92.0	15.0
27.0	35.5		62.0	2.2	93.0	13.2
29.0	5.1		63.0	5.5	94.0	3.4
38.0	3.3		64.0	2.1	103.0	9.2
39.0	40.1		65.0	19.5	104.0	7.4
40.0	10.0		66.0	15.7	105.0	48.6
41.0	55.5		67.0	25.8	106.0	55.3
42.0	3.8		68.0	16.4	107.0	6.4
50.0	8.8		69.0	2.5	115.0	4.4
51.0	21.8		77.0	46.7	117.0	12.5
52.0	13.8		78.0	13.1	118.0	2.9
53.0	73.2		79.0	54.9	119.0	82.9
54.0	9.9		80.0	8.1	120.0	8.7
			81.0	5.8	133.0	12.5
			91.0	100.0	134.0	1.9

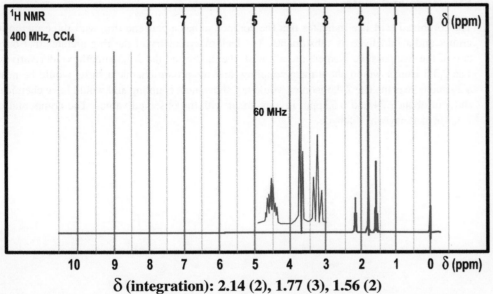

δ (integration): 2.14 (2), 1.77 (3), 1.56 (2)

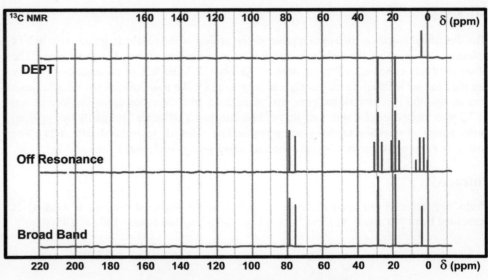

δ: 78.95, 75.63, 28.31, 18.38, 3.42

Exercise 071 (035)

Preliminary Observation

The IR spectrum gives no indication of the presence of a functional group.

Mass Spectrum

The peak at 133 is not the molecular ion; the ion at 119 is 14 units less, and there are no fragments of 14 that one could reasonably expect the molecular ion to lose. It is more likely that the molecular ion is 134, and that the 119 ion is the M−15 ion. The principal peaks are of odd mass: 27, 39, 41, 53, 55, 67, 79 and 91, which is normal for even-mass molecules. The peaks at 119, 105 and 91 (base peak) come from the loss of CH_3, CH_3CH_2 and $CH_3CH_2CH_2$, but this does not necessarily mean that these groups are part of the original structure because hydrogen migration occurs readily in unsaturated chains. The peaks at 39 $(C_3H_3)^+$ and 53 $(C_4H_5)^+$ are similar to three and four carbon alkyl chains (different from the saturated chains by four less hydrogens, suggesting a triple bond).

^{13}C NMR

Of the five signals in the ^{13}C NMR spectra, two are quaternary carbons (peaks disappear in DEPT), two are CH_2 carbons (peaks inverted in DEPT, and triplets in OR) and one is a CH_3 carbon (quartet in OR). The chemical shifts of the quaternary carbons suggest that they are sp carbons of an alkyne group. Note the chemical shift of the methyl group; the high field position (3.4 ppm) is attributed to the anisotropic shielding effect of the triple bond.

^{1}H NMR

Peaks are seen for one CH_3 and two CH_2 groups as expected from the ^{13}C NMR spectrum. The methyl peak at 1.77 ppm is split into a triplet with a small coupling constant.

This is due to five-bond long-range coupling of the CH_3 to a CH_2 through the triple bond (H–C–C≡C–C–H), $^5J_{HH} \approx 3$ Hz. The signal at 2.14 ppm represents the CH_2 to which the CH_3 is coupled. This CH_2 is coupled to another CH_2 ($^3J \approx 6$ Hz) as well as to the CH_3 group. The signal at 1.56 ppm represents the second CH_2 coupled to the first one by 6 Hz.

Infrared

The infrared spectrum looks very much like the spectrum of a straight chain alkane. There are two very small bands in the C≡C stretching region (2230 and 2055 cm^{-1}). The peak at 2055 cm^{-1} is at the precise location observed for triple bonds bearing a methyl group and another alkyl group (CH$_3$C≡CR). Because of the similar nature of the substituents, the dipole moment change on stretching is very small and the resulting peak is too weak to be diagnostically significant. No band is seen near 3300 cm^{-1} for a terminal alkyne (HC≡C–) near 3300 cm^{-1} because there is no H–C≡ bond.

Summary

The ^{1}H NMR coupling pattern shows that the molecule has the substructure: $CH_3C≡CCH_2CH_2-$. As there is no evidence for any other group, two of these need to be combined to complete the structure. The structure is $CH_3C≡CCH_2CH_2–CH_2CH_2C≡CCH_3$. This fits with the mass spectrum indication of a molecular mass of 134, $C_{10}H_{14}$. The compound is 2,8-decadiyne.

Exercise 072 (014)

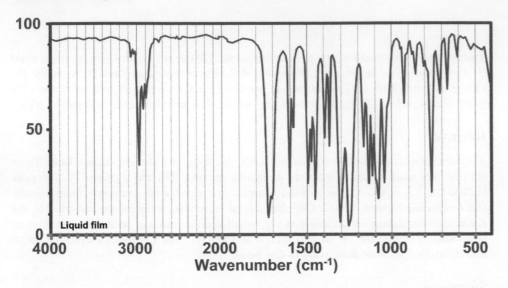

Liquid film

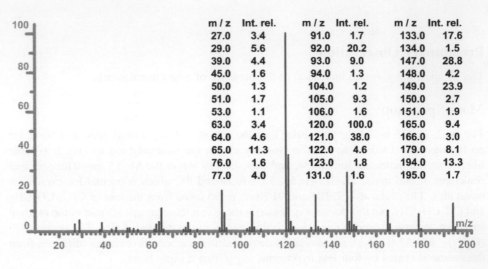

m / z	Int. rel.	m / z	Int. rel.	m / z	Int. rel.
27.0	3.4	91.0	1.7	133.0	17.6
29.0	5.6	92.0	20.2	134.0	1.5
39.0	4.4	93.0	9.0	147.0	28.8
45.0	1.6	94.0	1.3	148.0	4.2
50.0	1.3	104.0	1.2	149.0	23.9
51.0	1.7	105.0	9.3	150.0	2.7
53.0	1.1	106.0	1.6	151.0	1.9
63.0	3.4	120.0	100.0	165.0	9.4
64.0	4.6	121.0	38.0	166.0	3.0
65.0	11.3	122.0	4.6	179.0	8.1
76.0	1.6	123.0	1.8	194.0	13.3
77.0	4.0	131.0	1.6	195.0	1.7

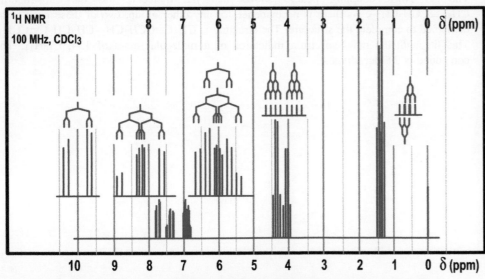

δ (integration): 7.76 (1), 7.39 (1), 6.94 (1), 6.92 (1), 4.34 (2), 4.06 (2), 1.43 (3), 1.36 (3)

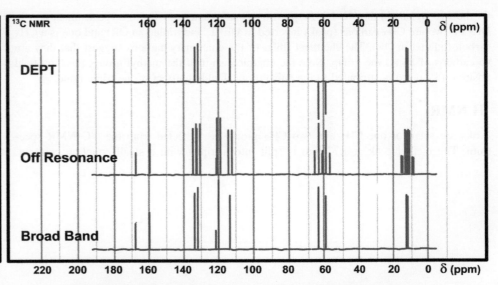

δ : 166.6, 158.5, 133.1, 131.4, 121.4, 120.2, 113.6, 64.7, 60.7, 14.8, 14.3

Exercise 072 (014)

Preliminary Observation

The IR spectrum shows evidence for an aromatic ring and an ester.

^{13}C NMR

The BB spectrum has peaks for 11 different types of carbon. In the off-resonance spectrum one can see two CH_3, two CH_2, four CH and three quaternary carbons. The four CH carbons and two of the quaternary carbons are in the aromatic region. The strongly deshielded (159 ppm) quaternary aromatic carbon bears an oxygen substituent. The quaternary carbon at 167 ppm belongs to the carbonyl of an ester group.

^{1}H NMR

The integration indicates the presence of 14 protons, four of which are on an aromatic ring. The peaks of the aromatic protons overlap, but the stick diagrams above the peaks clarify the observed patterns. The coupling pattern is derived from a larger ($\approx$ 8 Hz) ortho coupling and a smaller ($\approx$ 2 Hz) meta coupling. The proton at 7.76 is a doublet ($\approx$ 8 Hz) with a smaller meta coupling. The proton at 7.39 ppm is split by two large couplings and one smaller coupling, therefore it has two protons ortho to it and one proton meta to it. The ring is ortho disubstituted. The two triplet–quartet sets indicate that there are two ethyl groups bonded to oxygens.

Infrared

The band at 1725 cm^{-1} is at the appropriate position for an aromatic ester (aliphatic ester $\approx$ 1740, benzoate ester $\approx$ 1725). The ester is confirmed by the =C–O and O–C stretching bands between 1000 and 1250 cm^{-1}. The band at 1250 cm^{-1} is characteristic of conjugated ethers. The presence of bands at 1500 and 1600 cm^{-1} is characteristic of a benzene ring, and the band at 760 cm^{-1}, for the out-of-plane C–H bending is typical of the presence of four adjacent hydrogens.

Mass Spectrum

The molecular ion at 194 (13.3%) has an M+1 isotope peak with a relative intensity of 12.8%, which agrees with the ^{13}C NMR count of 11 carbons. The peak at 179 represents the loss of a CH_3 from the molecular ion, and the peak at 165 to the loss of an –CH_2CH_3 group. The fragment at 149 (M−45) represents the loss of the ethoxy radical (OCH_2CH_3) by α cleavage at the carbonyl group. A further loss of 28 (either carbon monoxide or ethylene) gives the ion at 121. Aromatic ethyl ethers typically lose ethylene (28) by a McLafferty-type rearrangement. The loss of ethylene from the molecular ion, followed by the loss of ethanol (46) gives the base peak at 120. The 120 ion is typically the base peak of ortho hydroxy benzoate esters, but not of the meta or para isomers.

Summary

The molecule has an ortho-substituted benzene ring, two different ethoxy groups and an ester carbonyl group. Combining these components gives $CH_3CH_2OC_6H_4COOCH_2CH_3$, ethyl o-ethoxybenzoate. The ortho substitution of the ring is established by the chemical shifts and splitting pattern of the aromatic protons. The IR spectrum confirms the ortho position of the groups by the deformation band at 760 cm^{-1}.

Exercise 073 (043)

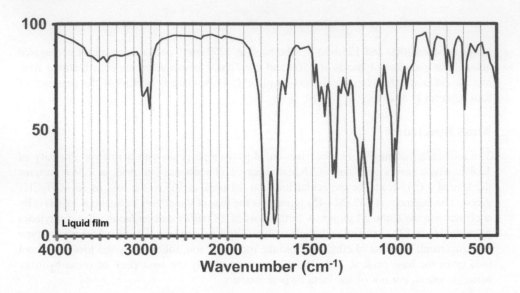

Liquid film

Wavenumber (cm⁻¹)

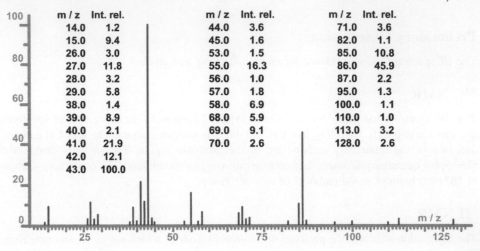

m / z	Int. rel.	m / z	Int. rel.	m / z	Int. rel.
14.0	1.2	44.0	3.6	71.0	3.6
15.0	9.4	45.0	1.6	82.0	1.1
26.0	3.0	53.0	1.5	85.0	10.8
27.0	11.8	55.0	16.3	86.0	45.9
28.0	3.2	56.0	1.0	87.0	2.2
29.0	5.8	57.0	1.8	95.0	1.3
38.0	1.4	58.0	6.9	100.0	1.1
39.0	8.9	68.0	5.9	110.0	1.0
40.0	2.1	69.0	9.1	113.0	3.2
41.0	21.9	70.0	2.6	128.0	2.6
42.0	12.1				
43.0	100.0				

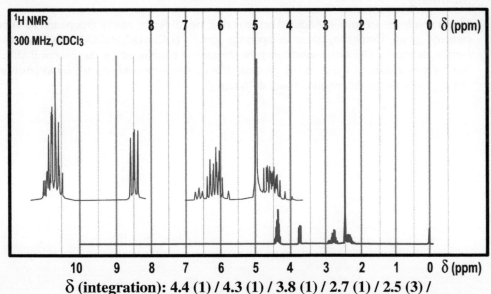

¹H NMR
300 MHz, CDCl₃

δ (integration): 4.4 (1) / 4.3 (1) / 3.8 (1) / 2.7 (1) / 2.5 (3) /
4.4 (1), 4.3 (1), 3.8 (1), 2.7 (1), 2.5 (3), 2.4 (1)

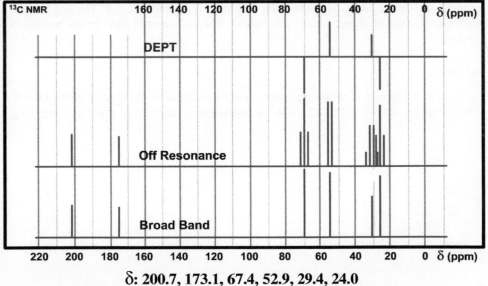

¹³C NMR

DEPT

Off Resonance

Broad Band

δ: 200.7, 173.1, 67.4, 52.9, 29.4, 24.0

Exercise 073 (043)

Infrared

The very strong peaks at 1770 and 1720 cm^{-1} indicate that there are two carbonyl groups in the molecule. The peak at 1770 is well above the 1740 cm^{-1} normal position of an ester. An increase in carbonyl frequency is usually due to a decrease in ring size or to substitution at the α carbon with an electronegative atom, such as fluorine or chlorine. The typical wavenumbers for the carbonyls of cyclic esters (lactones) are: six-membered, 1740; five-membered, 1775; four-membered, 1820. Substitution with an α halogen can also raise the frequency, but usually by only 10 to 20 cm^{-1}. The peak at 1770 cm^{-1} is most likely due to the presence of a γ lactone. The expected bands for the =C–O and O–C stretching are found between 1160 and 1000 cm^{-1}. The second peak in the carbonyl region (1720 cm^{-1}) is at the appropriate position for a ketone. These assignments should be confirmed by the ^{13}C NMR spectrum.

^{13}C NMR

The spectrum has peaks for six different carbons, two quaternary carbons at 201 and 173 ppm (disappear in DEPT), two CH$_2$ at 67 and 24 ppm (inverted in DEPT, triplet in OR), a CH at 53 ppm (doublet in OR) and one methyl carbon at 24 ppm (quartet in OR). The chemical shift of the quaternary carbon at 173 is in the range for the carbonyl of an ester, and that at 201 is in the ketone range. The CH$_2$ at 67 ppm is the alkoxy carbon (–O–CH$_2$) of the lactone, being strongly deshielded by the oxygen.

Mass Spectrum

Esters and ketones generally undergo easy fragmentation, so it is not surprising that the peak for the molecular ion at 128 is very weak. The fragment at 86 (49.5%) is due to the loss of ketene (reduce spacing) by a McLafferty rearrangement reaction with the carbonyl group of the lactone accepting the γ hydrogen (on the –CH$_3$).

The base peak at 43 is due to α cleavage at the ketone. These fragments indicate the presence of a methyl ketone. The six carbons, as indicated by the ^{13}C NMR spectrum, and three oxygens for the ketone and lactone groups, give a mass of 120, thus requiring eight hydrogens to complete the molecular formula, C$_6$H$_8$O$_3$.

^{1}H NMR

The integration ratios give a total of eight hydrogens, as expected from the molecular formula derived from the ^{13}C NMR and mass spectra. The singlet (3H) at 2.43 ppm is due to the CH$_3$ group of a methyl ketone. The low field multiplet (2 H) at 4.36 belong to a CH$_2$ group attached to the oxygen, (–CH$_2$–O–C=O). The protons have very complex splitting patterns except the proton at 3.78 ppm (1 H), which is a double doublet with J = 9.3 and 7.2 Hz. This is evidently a proton that is isolated except for being coupled to two protons with slightly different coupling constants. The other protons are parts of highly coupled splitting patterns.

Summary

The evidence indicates the presence of a γ lactone having a methyl ketone as a substituent. There are three positions on the ring where the substituent may be placed (α, β or γ) to the carbonyl group. The chemical shifts of the γ CH$_2$ group at 67 ppm (^{13}C) and at 4.3 ppm (^{1}H) indicate that it is the CH$_2$ attached to the oxygen (–CH$_2$–O–C=O), thus the γ carbon is unsubstituted. The high-field carbon (CH$_2$) at 24 ppm in the ^{13}C spectrum is the β carbon. It is also unsubstituted, therefore the substituent must be on the α carbon. The compound is α-acetyl-γ-butyrolactone. The hydrogen on the substituted α carbon is split by the adjacent *cis* and *trans* protons into a double doublet. Its chemical shift (3.78 ppm) is at the appropriate position for a proton between two carbonyl groups.

Exercise 074 (055)

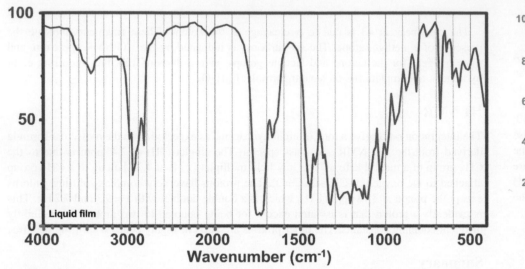

Liquid film

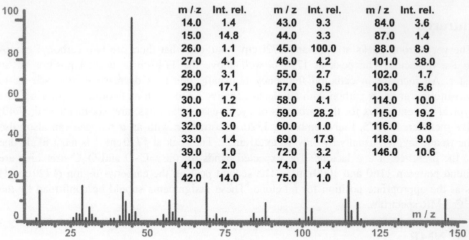

m / z	Int. rel.	m / z	Int. rel.	m / z	Int. rel.
14.0	1.4	43.0	9.3	84.0	3.6
15.0	14.8	44.0	3.3	87.0	1.4
26.0	1.1	45.0	100.0	88.0	8.9
27.0	4.1	46.0	4.2	101.0	38.0
28.0	3.1	55.0	2.7	102.0	1.7
29.0	17.1	57.0	9.5	103.0	5.6
30.0	1.2	58.0	4.1	114.0	10.0
31.0	6.7	59.0	28.2	115.0	19.2
32.0	3.0	60.0	1.0	116.0	4.8
33.0	1.4	69.0	17.9	118.0	9.0
39.0	1.0	72.0	3.2	146.0	10.6
41.0	2.0	74.0	1.4		
42.0	14.0	75.0	1.0		

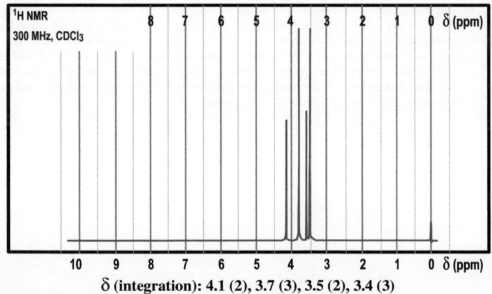

^{1}H NMR
300 MHz, CDCl$_3$

δ (integration): 4.1 (2), 3.7 (3), 3.5 (2), 3.4 (3)

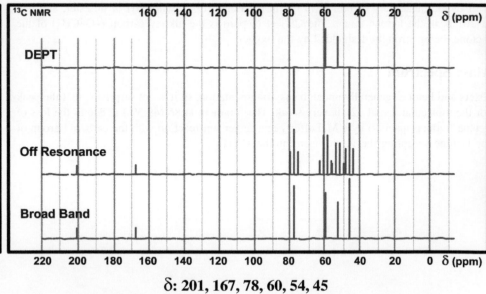

^{13}C NMR

DEPT

Off Resonance

Broad Band

δ: 201, 167, 78, 60, 54, 45

Exercise 074 (055)

^{13}C NMR

The spectrum shows six types of carbon, two CH_3, two CH_2 and two strongly deshielded quaternary carbons, at the appropriate shifts for a ketone (201 ppm) and an ester (167 ppm).

Infrared

The broad band between 1760 and 1720 cm^{-1} shows the presence of two carbonyls, an ester and a ketone. The overlapping bands between 1100 and 1020 cm^{-1} represent C–O stretching vibrations.

^{1}H NMR

The four singlets represent two methyl and two methylene groups, as indicated by the integration ratios (2:3:2:3). The lack of coupling indicates that they are isolated from one another. The chemical shifts indicate that the protons are strongly deshielded by the oxygens and carbonyl groups.

Mass Spectrum

The very weak peak for the molecular ion at 146 indicates that the molecule has an easily fragmented functional group. The weak ion at 131 comes from the loss of CH_3, and the ion at 115 from the loss of OCH_3 (α cleavage at the ester). The peak at 101 is due to the loss of 45 (loss of CH_2OCH_3, α cleavage at the ketone) and the base peak (45) is due to the $[CH_2=OCH_3]^+$ ion. The molecular formula is $C_6H_{10}O_4$.

Conclusion

The compound contains: CH_3OCH_2-, $>C=O$, CH_2- and $-COOCH_3$. Combining these substructures so that none of the protons is coupled, gives the structure: $CH_3OCH_2(C=O)CH_2COOCH_3$, methyl 4-methoxyacetoacetate.

Notes

Although the ketone and ester carbonyl bands are not clearly separated in the IR spectrum, the broadness of the carbonyl band shows that more than one carbonyl is present, and the two carbonyl groups are clearly identified in the ^{13}C NMR. The peak at 78 ppm in the ^{13}C spectrum is assigned to the CH_2 between the oxygen and the ketone; the 60 ppm peak to the CH_3 of the ether; the 54 ppm peak to the CH_3 of the ester; and the 45 ppm peak to the CH_2 between the two carbonyls.

Exercise 075 (059)

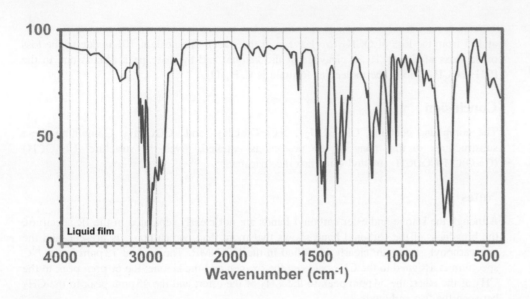

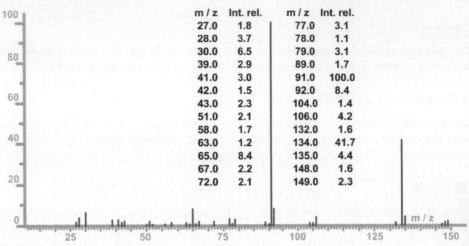

m/z	Int. rel.	m/z	Int. rel.
27.0	1.8	77.0	3.1
28.0	3.7	78.0	1.1
30.0	6.5	79.0	3.1
39.0	2.9	89.0	1.7
41.0	3.0	91.0	100.0
42.0	1.5	92.0	8.4
43.0	2.3	104.0	1.4
51.0	2.1	106.0	4.2
58.0	1.7	132.0	1.6
63.0	1.2	134.0	41.7
65.0	8.4	135.0	4.4
67.0	2.2	148.0	1.6
72.0	2.1	149.0	2.3

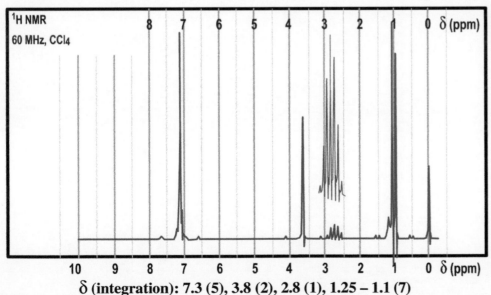

δ (integration): 7.3 (5), 3.8 (2), 2.8 (1), 1.25 – 1.1 (7)

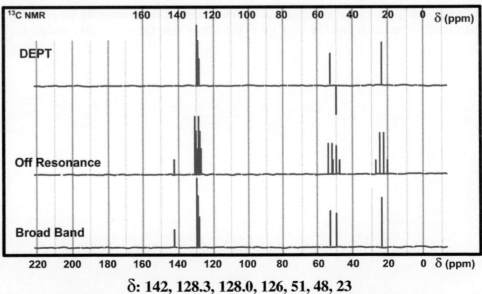

δ: 142, 128.3, 128.0, 126, 51, 48, 23

Exercise 075 (059)

Preliminary Observation

The IR spectrum has the noteworthy pattern of an aromatic ring in the overtone-combination region between 2000 and 1600 cm^{-1}. It also has the peaks expected in the other aromatic regions. The peak at 3300 cm^{-1} is possibly that of NH stretching. In the ^{1}H NMR spectrum there is a peak at the base of the high-field doublet which disappears in a 'D$_2$O shake' experiment.

^{13}C NMR

There are signals for seven types of carbon: one quaternary and three CH carbons in the aromatic region, and a CH$_3$, CH$_2$ and CH in the aliphatic region. The aromatic signals have the classical pattern for monosubstituted benzene rings, one weak signal for the quaternary carbon, a more intense signal for the para carbon, and even more intense peaks for the ortho and meta pairs. The peaks at 51 and 48 ppm represent, respectively, the CH and CH$_2$ carbons bonded to a nitrogen atom. Finally, the OR quartet represents two isochronous CH$_3$ groups.

Infrared

Peaks are found in all regions expected for a benzene ring. The broad band at 3300 cm^{-1} represents the weak N–H stretching of a secondary amine. The NH bending band of a secondary amine is found between 660 and 900 cm^{-1}, but it is very variable and not a reliable indicator for the NH group.

Mass Spectrum

The peak at 134 (41%) would be a candidate for the molecular ion, except for the fact that it is an even-mass ion and there appears to be nitrogen in the compound. There is also a weak peak at 149 (2.3%), which is 15 units higher than the 134 peak. This suggests that it is more likely to be a fragment arising from the loss of a methyl group. The molecule has two functions that direct the fragmentation, the phenyl ring and the amino group. Cleavage at the α position of the benzene ring gives C$_6$H$_5$CH$_2^+$, which is responsible for the base peak at 91 (100%). The M−15 peak at 134 (41%) is due to α cleavage at the nitrogen. The benzyl group and nitrogen (C$_6$H$_5$CH$_2$N) contribute a mass of 105, which requires another 44 (C$_3$H$_8$) to complete the molecular formula of C$_{10}$H$_{15}$N.

^{1}H NMR

The single peak at 7.3 ppm represents the five protons of a monosubstituted benzene ring. Although there are three different types of proton on such a ring, the chemical shifts of these protons are so close that they appear as one peak in a low–field spectrometer (60 MHz). In high–field spectrometers (400 MHz) separate peaks may be observed. The singlet at 3.7 ppm represents a CH$_2$ bonded to an aromatic ring and to a nitrogen atom. The septet at 2.8 ppm represents the CH of an isopropyl group, and the doublet at 1.1 ppm represents its two CH$_3$ groups. The CH chemical shift indicates that it is bonded to nitrogen. The broad signal at 1.25 ppm represents the NH of a secondary amine (this peak disappears when the sample is shaken with one or two drops of D$_2$O).

Conclusion

The identified substructures are: a phenyl ring, a methylene group, >NH and an isopropyl group. This gives the structure, C$_6$H$_5$CH$_2$NHCH(CH$_3$)$_2$, isopropylbenzylamine.

Exercise 076 (064)

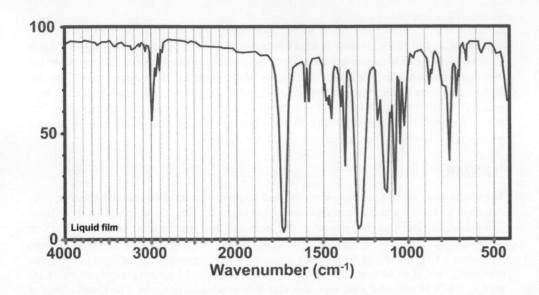

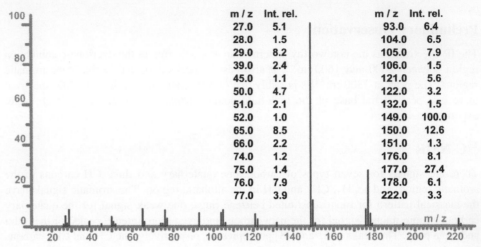

m / z	Int. rel.		m / z	Int. rel.
27.0	5.1		93.0	6.4
28.0	1.5		104.0	6.5
29.0	8.2		105.0	7.9
39.0	2.4		106.0	1.5
45.0	1.1		121.0	5.6
50.0	4.7		122.0	3.2
51.0	2.1		132.0	1.5
52.0	1.0		149.0	100.0
65.0	8.5		150.0	12.6
66.0	2.2		151.0	1.3
74.0	1.2		176.0	8.1
75.0	1.4		177.0	27.4
76.0	7.9		178.0	6.1
77.0	3.9		222.0	3.3

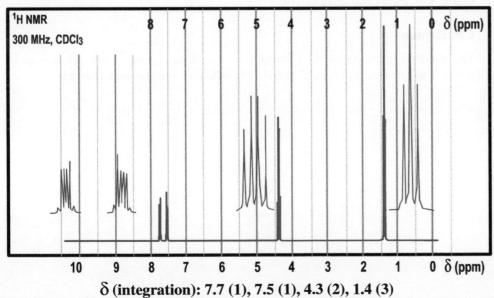

δ (integration): 7.7 (1), 7.5 (1), 4.3 (2), 1.4 (3)

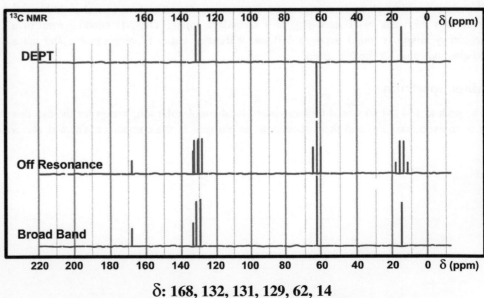

δ: 168, 132, 131, 129, 62, 14

Exercise 076 (064)

Preliminary Observation

The IR spectrum has strong bands at 1730 and 1290 cm^{-1} which indicate that the compound is an ester. The symmetrical splitting pattern of the aromatic protons in the ^{1}H NMR spectrum shows that there is a symmetrically substituted benzene ring.

^{13}C NMR

Because of the symmetrical substitution, there are only three peaks in the aromatic region, two CH and one quaternary carbon. The carbonyl of the ester gives a peak at 168 ppm. The triplet at 62 ppm corresponds to a CH$_2$ attached to the oxygen of an ester. Finally, the signal at 14 ppm corresponds to a CH$_3$ group.

Mass Spectrum

The molecular ion shows a weak peak at 222 (3.3%). The fragment at 177 (27.4%) represents the loss of 45 (CH$_3$CH$_2$O) which suggests that the ester is an ethyl ester. A further loss of CO gives the base peak at 149. The weak peak at 194 is due to the loss of ethylene by a McLafferty rearrangement of the ethyl ester. The ions at 39, 50, 51, 76 and 77 are typical of benzene rings.

^{1}H NMR

The typical symmetrical AA'BB' splitting pattern of a benzene ring with two identical ortho substituents is seen in the aromatic region. The peaks are shifted downfield by the electron-withdrawing substituents. The quartet at 4.3 ppm and triplet at 1.4 ppm together represent the ethyl group of the ester.

Infrared

The band at 1730 cm^{-1} is at the high end of the range for a conjugated ester. The other typical band for the =C–O and O–C stretching of the ester is seen at 1290 cm^{-1}. The bands between 1500 and 1600 cm^{-1} show the presence of an aromatic ring (>C=C<). The band near 750 cm^{-1} represents the deformations corresponding to four adjacent hydrogens (ortho-disubstituted ring).

Conclusion

The components of the structure identified from the spectra include an ortho-disubstituted benzene ring and an ethyl ester. The mass spectrum requires two of the ethyl ester groups to arrive at the molecular formula of C$_{12}$H$_{14}$O$_4$. Assembly of these components gives the compound: diethyl phthalate, C$_6$H$_4$(COOCH$_2$CH$_3$)$_2$.

Exercise 077 (071)

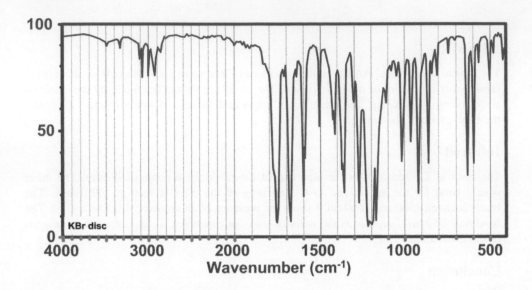

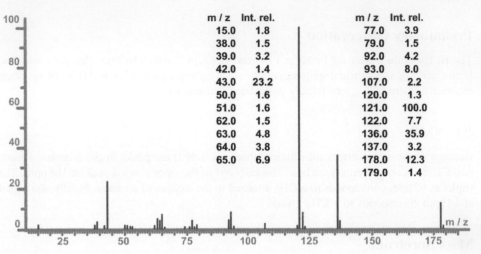

m / z	Int. rel.		m / z	Int. rel.
15.0	1.8		77.0	3.9
38.0	1.5		79.0	1.5
39.0	3.2		92.0	4.2
42.0	1.4		93.0	8.0
43.0	23.2		107.0	2.2
50.0	1.6		120.0	1.3
51.0	1.6		121.0	100.0
62.0	1.5		122.0	7.7
63.0	4.8		136.0	35.9
64.0	3.8		137.0	3.2
65.0	6.9		178.0	12.3
			179.0	1.4

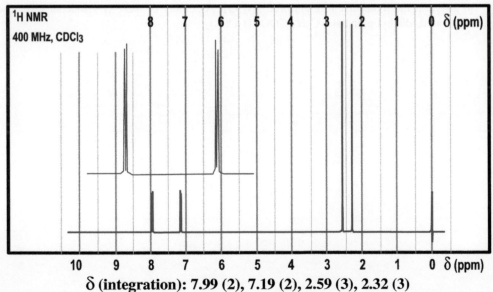

δ (integration): 7.99 (2), 7.19 (2), 2.59 (3), 2.32 (3)

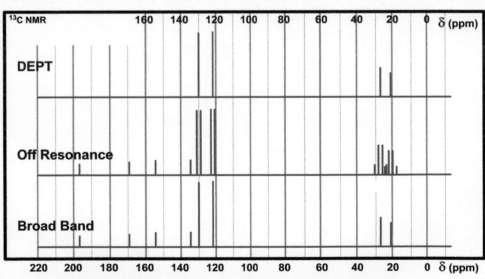

δ: 196.8, 168.8, 154.4, 134.8, 129.9, 121.8, 26.6, 21.1

Exercise 077 (071)

Preliminary Observation

The IR spectrum has two strong carbonyl peaks, one ($1755\,cm^{-1}$) in the ester range, and the other ($1675\,cm^{-1}$) in the conjugated ketone region. The IR and NMR spectra have peaks that indicate the presence of an aromatic ring. Note that the peaks at 7.99 and 7.19 ppm in the proton NMR are doublets with a splitting of 8.9 Hz.

Mass Spectrum

The molecular ion appears at 178 (12.3%) with an M+1 isotope peak at 179 (1.4%) indicating a carbon count of 10. The first fragment ion at 136 (35.9%) comes from a rearrangement reaction in which ketene ($CH_2=C=O$) is lost, a common reaction of aryl acetates. The base peak, 121 (100%) arises from the further loss of a methyl group. Its isotope peak at 122 (7.7%) indicates that the fragment has seven carbons, which is in agreement with a total of 10 carbons in the molecule.

^{13}C NMR

The ^{13}C spectrum shows eight peaks for 10 carbons. The more intense peaks (doublets in OR) at 121.8 and 129.9 ppm appear to be each representing a pair of isochronous CH carbons. As well as these two peaks, there are two other peaks in the aromatic region: quaternary carbons at 154.4 and 134.8 ppm. The peak at 154.4 ppm is so far downfield that it must be bonded to an oxygen. There are two other quaternary carbons that can be assigned to carbonyl groups,

one (196.8) to a ketone and the other (168.8) to an ester. The last two peaks (26.6 and 23.2) belong to two methyl groups.

^{1}H NMR

The proton NMR spectrum is quite simple. It shows two methyl singlets, each in the region of methyl groups attached to carbonyls, and two doublets in the aromatic region. The size of the observed splitting (8.9 Hz) is that expected for ortho coupling. The downfield protons at 7.99 ppm are deshielded by an ortho carbonyl, while the upfield protons at 7.19 ppm are shielded by an ortho oxygen atom.

Infrared

As well as the carbonyl bands at 1755 and $1675\,cm^{-1}$, there are peaks attributable to a benzene ring between 3000 and $3100\,cm^{-1}$, and at $1600\,cm^{-1}$, $1500\,cm^{-1}$, and $855\,cm^{-1}$.

Structure

The accumulated evidence establishes the presence of a benzene ring with two substituents, a ketone and an ester. The ester substituent is an acetoxy group, $CH_3(C=O)O-$, and the ketone substituent is an acetyl group, $CH_3C=O$. The equivalence of the ortho positions in the NMR spectra and the ortho coupling of the ring protons establish the substitution as para. The compound is 4-acetoxyacetophenone, $CH_3(C=O)C_6H_4O(C=O)CH_3$.

Exercise 078 (081)

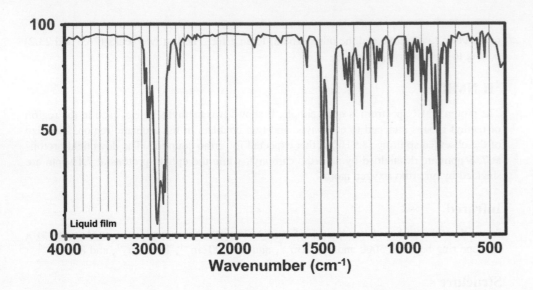

Liquid film

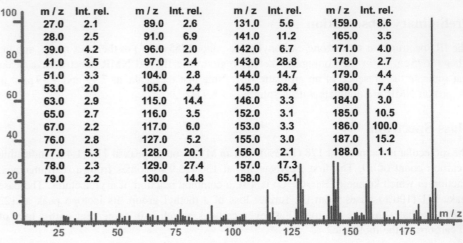

m/z	Int. rel.	m/z	Int. rel.	m/z	Int. rel.	m/z	Int. rel.
27.0	2.1	89.0	2.6	131.0	5.6	159.0	8.6
28.0	2.5	91.0	6.9	141.0	11.2	165.0	3.5
39.0	4.2	96.0	2.0	142.0	6.7	171.0	4.0
41.0	3.5	97.0	2.4	143.0	28.8	178.0	2.7
51.0	3.3	104.0	2.8	144.0	14.7	179.0	4.4
53.0	2.0	105.0	2.4	145.0	28.4	180.0	7.4
63.0	2.9	115.0	14.4	146.0	3.3	184.0	3.0
65.0	2.7	116.0	3.5	152.0	3.1	185.0	10.5
67.0	2.2	117.0	6.0	153.0	3.3	186.0	100.0
76.0	2.8	127.0	5.2	155.0	3.0	187.0	15.2
77.0	5.3	128.0	20.1	156.0	2.1	188.0	1.1
78.0	2.3	129.0	27.4	157.0	17.3		
79.0	2.2	130.0	14.8	158.0	65.1		

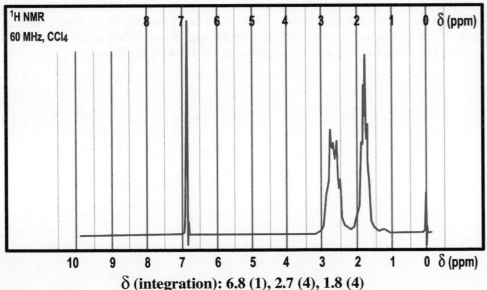

¹H NMR

60 MHz, CCl₄

δ (integration): 6.8 (1), 2.7 (4), 1.8 (4)

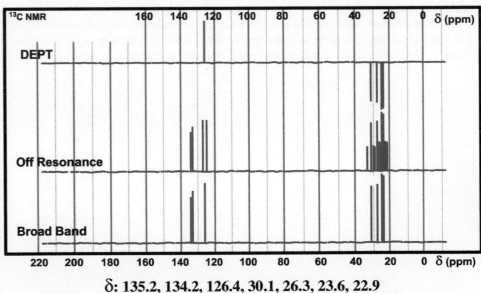

¹³C NMR

DEPT

Off Resonance

Broad Band

δ: 135.2, 134.2, 126.4, 30.1, 26.3, 23.6, 22.9

Exercise 078 (081)

Mass Spectrum

The molecular ion at 186 (100%) is very stable, forming the base peak of the spectrum. The M+1 isotope peak at 187 (15.2%) indicates that there are 14 carbons in the molecule. The total mass of 14 carbons is 168, that is 18 less than the molecular mass. The most likely molecular formula of the compound is $C_{14}H_{18}$. A very stable molecular ion is usually associated with a cyclic π system. The next most intense ion is the M−28 ion at 158 (65.1%). It is generated by a retro Diels–Alder reaction, a common mechanism for six-membered rings containing a π bond. The ion at 130 is most likely derived from the loss of a second ethylene molecule (28) from the 158 by a second retro Diels–Alder reaction.

^{13}C NMR

There are three peaks in the aromatic region, two quaternary (135.2, 134.2 ppm) and one CH carbon (126.4 ppm). The other peaks are all methylene carbons occurring in the aliphatic region (30.1, 26.3, 23.6, 22.9 ppm). With 14 carbons in the molecule, and only seven carbons seen in the spectrum, the molecule must have a symmetry element that divides the molecule into equivalent halves.

^{1}H NMR

The single peak in the aromatic region indicates only one type of aromatic proton, as expected from the observation of a single CH in the ^{13}C NMR spectrum. The relative shielding of these protons (6.8 ppm) is due to the inductive effect of the CH₂ groups. The multiplets centred at about 2.7 and 1.8 ppm represent the CH₂ groups. The lower field peaks are those of the CH₂ bonded to the aromatic ring, and the upper set are those of the CH₂ one further removed from the ring.

Infrared

The aromatic ring is evidenced by the bands just above 3000 cm⁻¹, the bands at 1500 and 1580 cm⁻¹, and the out-of-plane bending band at 800 cm⁻¹ corresponding to two adjacent protons on the aromatic ring.

Summary

The proposed formula ($C_{14}H_{18}$) has six centres of unsaturation and the benzene ring is responsible for four of these. That leaves two to be accounted for, in the form of rings or double bonds. Other than the sp² carbons of the benzene ring there are only sp³ methylene carbons; therefore the unsaturations are accounted for by two rings. The rings must be created by adding chains of four methylene groups to ortho carbons of the benzene ring. Two such rings can be added to the benzene ring to make either an octahydrophenanthrene ring or an octahydroanthracene ring. The octahydrophenanthrene ring has the correct symmetry to give the observed number of different types of methylene groups (four), whereas the octahydroanthracene ring gives only two types of methylene groups.

The resulting compound is 1,2,3,4,5,6,7,8-octahydro-phenanthrene. A plane of symmetry bisects the molecule through the centre ring with identical atoms on either side of the plane.

Exercise 079 (089)

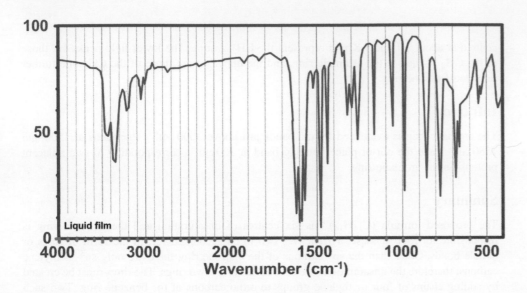

Liquid film

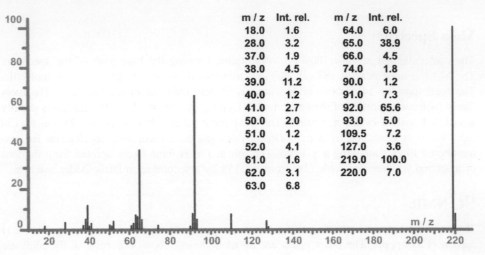

m / z	Int. rel.	m / z	Int. rel.
18.0	1.6	64.0	6.0
28.0	3.2	65.0	38.9
37.0	1.9	66.0	4.5
38.0	4.5	74.0	1.8
39.0	11.2	90.0	1.2
40.0	1.2	91.0	7.3
41.0	2.7	92.0	65.6
50.0	2.0	93.0	5.0
51.0	1.2	109.5	7.2
52.0	4.1	127.0	3.6
61.0	1.6	219.0	100.0
62.0	3.1	220.0	7.0
63.0	6.8		

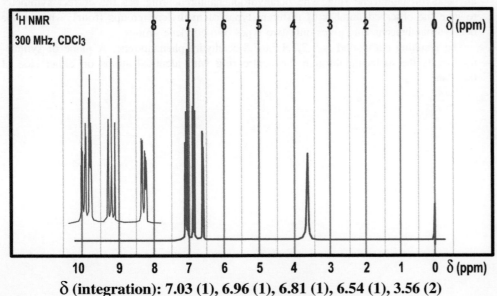

δ (integration): 7.03 (1), 6.96 (1), 6.81 (1), 6.54 (1), 3.56 (2)

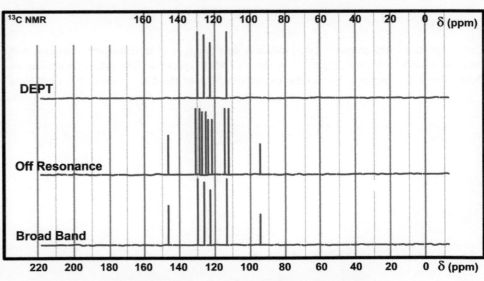

δ: 147.6, 130.7 127.2, 123.6, 114.2, 94.9

Exercise 079 (089)

Infrared

An aromatic ring is indicated by the bands between 3000 and 3050 cm^{-1}, by the strong band near 1600 cm^{-1} (strongly polarized >C=C<) and by the band at 1490 cm^{-1}. The peaks at 770 and 680 cm^{-1} represent the out-of-plane bending, corresponding to a meta-substituted benzene ring. The two bands at 3380 and 3450 cm^{-1} are characteristic symmetric and asymmetric N–H stretching of a primary amine. One of the strong bands at 1600 cm^{-1} corresponds to the scissoring deformation of the NH$_2$, while the medium broad band at 650 cm^{-1} represents wagging deformation bands.

^{13}C NMR

There are signals for six aromatic carbons, two quaternary carbons (147.6 and 94.9 ppm) and four CH carbons (130.7, 127.2, 123.6 and 114.2 ppm). The carbon at 147.6 ppm is deshielded relative to benzene (128.5 ppm) by 19.1 ppm. This is in the range expected for the effect of an NH$_2$ group (+18.2 ppm) on the carbon it is attached to (ipso carbon). The other substituent has a strong shielding effect on its carbon (−33.4 ppm). The only substituent that has such a large ipso shielding effect is iodine (−34.1). The shielding effect of the large electron cloud of the iodine atom is known as the 'heavy atom effect'. Estimation of the chemical shifts for a benzene ring with an iodine and NH$_2$ as meta substituents gives calculated values very close to those for this compound.

Mass Spectrum

The spectrum has a peak for the molecular ion at 219 (100%) and a strong peak for the loss of iodine (127) at 92 (65.6%). The molecular ion is very stable, having a large electron source in the ring π system, and the non-bonding electrons of the nitrogen and the iodine. The system is so electron rich that it will even support the loss of two electrons, as evidenced by the presence of the doubly charged molecular ion at 109.5 (7.2%). The peak at 65 (39%) is formed from the 92 ion by the loss of HCN (27).

^{1}H NMR

The NH$_2$ protons are found in the slightly broadened singlet at 3.56 ppm. There are four signals for the aromatic protons, which are split by $^3J_{ortho}$ couplings into the following patterns: 7.03 (doublet), 6.96 (unsplit), 6.81 (triplet) and 6.54 (doublet). The peaks at 7.03, 6.96 and 6.54 ppm also show smaller $^4J_{meta}$ coupling. This ^3J splitting pattern is compatible with meta substitution only; the peak at 6.96 ppm, which shows only $^4J_{meta}$ coupling, is between the two substituents. Para substitution would have a symmetrical AA'XX' pattern, and ortho substitution would have two triplets and two doublets from the ^3J couplings. The assignments of the chemical shifts are: 6.54, ortho NH$_2$, para I; 6.81, meta NH$_2$, meta I; 6.96, ortho NH$_2$, ortho I; and 7.03, para NH$_2$, ortho I. These are compatible with the calculated values from the substituent tables.

Summary

The accumulated evidence indicates that the molecule has an aromatic ring with a primary amino group and an iodine as substituents. The lack of symmetry indicates that the ring cannot be para substituted. The out-of-plane deformation bands in the IR spectrum are compatible with meta substitution but do not provide conclusive evidence. The splitting pattern in ^{1}H NMR, and the chemical shifts in the ^{1}H and ^{13}C spectra provide conclusive proof that the ring is meta substituted. The compound is 3-iodoaniline.

Exercise 080 (090)

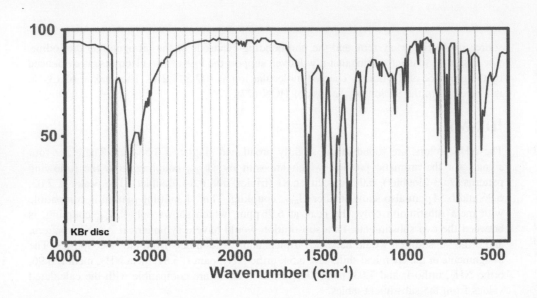

High resolution mass = 228.0721

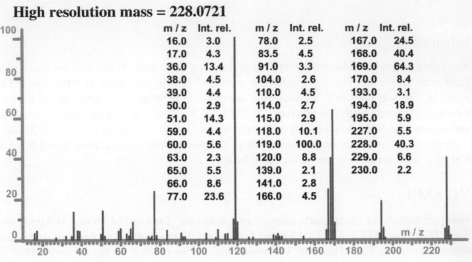

m / z	Int. rel.	m / z	Int. rel.	m / z	Int. rel.
16.0	3.0	78.0	2.5	167.0	24.5
17.0	4.3	83.5	4.5	168.0	40.4
36.0	13.4	91.0	3.3	169.0	64.3
38.0	4.5	104.0	2.6	170.0	8.4
39.0	4.4	110.0	4.5	193.0	3.1
50.0	2.9	114.0	2.7	194.0	18.9
51.0	14.3	115.0	2.9	195.0	5.9
59.0	4.4	118.0	10.1	227.0	5.5
60.0	5.6	119.0	100.0	228.0	40.3
63.0	2.3	120.0	8.8	229.0	6.6
65.0	5.5	139.0	2.1	230.0	2.2
66.0	8.6	141.0	2.8		
77.0	23.6	166.0	4.5		

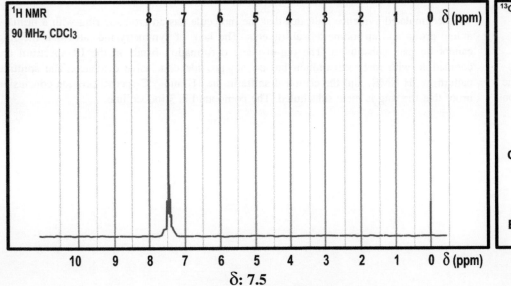

δ: 7.5

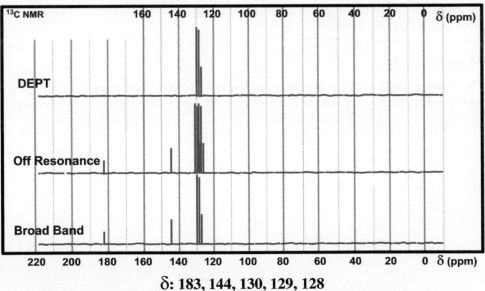

δ: 183, 144, 130, 129, 128

Exercise 080 (090)

Preliminary Observation

The IR spectrum indicates the presence of an aromatic ring and a primary amino group (3280 and $3460\,cm^{-1}$).

Mass Spectrum

The molecular peak at 228 (40.3%) has isotope peaks for $M+1$ at 229 (6.6%) and $M+2$ at 230 (2.2%), which gives relative percentages of $M+1 = 16.3\%$ and $M+2 = 5.4\%$. Sulfur or silicon could give an $M+2$ in this range, but silicon would have a stronger $M+1$. Thus, we have a compound containing sulfur, and since the natural abundance of ^{34}S is 4.4%, the observed intensity of 5.4% for $M+2$ indicates that there is one S in the molecule. Calculation of the number of carbons from the intensity of $M+1$ (16.3%) has to take into account the contribution of ^{33}S and ^{15}N. For example, one S (0.8%) and two N ($2 \times 0.36\%$) contribute 1.5% to $M+1$, thus the contribution of ^{13}C to $M+1$ would be 14.8%, ($16.3 - 1.5$), indicating the presence of 13 carbons in the molecule. The base peak at 119 does not contain any sulfur, as evidenced by the lack of a peak at 121. It represents the loss of 109, consisting of a phenyl group (77) and a sulfur atom (32), the thiophenoxy radical. This is a typical fragmentation in compounds containing a phenyl ring and sulfur.

^{13}C NMR

There are four carbons in the aromatic region: two CH peaks of twice the intensity of a third CH peak, and a quaternary carbon. This is the typical pattern of a monosubstituted benzene ring. The final peak at $182.7\,ppm$ is a strongly deshielded quaternary carbon. This is in the region where one finds sp^2 carbons bonded to heteroatoms, such as in a carbonyl group.

However, there is no evidence of a carbonyl group in the IR spectrum, but there is a sulfur atom in the compound, so this could be a thiocarbonyl group, $>C=S$.

^1H NMR

The ^1H NMR spectrum has only one signal ($7.5\,ppm$) in the aromatic region. The lack of separation of the ring protons indicates that it is substituted with a non-polar group. The occurrence of the NH protons at the same chemical shift is accidental.

Infrared

The aromatic ring is evidenced by the C–H bands between 3000 and $3500\,cm^{-1}$, the $>C=C<$ stretching bands near 1600 and $1500\,cm^{-1}$, and the out-of-plane bending bands at $700\,cm^{-1}$ and $760\,cm^{-1}$ for five adjacent ring protons. The two bands at 3290 and $3470\,cm^{-1}$ represent a primary amine (N–H symmetric and asymmetric stretching). One of the strong bands at $1600\,cm^{-1}$ represents the scissoring deformation of the $-NH_2$. The $>C=S-$ stretching band is probably at $1440\,cm^{-1}$.

Summary

The only carbons in the molecule other than the quaternary carbon of the functional group are the carbons of a phenyl ring. As the molecule has 13 carbons, there must be two such phenyl rings. The quaternary carbon of the functional group bears the sulfur atom, the primary amino group and the nitrogen with its two phenyl rings. The molecular formula is $C_{13}H_{12}N_2S$. The molecule is created by attaching the two phenyl rings to a N which forms part of a thiourea $(C_6H_5)_2N-(C=S)-$. Still to be placed is the NH_2 group. The completed structure is N,N-diphenylthiourea, $(C_6H_5)_2N(C=S)NH_2$.

Exercise 081 (100)

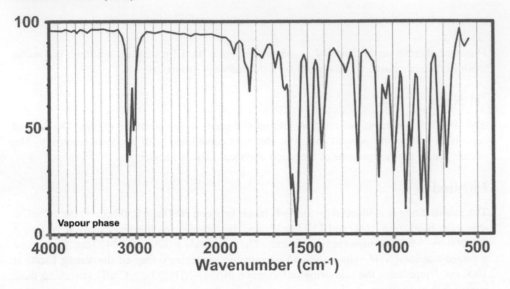

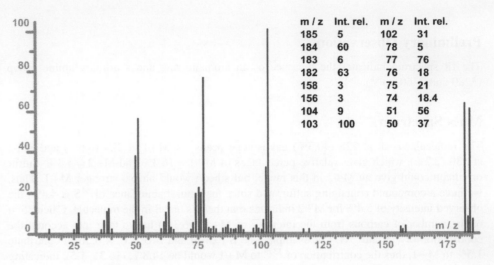

m / z	Int. rel.	m / z	Int. rel.
185	5	102	31
184	60	78	5
183	6	77	76
182	63	76	18
158	3	75	21
156	3	74	18.4
104	9	51	56
103	100	50	37

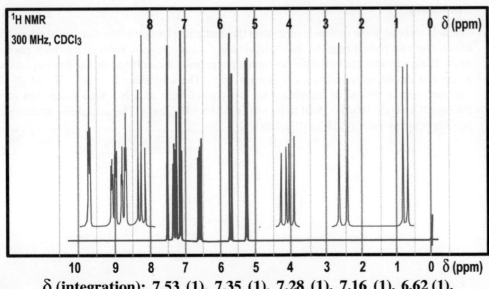

δ (integration): 7.53 (1), 7.35 (1), 7.28 (1), 7.16 (1), 6.62 (1), 5.73 (1), 5.27 (1)

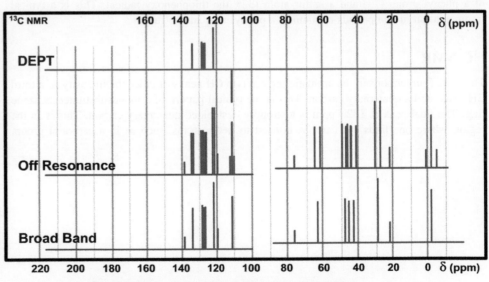

δ: 139, 135, 132, 131, 130, 125, 123, 115

Exercise 081 (100)

Preliminary Observation

The verbal descriptions given in these boxes show accepted formats for presenting spectral data. The peaks around $3500\,\text{cm}^{-1}$ are overtone bands.

Infrared

IR (gas) 3094, 3070, 3016, 1589, 1562, 1475, 1412, 1199, 1074, 989, 916, 881, 821, 785, 706, 666 cm^{-1}.

Mass Spectrum

EIMS (70 eV) m/z (relative intensity) 185 (5), 184 (60), 183 (6), 182 (63), 158 (3), 156 (3), 104 (9), 103 (100), 102 (31), 78 (5), 77 (76), 76 (18), 75 (21), 74 (18.4), 51 (56), 50 (37).

Note the peaks at 182 and 184.

^{13}C NMR

^{13}C NMR (300 MHz, CDCl$_3$) δ 139, 135, 132, 131, 130, 125, 123, 115. Note that all of the peaks are in the sp^2 carbon region.

^{1}H NMR

^{1}H NMR (300 MHz, CDCl$_3$) δ 7.53 (t, J = 1.7 Hz, 1H), 7.35 (dt, J = 7.8, 1.7 Hz, 1H), 7.28 (dt, J = 7.8, 1.7 Hz, 1H), 7.16 (t, J = 7.8 Hz, 1H), 6.62 (dd, J = 17.5, 10.8 Hz, 1H), 5.73 (dd, J = 17.5, 0.7 Hz, 1H), 5.27 (dd, J = 10.8, 0.7, 1H).

There is a lot of information in the splitting patterns. Assign all of the couplings. The triplet (J = 7.8 Hz) is significant. Note the very large coupling of 17.5 Hz.

Structure

Exercise 082 (079)

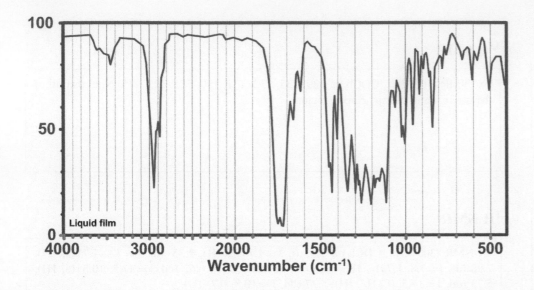

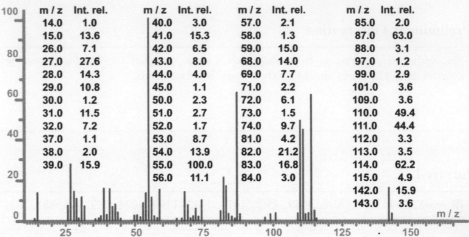

m / z	Int. rel.	m / z	Int. rel.	m / z	Int. rel.	m / z	Int. rel.
14.0	1.0	40.0	3.0	57.0	2.1	85.0	2.0
15.0	13.6	41.0	15.3	58.0	1.3	87.0	63.0
26.0	7.1	42.0	6.5	59.0	15.0	88.0	3.1
27.0	27.6	43.0	8.0	68.0	14.0	97.0	1.2
28.0	14.3	44.0	4.0	69.0	7.7	99.0	2.9
29.0	10.8	45.0	1.1	71.0	2.2	101.0	3.6
30.0	1.2	50.0	2.3	72.0	6.1	109.0	3.6
31.0	11.5	51.0	2.7	73.0	1.5	110.0	49.4
32.0	7.2	52.0	1.7	74.0	9.7	111.0	44.4
37.0	1.1	53.0	8.7	81.0	4.2	112.0	3.3
38.0	2.0	54.0	13.9	82.0	21.2	113.0	3.5
39.0	15.9	55.0	100.0	83.0	16.8	114.0	62.2
		56.0	11.1	84.0	3.0	115.0	4.9
						142.0	15.9
						143.0	3.6

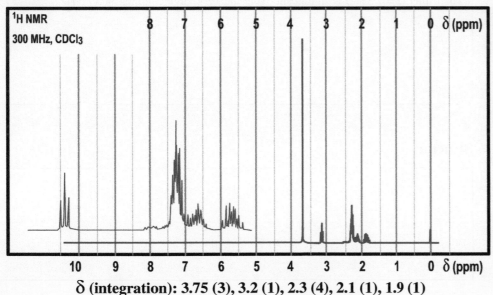

δ (integration): 3.75 (3), 3.2 (1), 2.3 (4), 2.1 (1), 1.9 (1)

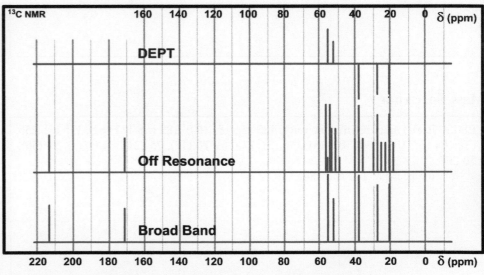

δ: 212.2, 169.9, 54.7, 52.3, 38.0, 27.5, 21.0

Exercise 082 (079)

Preliminary Observation

Infrared

The two peaks at 1766 and 1728 cm^{-1} indicate the presence of two functional groups. This can be confirmed by the carbon NMR. Take careful note their IR frequencies.

Mass Spectrum

The peak at 142 is that of the molecular ion.

^{13}C NMR

The DEPT spectrum has information that can help with the interpretation of the proton NMR.

^{1}H NMR

Note the integration ratios, and check the chemical shifts of the singlet and the triplet. Compare the integration ratios with the data in the off-resonance ^{13}C NMR spectrum. This may help to explain the complexity of the signals between 1.9 and 2.1 ppm.

Structure

Exercise 083 (075)

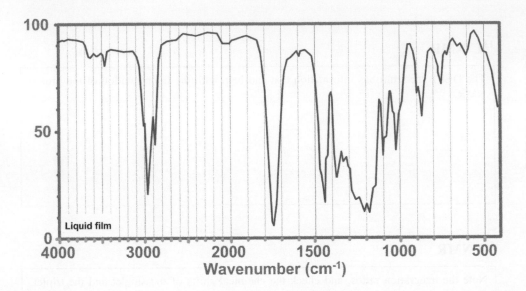

Liquid film

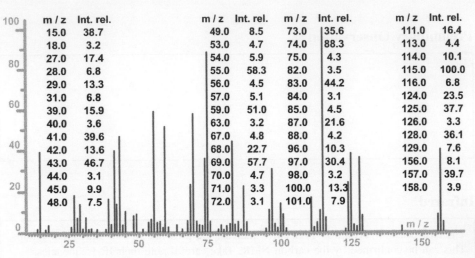

m / z	Int. rel.		m / z	Int. rel.	m / z	Int. rel.		m / z	Int. rel.
15.0	38.7		49.0	8.5	73.0	35.6		111.0	16.4
18.0	3.2		53.0	4.7	74.0	88.3		113.0	4.4
27.0	17.4		54.0	5.9	75.0	4.3		114.0	10.1
28.0	6.8		55.0	58.3	82.0	3.5		115.0	100.0
29.0	13.3		56.0	4.5	83.0	44.2		116.0	6.8
31.0	6.8		57.0	5.1	84.0	3.1		124.0	23.5
39.0	15.9		59.0	51.0	85.0	4.5		125.0	37.7
40.0	3.6		63.0	3.2	87.0	21.6		126.0	3.3
41.0	39.6		67.0	4.8	88.0	4.2		128.0	36.1
42.0	13.6		68.0	22.7	96.0	10.3		129.0	7.6
43.0	46.7		69.0	57.7	97.0	30.4		156.0	8.1
44.0	3.1		70.0	4.7	98.0	3.2		157.0	39.7
45.0	9.9		71.0	3.3	100.0	13.3		158.0	3.9
48.0	7.5		72.0	3.1	101.0	7.9			

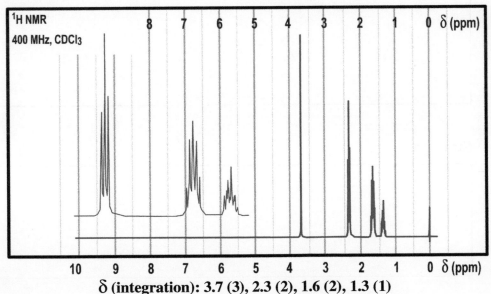

¹H NMR
400 MHz, CDCl₃

δ (integration): 3.7 (3), 2.3 (2), 1.6 (2), 1.3 (1)

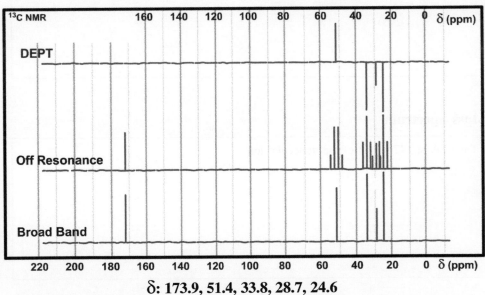

¹³C NMR
DEPT
Off Resonance
Broad Band

δ: 173.9, 51.4, 33.8, 28.7, 24.6

Exercise 083 (075)

Preliminary Observation

Infrared

The functional group can be identified and confirmed by the ^{13}C NMR.

Mass Spectrum

There is no molecular peak in this spectrum, a typical situation for compounds with this functional group.

^{13}C NMR

^{1}H NMR

^{1}H NMR (400 MHz, CDCl$_3$) δ 3.67 (s, 3H), 2.32 (t, J = 7.5 Hz, 2H), 1.66 (quintet, J = 7.5 Hz, 2H), 1.36 (m, 1H).

Compare the integration ratios with the data in the off-resonance spectrum to help with the construction of the molecule.

Structure

Exercise 084 (061)

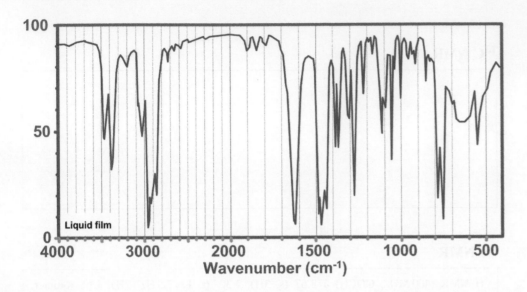

Liquid film

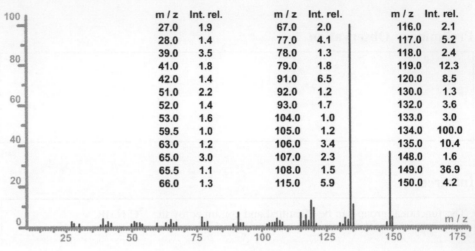

m / z	Int. rel.	m / z	Int. rel.	m / z	Int. rel.
27.0	1.9	67.0	2.0	116.0	2.1
28.0	1.4	77.0	4.1	117.0	5.2
39.0	3.5	78.0	1.3	118.0	2.4
41.0	1.8	79.0	1.8	119.0	12.3
42.0	1.4	91.0	6.5	120.0	8.5
51.0	2.2	92.0	1.2	130.0	1.3
52.0	1.4	93.0	1.7	132.0	3.6
53.0	1.6	104.0	1.0	133.0	3.0
59.5	1.0	105.0	1.2	134.0	100.0
63.0	1.2	106.0	3.4	135.0	10.4
65.0	3.0	107.0	2.3	148.0	1.6
65.5	1.1	108.0	1.5	149.0	36.9
66.0	1.3	115.0	5.9	150.0	4.2

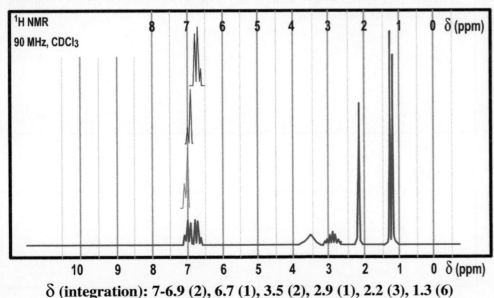

δ (integration): 7-6.9 (2), 6.7 (1), 3.5 (2), 2.9 (1), 2.2 (3), 1.3 (6)

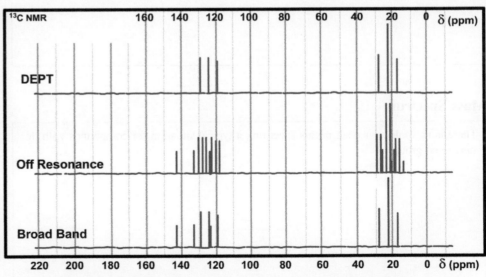

δ: 141.5, 131.8, 127.9, 123.1, 122.1, 118.2, 27.8, 22.3, 17.9

Exercise 084 (061)

Preliminary Observations

Infrared

Look carefully at the frequency of the strong peak just above $1600\,cm^{-1}$, and don't forget to confirm any proposed functional group with the ^{13}C data. The out-of-plane deformation bands give an indication of the substitution pattern of the ring.

Mass Spectrum

The peak of the molecular ion is at 149.

^{13}C NMR

The DEPT and OR spectra indicate how many substituents are present on the ring.

^1H NMR

^1H NMR (90 MHz, CDCl$_3$) δ 7.04 (d, J = 8 Hz, 1H), 6.95 (d, J = 8 Hz, 1H), 6.73 (t, J = 8 Hz, 1H), 3.5 (br, 2H), 2.92 (septet, J = 6.8 Hz, 1H), 2.18 (s, 3H), 1.26 (d, J = 6.8 Hz, 6H). The chemical shift of the aromatic triplet gives important information.

Structure

Exercise 085 (052)

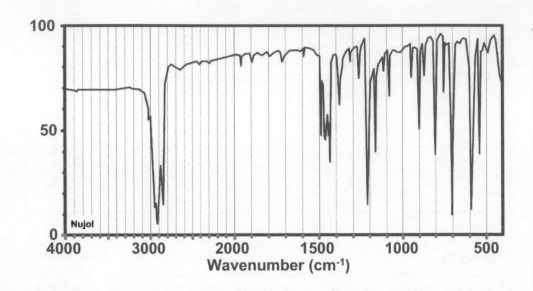

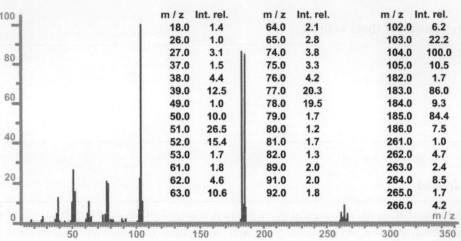

m / z	Int. rel.	m / z	Int. rel.	m / z	Int. rel.
18.0	1.4	64.0	2.1	102.0	6.2
26.0	1.0	65.0	2.8	103.0	22.2
27.0	3.1	74.0	3.8	104.0	100.0
37.0	1.5	75.0	3.3	105.0	10.5
38.0	4.4	76.0	4.2	182.0	1.7
39.0	12.5	77.0	20.3	183.0	86.0
49.0	1.0	78.0	19.5	184.0	9.3
50.0	10.0	79.0	1.7	185.0	84.4
51.0	26.5	80.0	1.2	186.0	7.5
52.0	15.4	81.0	1.7	261.0	1.0
53.0	1.7	82.0	1.3	262.0	4.7
61.0	1.8	89.0	2.0	263.0	2.4
62.0	4.6	91.0	2.0	264.0	8.5
63.0	10.6	92.0	1.8	265.0	1.7
				266.0	4.2

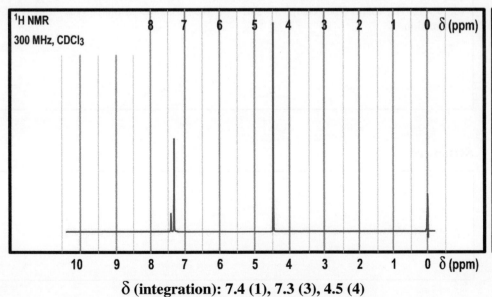

δ (integration): 7.4 (1), 7.3 (3), 4.5 (4)

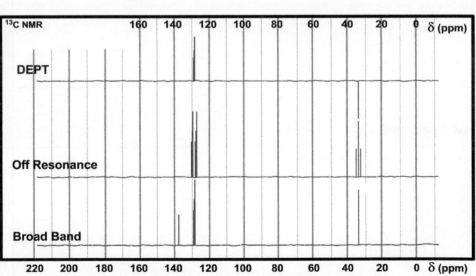

δ: 138.4, 129.5, 129.2, 129.1, 32.8

Exercise 085 (052)

Preliminary Observation

Infrared

Mass Spectrum

The peaks at 262, 264 and 266 form a particular isotope pattern.

^{13}C NMR

The number of carbons observed in the aromatic region should help to determine the substitution pattern.

^{1}H NMR

The chemical shift of one of the aromatic protons is slightly different from the others.

Structure

Exercise 086 (069)

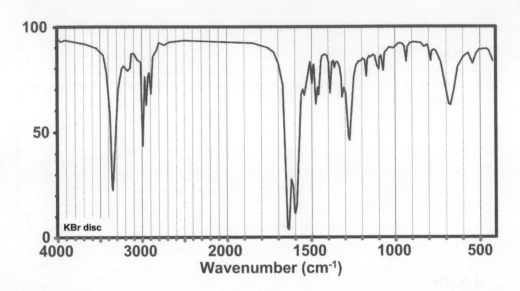

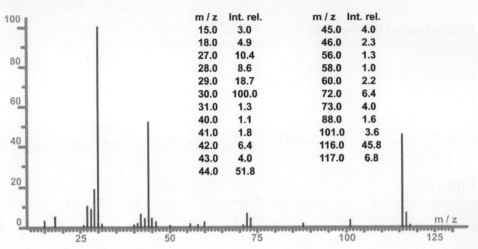

m / z	Int. rel.	m / z	Int. rel.
15.0	3.0	45.0	4.0
18.0	4.9	46.0	2.3
27.0	10.4	56.0	1.3
28.0	8.6	58.0	1.0
29.0	18.7	60.0	2.2
30.0	100.0	72.0	6.4
31.0	1.3	73.0	4.0
40.0	1.1	88.0	1.6
41.0	1.8	101.0	3.6
42.0	6.4	116.0	45.8
43.0	4.0	117.0	6.8
44.0	51.8		

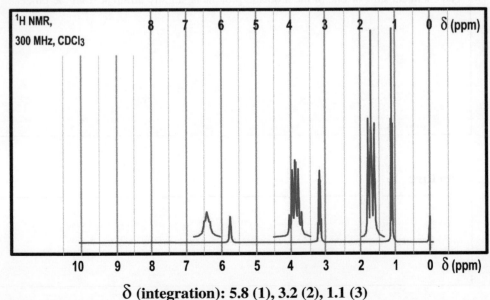

δ (integration): 5.8 (1), 3.2 (2), 1.1 (3)

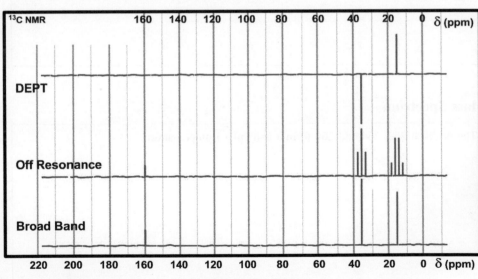

δ: 161.4, 35.7, 15.3

Exercise 086 (069)

Preliminary Observation

Infrared

A careful analysis of this spectrum should reveal bands that give a strong indication of the functional group of the molecule.

Mass Spectrum

The peak at 116 is the molecular ion, although in view of the presence of the nitrogen-containing even-mass fragment at 30 (100%), one might have been looking for an odd-mass molecular ion.

^{13}C NMR

^{1}H NMR

^{1}H NMR (300 MHz, CDCl$_3$) t 5.8 (br, 2H, NH), 3.2, (dt, J = 5.6, 7.3 Hz, 4H), 1.1 (t, J = 7.3 Hz, 6H). The alkyl group is easily identified. Note the additional coupling in the CH$_2$ group.

Structure

You must take into account the even mass and any apparent conflict with a possible functional group. Note the simplicity of the ^{13}C NMR spectrum.

Exercise 087 (062)

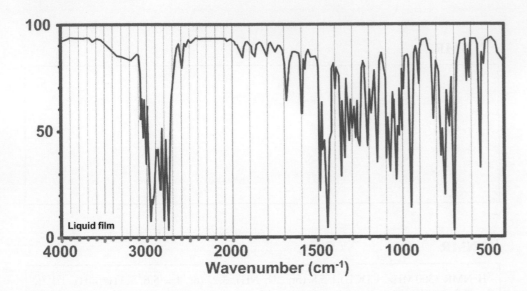

Liquid film

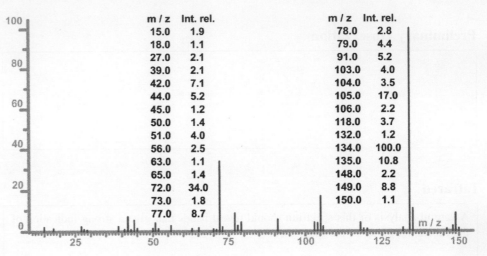

m / z	Int. rel.	m / z	Int. rel.
15.0	1.9	78.0	2.8
18.0	1.1	79.0	4.4
27.0	2.1	91.0	5.2
39.0	2.1	103.0	4.0
42.0	7.1	104.0	3.5
44.0	5.2	105.0	17.0
45.0	1.2	106.0	2.2
50.0	1.4	118.0	3.7
51.0	4.0	132.0	1.2
56.0	2.5	134.0	100.0
63.0	1.1	135.0	10.8
65.0	1.4	148.0	2.2
72.0	34.0	149.0	8.8
73.0	1.8	150.0	1.1
77.0	8.7		

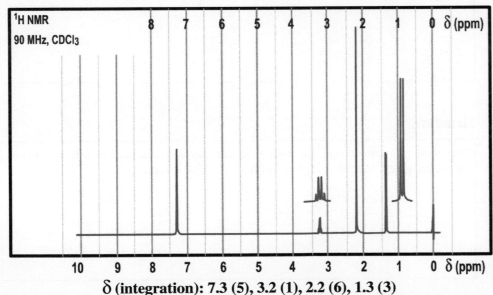

δ (integration): 7.3 (5), 3.2 (1), 2.2 (6), 1.3 (3)

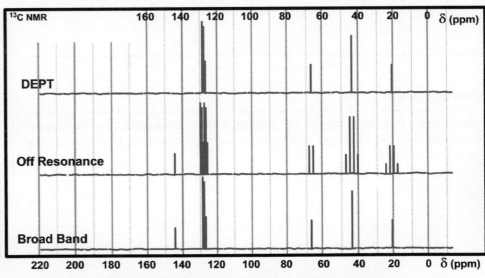

δ: 144.2, 128.2, 127.5, 126.9, 66, 43.2, 20.2

Exercise 087 (062)

Preliminary Observation

Infrared

There is a functional group present that is difficult to detect in the IR spectrum.

Mass Spectrum

The parent peak is at 149.

^{13}C NMR

Sometimes in ^{13}C NMR, it is possible to compare intensities of the signals for carbons of same kind. In this case comparisons of the CH signals and of the CH$_3$ signals are useful.

^{1}H NMR

^{1}H NMR (90 MHz, CDCl$_3$) δ 7.3 (s, 5H, Ar), 3.2 (q, J = 8 Hz, 1H), 2.2 (s, 6H), 1.3 (d, J = 8 Hz, 3H).

Structure

Exercise 088 (072)

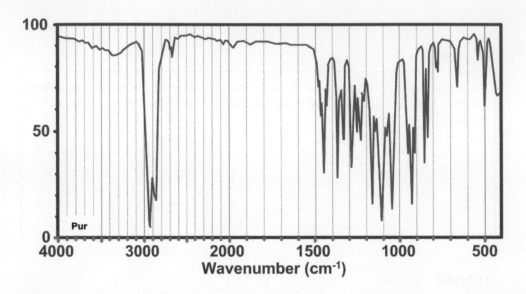

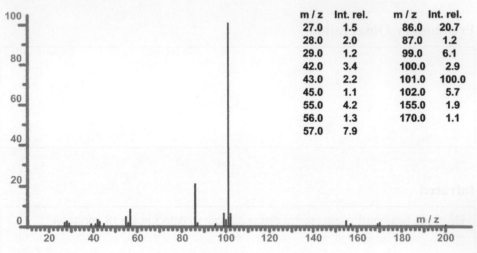

m / z	Int. rel.	m / z	Int. rel.
27.0	1.5	86.0	20.7
28.0	2.0	87.0	1.2
29.0	1.2	99.0	6.1
42.0	3.4	100.0	2.9
43.0	2.2	101.0	100.0
45.0	1.1	102.0	5.7
55.0	4.2	155.0	1.9
56.0	1.3	170.0	1.1
57.0	7.9		

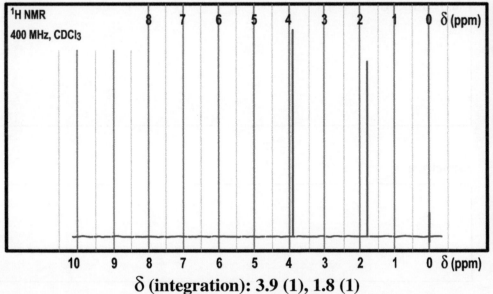

δ (integration): 3.9 (1), 1.8 (1)

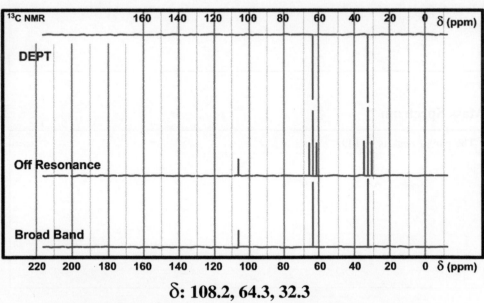

δ: 108.2, 64.3, 32.3

Exercise 088 (072)

Preliminary Observation

Infrared

It is possible to eliminate many functional groups with this spectrum.

Mass Spectrum

Take note of the very small peak at 200; although its numerical value is not recorded on the chart it is an important ion, the molecular ion.

^{13}C NMR

Considering the mass of the compound, there are very few peaks. The IR spectrum should be taken into consideration when trying to explain the chemical shift of the peak at 108 ppm.

^1H NMR

It is noteworthy that the alkyl groups show no coupling although the ^{13}C spectrum shows them to be methylene groups.

Structure

The compound obviously has some elements of symmetry that account for the equivalence of its structural components.

Exercise 089 (053)

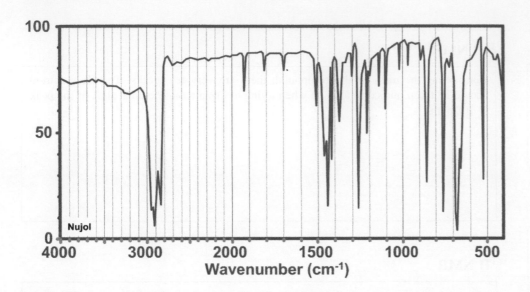

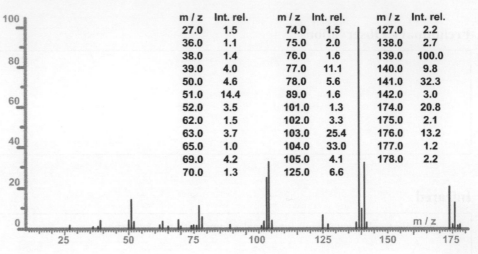

m/z	Int. rel.	m/z	Int. rel.	m/z	Int. rel.
27.0	1.5	74.0	1.5	127.0	2.2
36.0	1.1	75.0	2.0	138.0	2.7
38.0	1.4	76.0	1.6	139.0	100.0
39.0	4.0	77.0	11.1	140.0	9.8
50.0	4.6	78.0	5.6	141.0	32.3
51.0	14.4	89.0	1.6	142.0	3.0
52.0	3.5	101.0	1.3	174.0	20.8
62.0	1.5	102.0	3.3	175.0	2.1
63.0	3.7	103.0	25.4	176.0	13.2
65.0	1.0	104.0	33.0	177.0	1.2
69.0	4.2	105.0	4.1	178.0	2.2
70.0	1.3	125.0	6.6		

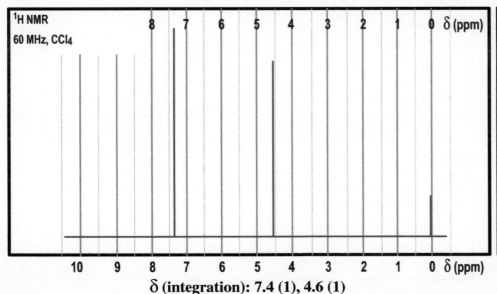

δ (integration): 7.4 (1), 4.6 (1)

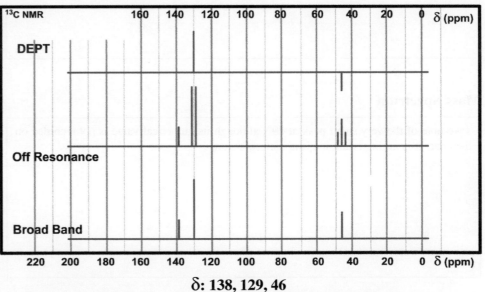

δ: 138, 129, 46

Exercise 089 (053)

Preliminary Observation

Infrared

Although the overtone-combination bands between 2000 and 1600 cm^{-1} are particularly sharp, and the out-of-plane bands between 700 and 900 cm^{-1} are difficult to assign, it appears that the compound may have an aromatic ring.

Mass Spectrum

The molecular peak is at 174. There are very significant isotope peaks in the spectrum.

^{13}C NMR

With only three peaks and such a large molecular mass, there have to be some elements of symmetry in the compound. The two signals at 128 and 139 ppm are characteristic of a particular aromatic substitution pattern.

^{1}H NMR

The lack of coupling and the chemical shifts of the protons (considering the mass spectrum) give critical information about the compound.

Summary

Exercise 090 (056)

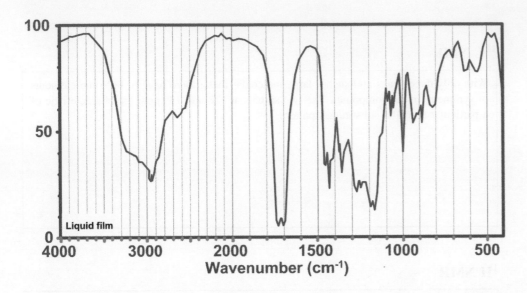

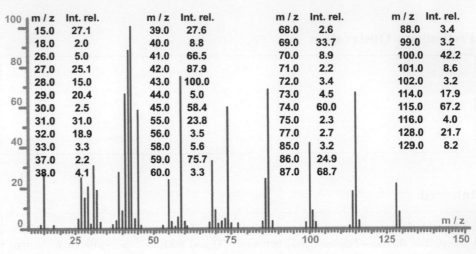

m / z	Int. rel.	m / z	Int. rel.	m / z	Int. rel.	m / z	Int. rel.
15.0	27.1	39.0	27.6	68.0	2.6	88.0	3.4
18.0	2.0	40.0	8.8	69.0	33.7	99.0	3.2
26.0	5.0	41.0	66.5	70.0	8.9	100.0	42.2
27.0	25.1	42.0	87.9	71.0	2.2	101.0	8.6
28.0	15.0	43.0	100.0	72.0	3.4	102.0	3.2
29.0	20.4	44.0	5.0	73.0	4.5	114.0	17.9
30.0	2.5	45.0	58.4	74.0	60.0	115.0	67.2
31.0	31.0	55.0	23.8	75.0	2.3	116.0	4.0
32.0	18.9	56.0	3.5	77.0	2.7	128.0	21.7
33.0	3.3	58.0	5.6	85.0	3.2	129.0	8.2
37.0	2.2	59.0	75.7	86.0	24.9		
38.0	4.1	60.0	3.3	87.0	68.7		

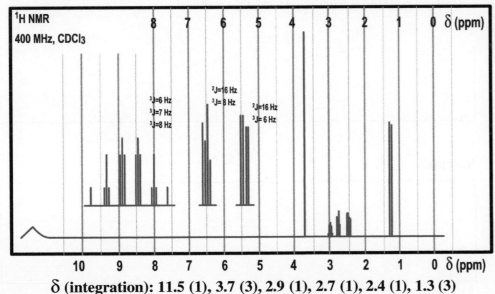

δ (integration): 11.5 (1), 3.7 (3), 2.9 (1), 2.7 (1), 2.4 (1), 1.3 (3)

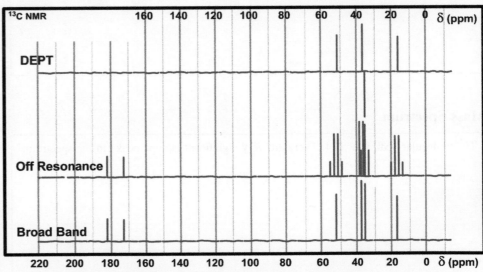

δ: 181.5, 172.3, 51.9, 37.1, 35.73, 16.8

Exercise 090 (056)

Preliminary Observation

Infrared

A functional group is easily identified, but note that there are two peaks in the signal near $1700\,\text{cm}^{-1}$.

^{13}C NMR

The presence of two functional groups is verified by this spectrum.

^{1}H NMR

^{1}H NMR (400 MHz, CDCl$_3$) δ 11.5 (br, 1H), 3.7 (s, 3H), 2.97 (ddq, J = 8, 6, 7 Hz, 1H), 2.75 (dd, J = 16, 8, 1H), 2.4 (dd, J = 16, 6 Hz, 1H), 1.3 (d, J = 7 Hz, 3H).

The presence of three CH signals with integrations of 1 seems to be in contradiction with the data of the ^{13}C NMR off-resonance spectrum. This apparent conflict is resolved by stereochemical considerations.

Mass Spectrum

The peak at 128 is not the molecular ion. The functional groups are easily cleaved. An analysis of the possible McLafferty rearrangements may be essential in distinguishing between two possible structures.

Summary

Using the two functional groups and the identified alkyl groups, construction of the molecule leads to two possible structures. They may be distinguished by careful analysis of the mass spectrum.

Exercise 091 (058)

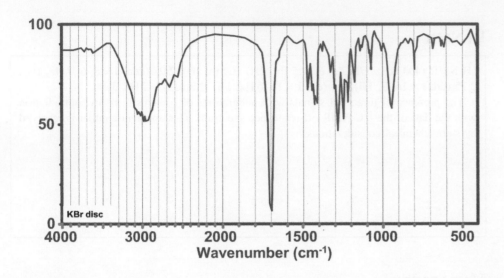

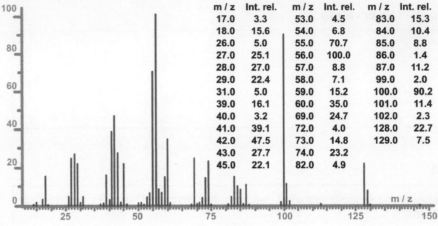

m / z	Int. rel.	m / z	Int. rel.	m / z	Int. rel.
17.0	3.3	53.0	4.5	83.0	15.3
18.0	15.6	54.0	6.8	84.0	10.4
26.0	5.0	55.0	70.7	85.0	8.8
27.0	25.1	56.0	100.0	86.0	1.4
28.0	27.0	57.0	8.8	87.0	11.2
29.0	22.4	58.0	7.1	99.0	2.0
31.0	5.0	59.0	15.2	100.0	90.2
39.0	16.1	60.0	35.0	101.0	11.4
40.0	3.2	69.0	24.7	102.0	2.3
41.0	39.1	72.0	4.0	128.0	22.7
42.0	47.5	73.0	14.8	129.0	7.5
43.0	27.7	74.0	23.2		
45.0	22.1	82.0	4.9		

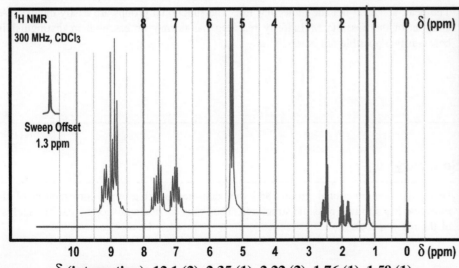

δ (integration): 12.1 (2), 2.35 (1), 2.22 (2), 1.76 (1), 1.58 (1), 1.06 (3)

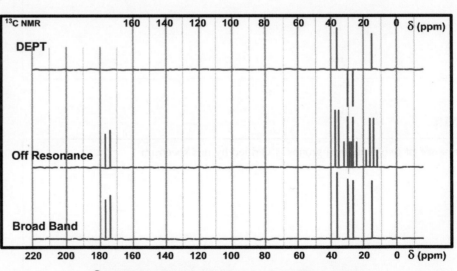

δ: 176.89, 173.97, 37.95, 31.35, 28.29, 16.73

Exercise 091 (058)

Preliminary Observation

Infrared

Mass Spectrum

Although not shown on the peak list, there are small peaks at 146 (0.7%) and 147 (0.3%). The large number of even-mass fragments suggests rearrangements that involve the loss of small molecules.

^{13}C NMR

^{1}H NMR

^{1}H NMR (300 MHz, CDCl$_3$) δ 12.1 (s, 2H), 2.56 (sextet, J = 7 Hz, 1H), 2.45 (t, J = 7 Hz, 2H), 2.01 (dq, J = 14, 7 Hz, 1H), 1.82 (dq, J = 14, 7 Hz, 1H), 1.23 (d, J = 7 Hz, 3H).

Structure

Exercise 092 (073)

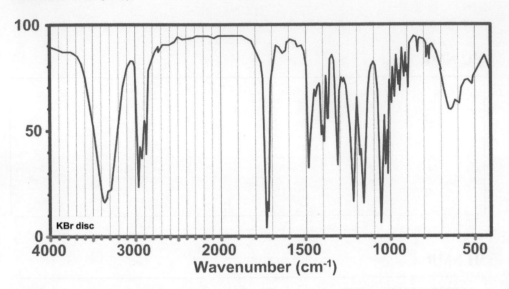

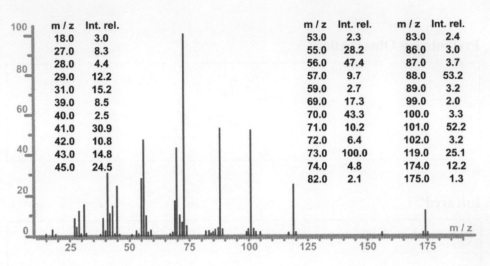

m / z	Int. rel.
18.0	3.0
27.0	8.3
28.0	4.4
29.0	12.2
31.0	15.2
39.0	8.5
40.0	2.5
41.0	30.9
42.0	10.8
43.0	14.8
45.0	24.5

m / z	Int. rel.
53.0	2.3
55.0	28.2
56.0	47.4
57.0	9.7
59.0	2.7
69.0	17.3
70.0	43.3
71.0	10.2
72.0	6.4
73.0	100.0
74.0	4.8
82.0	2.1

m / z	Int. rel.
83.0	2.4
86.0	3.0
87.0	3.7
88.0	53.2
89.0	3.2
99.0	2.0
100.0	3.3
101.0	52.2
102.0	3.2
119.0	25.1
174.0	12.2
175.0	1.3

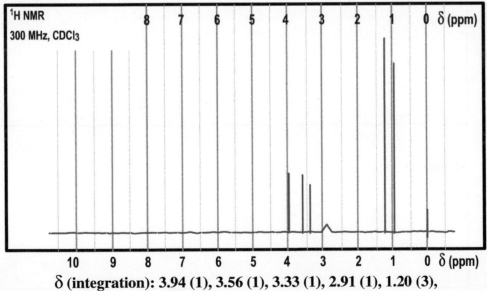

δ (integration): 3.94 (1), 3.56 (1), 3.33 (1), 2.91 (1), 1.20 (3), 0.93 (3)

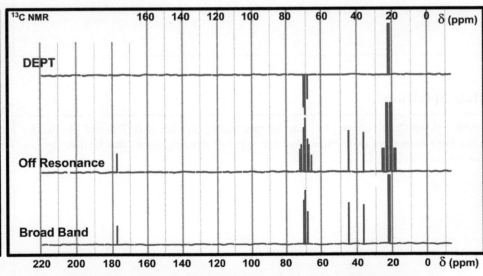

δ: 177.51, 69.81, 69.72, 68.28, 44.78, 36.23, 22.10, 21.59

Exercise 092 (073)

Preliminary Observation

Infrared

The indications of functional groups in this spectrum can be verified by analysis of the NMR spectra.

Mass Spectrum

As there is a carbonyl in the compound, a peak for the molecular ion is not expected because of prevalent α cleavage, and possible β cleavage if a γ hydrogen is present.

^{13}C NMR

Don't neglect the three peaks between 70 and 65 ppm.

^{1}H NMR

The lack of coupling limits the possibilities of construction of the molecule. Look at the integration and compare it with the ^{13}C NMR data.

Structure

Exercise 093 (076)

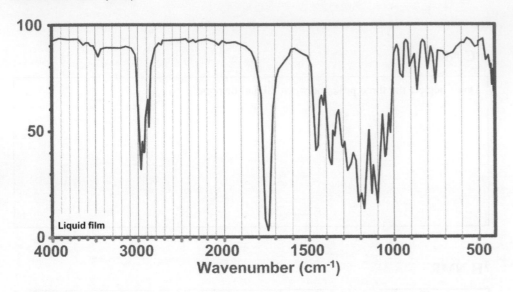

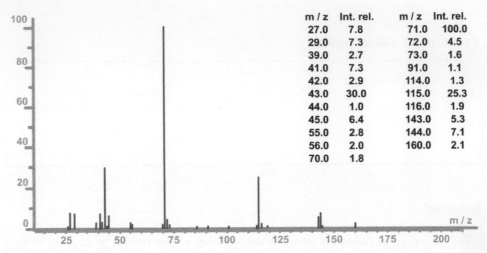

m / z	Int. rel.	m / z	Int. rel.
27.0	7.8	71.0	100.0
29.0	7.3	72.0	4.5
39.0	2.7	73.0	1.6
41.0	7.3	91.0	1.1
42.0	2.9	114.0	1.3
43.0	30.0	115.0	25.3
44.0	1.0	116.0	1.9
45.0	6.4	143.0	5.3
55.0	2.8	144.0	7.1
56.0	2.0	160.0	2.1
70.0	1.8		

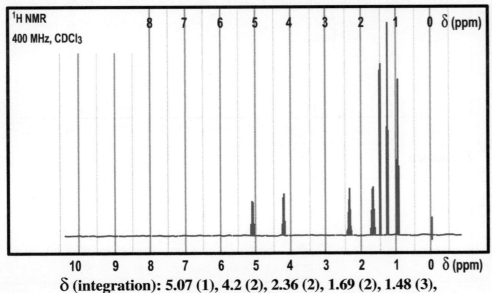

δ (integration): 5.07 (1), 4.2 (2), 2.36 (2), 1.69 (2), 1.48 (3), 1.27 (3), 0.98 (3)

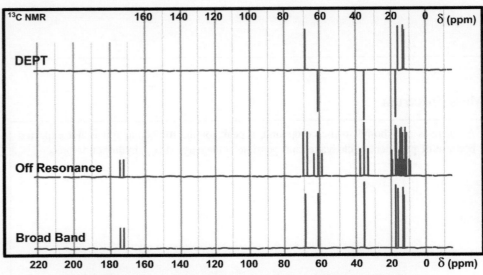

δ: 173.02, 170.93, 68.46, 61.29, 35.89, 18.38, 16.96, 14.11, 13.59

Exercise 093 (076)

Preliminary Observation

Infrared

IR (liquid film) 1744 cm^{-1}.

Mass Spectrum

There is no peak showing for the molecular ion.

^{13}C NMR

This spectrum gives information on the functional groups in addition to that provided by the infrared spectrum.

^{1}H NMR

^{1}H NMR (400 MHz, CDCl$_3$) δ 5.07 (q, J = 7.1 Hz, 1H), 4.201 (q, J = 7.1 Hz, 1H), 4.199 (q, J = 7.1 Hz, 1H), 2.36 (t, J = 7.4 Hz, 1H), 2.37 (t, J = 7.4 Hz, 1H), 1.69 (sextet, J = 7.4 Hz, 2H), 1.48 (d, J = 7.1 Hz, 3H), 1.28 (t, J = 7.1 Hz, 3H), 0.98 (t, J = 7.4 Hz, 3H).

Although the signal at 4.20 appears to be a triplet, close examination shows that it is more complex. Consider the effect of a chirality centre on the methylene protons.

Structure

Exercise 094 (083)

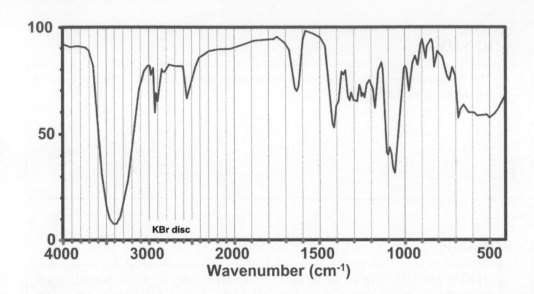

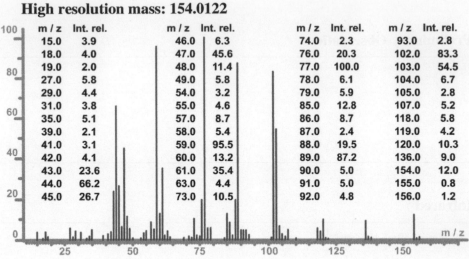

High resolution mass: 154.0122

m/z	Int. rel.	m/z	Int. rel.	m/z	Int. rel.	m/z	Int. rel.
15.0	3.9	46.0	6.3	74.0	2.3	93.0	2.8
18.0	4.0	47.0	45.6	76.0	20.3	102.0	83.3
19.0	2.0	48.0	11.4	77.0	100.0	103.0	54.5
27.0	5.8	49.0	5.8	78.0	6.1	104.0	6.7
29.0	4.4	54.0	3.2	79.0	5.9	105.0	2.8
31.0	3.8	55.0	4.6	85.0	12.8	107.0	5.2
35.0	5.1	57.0	8.7	86.0	8.7	118.0	5.8
39.0	2.1	58.0	5.4	87.0	2.4	119.0	4.2
41.0	3.1	59.0	95.5	88.0	19.5	120.0	10.3
42.0	4.1	60.0	13.2	89.0	87.2	136.0	9.0
43.0	23.6	61.0	35.4	90.0	5.0	154.0	12.0
44.0	66.2	63.0	4.4	91.0	5.0	155.0	0.8
45.0	26.7	73.0	10.5	92.0	4.8	156.0	1.2

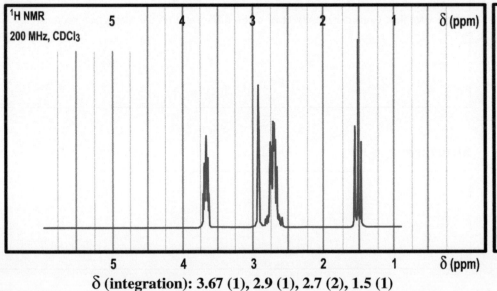

δ (integration): 3.67 (1), 2.9 (1), 2.7 (2), 1.5 (1)

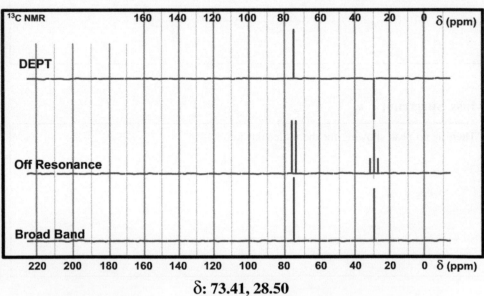

δ: 73.41, 28.50

Exercise 094 (083)

Preliminary Observation

Infrared

One functional group is evident from the broad band centred near $3400\,cm^{-1}$, and a second by the band at $2560\,cm^{-1}$.

Mass Spectrum

The molecular ion is at 154, and has isotope peaks at M+1 and M+2 that contain significant information.

^{13}C NMR

The great simplicity of the spectrum and the high molecular mass indicates the presence of equivalent carbons.

^{1}H NMR

The integration ratios give a total of 5H, which indicates a doubling of the number for an even-mass compound.

Structure

Exercise 095 (085)

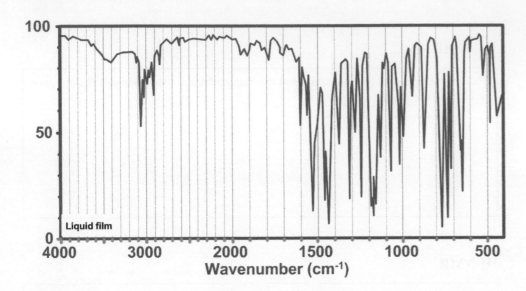

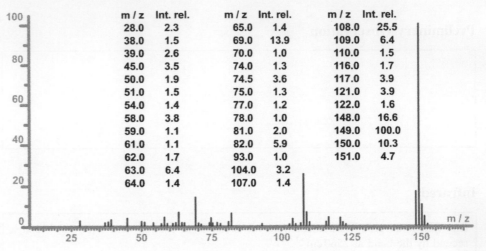

m / z	Int. rel.	m / z	Int. rel.	m / z	Int. rel.
28.0	2.3	65.0	1.4	108.0	25.5
38.0	1.5	69.0	13.9	109.0	6.4
39.0	2.6	70.0	1.0	110.0	1.5
45.0	3.5	74.0	1.3	116.0	1.7
50.0	1.9	74.5	3.6	117.0	3.9
51.0	1.5	75.0	1.3	121.0	3.9
54.0	1.4	77.0	1.2	122.0	1.6
58.0	3.8	78.0	1.0	148.0	16.6
59.0	1.1	81.0	2.0	149.0	100.0
61.0	1.1	82.0	5.9	150.0	10.3
62.0	1.7	93.0	1.0	151.0	4.7
63.0	6.4	104.0	3.2		
64.0	1.4	107.0	1.4		

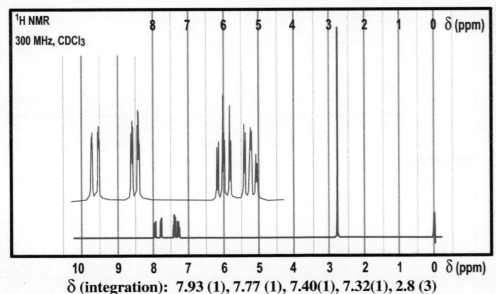

δ (integration): 7.93 (1), 7.77 (1), 7.40(1), 7.32(1), 2.8 (3)

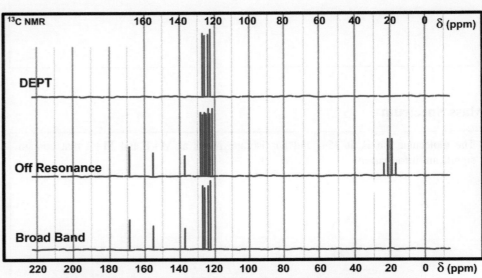

δ: 166.6, 153.4, 135.6, 125.8, 124.6, 122.3, 121.3, 19.9

Exercise 095 (085)

Preliminary Observation

Infrared

The spectrum allows the elimination of several functional groups.

Mass Spectrum

The molecular ion is at 149 and has two revealing isotope peaks.

^{13}C NMR

^{1}H NMR

^{1}H NMR (300 MHz, CDCl$_3$) δ 7.93 (dd, J = 8.1, 1.1 Hz, 1H), 7.77 (dd, J = 8.2, 1.1 Hz, 1H), 7.40 (ddd, J = 8.2, 7.2, 1.1 Hz, 1H), 7.32 (ddd, J = 8.1, 7.2, 1.1 Hz, 1H), 2.79 (s, 3H).

Structure

Exercise 096 (088)

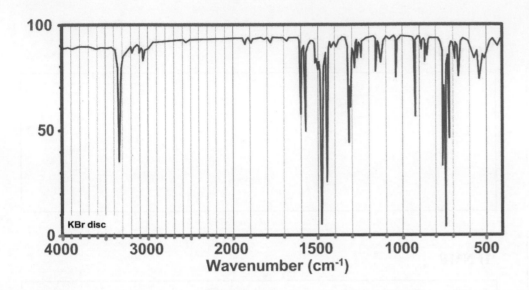

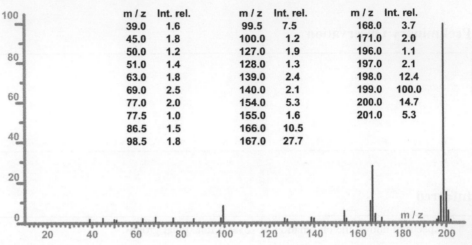

m / z	Int. rel.	m / z	Int. rel.	m / z	Int. rel.
39.0	1.6	99.5	7.5	168.0	3.7
45.0	1.8	100.0	1.2	171.0	2.0
50.0	1.2	127.0	1.9	196.0	1.1
51.0	1.4	128.0	1.3	197.0	2.1
63.0	1.8	139.0	2.4	198.0	12.4
69.0	2.5	140.0	2.1	199.0	100.0
77.0	2.0	154.0	5.3	200.0	14.7
77.5	1.0	155.0	1.6	201.0	5.3
86.5	1.5	166.0	10.5		
98.5	1.8	167.0	27.7		

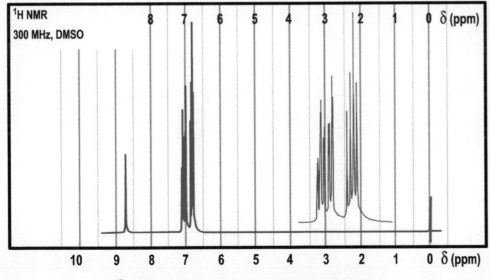

δ (integration): 8.7 (1), 7.05 (2), 6.95 (2)

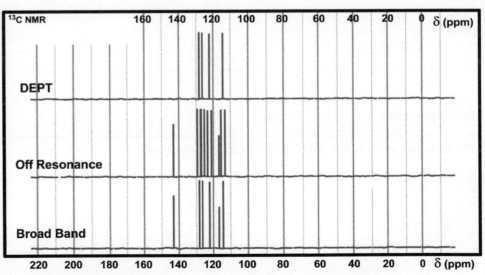

δ: 142.06, 127.41, 126.15, 121.67, 116.35, 114.36

Exercise 096 (088)

Preliminary Observation

Infrared

The ring system is apparent from the IR and NMR spectra. Take note of the sharp peak at $3320\,cm^{-1}$.

Mass Spectrum

The molecular ion is at 199, and there are significant isotope peaks for the M+1 and M+2 ions.

^{13}C NMR

^{1}H NMR

^{1}H NMR (300 MHz, CDCl$_3$) δ 8.7 (s, 1H), 6.98 (td, J = 7, 0.5 Hz, 1H), 6.90 (dd, J = 7, 0.5 Hz, 1H), 6.78 (td, J = 7, 0.5 Hz, 1H), 6.72 (dd, J = 7, 0.5 Hz, 1H).

Structure

Exercise 097 (095)

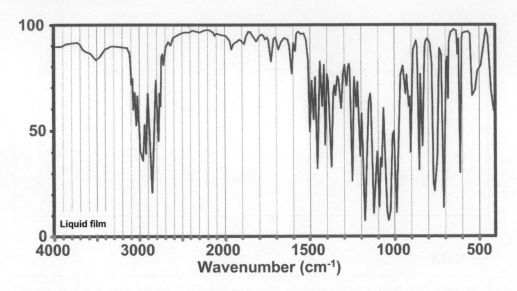

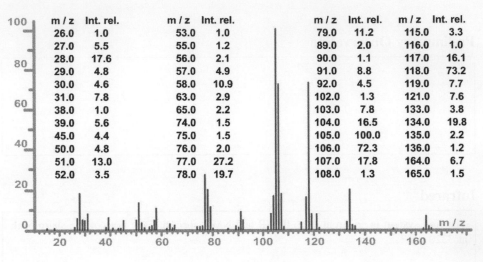

m / z	Int. rel.	m / z	Int. rel.	m / z	Int. rel.	m / z	Int. rel.
26.0	1.0	53.0	1.0	79.0	11.2	115.0	3.3
27.0	5.5	55.0	1.2	89.0	2.0	116.0	1.0
28.0	17.6	56.0	2.1	90.0	1.1	117.0	16.1
29.0	4.8	57.0	4.9	91.0	8.8	118.0	73.2
30.0	4.6	58.0	10.9	92.0	4.5	119.0	7.7
31.0	7.8	63.0	2.9	102.0	1.3	121.0	7.6
38.0	1.0	65.0	2.2	103.0	7.8	133.0	3.8
39.0	5.6	74.0	1.5	104.0	16.5	134.0	19.8
45.0	4.4	75.0	1.5	105.0	100.0	135.0	2.2
50.0	4.8	76.0	2.0	106.0	72.3	136.0	1.2
51.0	13.0	77.0	27.2	107.0	17.8	164.0	6.7
52.0	3.5	78.0	19.7	108.0	1.3	165.0	1.5

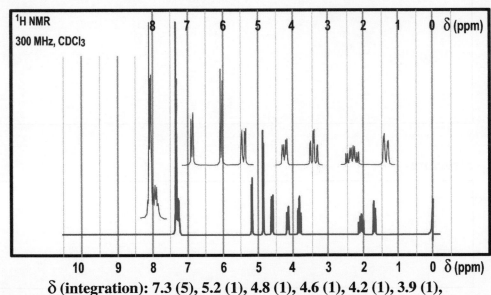

δ (integration): 7.3 (5), 5.2 (1), 4.8 (1), 4.6 (1), 4.2 (1), 3.9 (1), 2.1 (1), 1.7 (1)

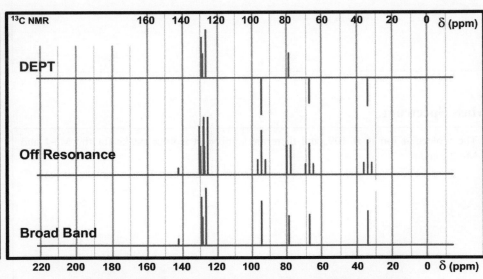

δ: 141.6, 128.4, 127.7, 125.7, 94.1, 78.6, 66.8, 34.0

Exercise 097 (095)

Preliminary Observation

Infrared

¹H NMR Spectrum

^{1}H NMR (300 MHz, CDCl$_3$) 7.3 (m, 5H, Ar), 5.20 (d, J = 6.2 Hz, 1H), 4.88 (d, J = 6.2 Hz, 1H), 4.63 (ddbr, J = 10.8, 2.8 Hz, 1H), 4.17 (ddbr, J = 11.4, 6.0 Hz, 1H), 3.84 (td, J = 11.4, 2.8 Hz, 1H), 2.06 (dddd, 13.5, 11.4, 10.8, 6.0 Hz, 1H), 1.69 (dbr, J = 13.5 Hz, 1H). The peaks at 4.17 and 1.69 are slightly broadened due to additional small coupling.

^{13}C NMR

Look carefully at the chemical shifts of the methylene groups.

Mass Spectrum

The molecular ion is at 164.

Structure

Draw the preferred conformation, and assign the coupling constants.

Exercise 098 (078)

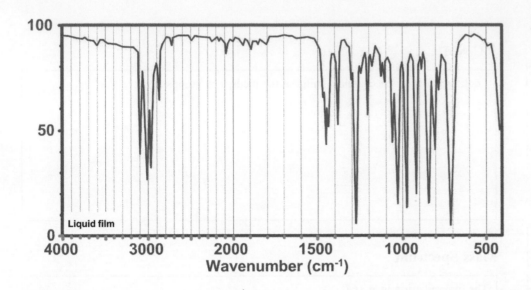

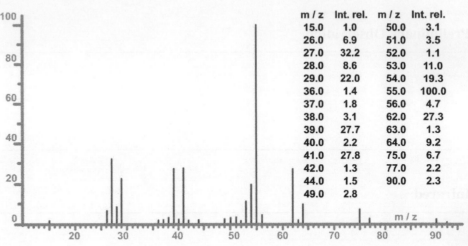

m / z	Int. rel.	m / z	Int. rel.
15.0	1.0	50.0	3.1
26.0	6.9	51.0	3.5
27.0	32.2	52.0	1.1
28.0	8.6	53.0	11.0
29.0	22.0	54.0	19.3
36.0	1.4	55.0	100.0
37.0	1.8	56.0	4.7
38.0	3.1	62.0	27.3
39.0	27.7	63.0	1.3
40.0	2.2	64.0	9.2
41.0	27.8	75.0	6.7
42.0	1.3	77.0	2.2
44.0	1.5	90.0	2.3
49.0	2.8		

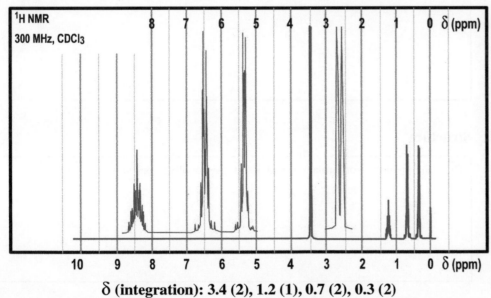

δ (integration): 3.4 (2), 1.2 (1), 0.7 (2), 0.3 (2)

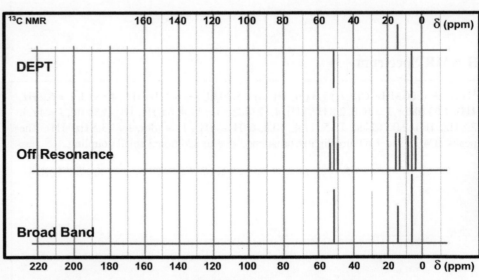

δ: 50.3, 13.9, 5.8

Exercise 098 (078)

Preliminary Observation

Infrared

There are peaks above $3000\,cm^{-1}$ in the region of H–C= stretching, but can this be confirmed by the NMR spectra?

Mass Spectrum

The molecular ion has a small peak at 90, and a barely discernible peak at 92. The first fragment ions give a better indication of the intensity of the M+2 isotope peak.

^{13}C NMR

Look carefully at the chemical shifts.

^{1}H NMR

^{1}H NMR (300 MHz, CDCl$_3$) 3.44 (d, J = 7.5, 2H), 1.23 (ttt, J = 8.1, 7.5, 4.9 Hz, 1H), 0.67 (m, 2H), 0.35 (m, 2H).

The odd number of protons indicates the presence of an atom other than C or O. Note the very low chemical shift of two of the signals.

Structure

Exercise 099 (065)

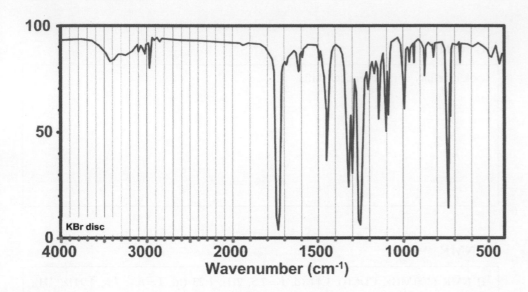

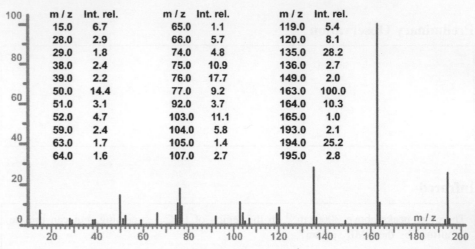

m / z	Int. rel.	m / z	Int. rel.	m / z	Int. rel.
15.0	6.7	65.0	1.1	119.0	5.4
28.0	2.9	66.0	5.7	120.0	8.1
29.0	1.8	74.0	4.8	135.0	28.2
38.0	2.4	75.0	10.9	136.0	2.7
39.0	2.2	76.0	17.7	149.0	2.0
50.0	14.4	77.0	9.2	163.0	100.0
51.0	3.1	92.0	3.7	164.0	10.3
52.0	4.7	103.0	11.1	165.0	1.0
59.0	2.4	104.0	5.8	193.0	2.1
63.0	1.7	105.0	1.4	194.0	25.2
64.0	1.6	107.0	2.7	195.0	2.8

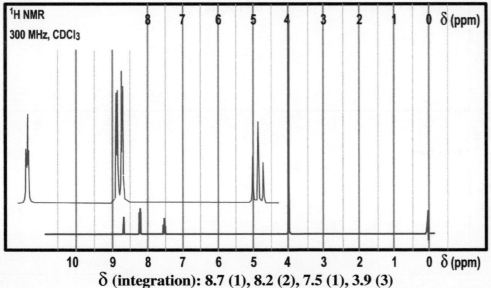

δ (integration): 8.7 (1), 8.2 (2), 7.5 (1), 3.9 (3)

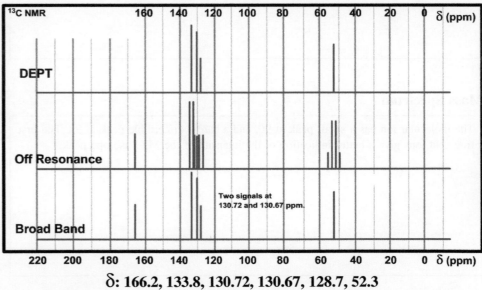

Two signals at 130.72 and 130.67 ppm.

δ: 166.2, 133.8, 130.72, 130.67, 128.7, 52.3

Exercise 099 (065)

Preliminary Observation

Infrared

Mass Spectrum

^{13}C NMR

Don't miss the singlet at 130.67 ppm in the OR spectrum.

^{1}H NMR

^{1}H NMR (300 MHz, CDCl$_3$) 8.7 (t, J = 1.5 Hz, 1H), 8.2 (dd, J = 6.5, 1.5 Hz, 2H), 7.5 (t, J = 6.5 Hz, 1H), 3.9 (s,3H).

Summary

Exercise 100 (054)

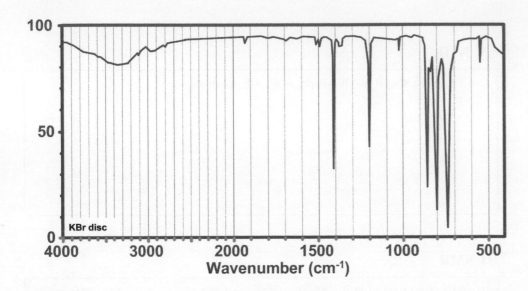

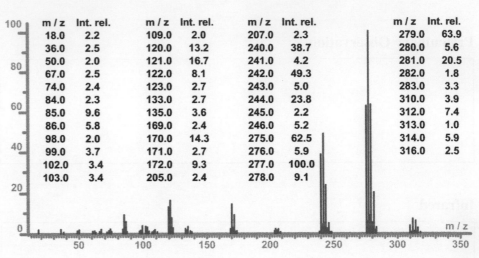

m/z	Int. rel.	m/z	Int. rel.	m/z	Int. rel.	m/z	Int. rel.
18.0	2.2	109.0	2.0	207.0	2.3	279.0	63.9
36.0	2.5	120.0	13.2	240.0	38.7	280.0	5.6
50.0	2.0	121.0	16.7	241.0	4.2	281.0	20.5
67.0	2.5	122.0	8.1	242.0	49.3	282.0	1.8
74.0	2.4	123.0	2.7	243.0	5.0	283.0	3.3
84.0	2.3	133.0	2.7	244.0	23.8	310.0	3.9
85.0	9.6	135.0	3.6	245.0	2.2	312.0	7.4
86.0	5.8	169.0	2.4	246.0	5.2	313.0	1.0
98.0	2.0	170.0	14.3	275.0	62.5	314.0	5.9
99.0	3.7	171.0	2.7	276.0	5.9	316.0	2.5
102.0	3.4	172.0	9.3	277.0	100.0		
103.0	3.4	205.0	2.4	278.0	9.1		

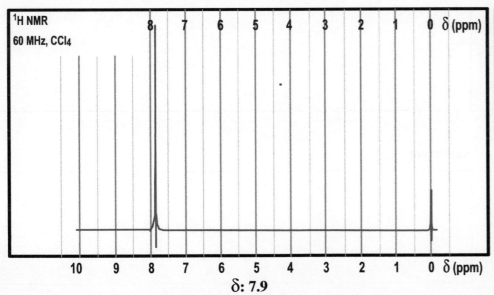

δ: 7.9

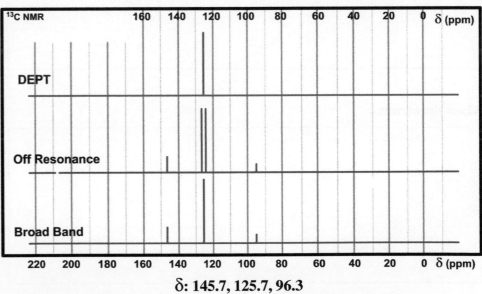

δ: 145.7, 125.7, 96.3

Exercise 100 (054)

Preliminary Observation

Infrared

The absence of absorption in regions where it might be expected is due to the symmetry of the molecule. No band is observed for vibrations that have no change in dipole moment.

Mass Spectrum

This mass spectrum has important information in the isotope patterns. The M^+ at 310 as well as several fragment ions have interesting isotope patterns.

^{13}C NMR

^{1}H NMR

Summary

Once the main skeleton is found, the substituent groups are easily determined when the molecular formula is taken into account.

Infrared Data

IR Frequencies: Display of Infrared Database by Frequency

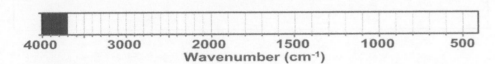

4000 - 3700 cm⁻¹

No significant bands are observed in this region...

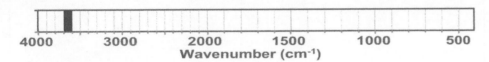

3700 - 3600 cm⁻¹

Stretching, O-H free (3600-3645 cm⁻¹).

A sharp peak is observed in this region for very dilute alcohol samples.

Stretching, O-H intra-molecular (3600-3450 cm⁻¹),
characteristically symmetric (gaussian) shape.

Stretching, O-H phenols (3125-3700 cm⁻¹),
strong broad band due to hydrogen bonding,

Sharp band (3640-3650 cm⁻¹) for free OH in dilute solutions, or
where large substituents (ortho to the OH) inhibit hydrogen bonding.

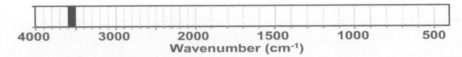

3600 - 3500 cm⁻¹

Stretching, O-H intra-molecular (3450-3600 cm⁻¹),
characteristic symmetric (gaussian) shape.

Stretching, O-H (3125-3700 cm⁻¹) observed for phenols.

Strong broad band due to hydrogen bonding, but a sharp peak (near 3645 cm⁻¹)
if large substituents (ortho to the OH) inhibit hydrogen bonding.

Stretching, O-H free (3600-3645 cm⁻¹).

A sharp peak is observed in this region for very dilute alcohol samples.

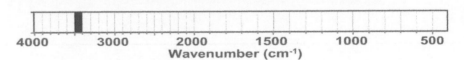

3500 - 3400 cm⁻¹

Stretching O-H inter-molecular (3200-3400 cm⁻¹).

A band characterized by its symmetric (gaussian) shape, quite different from that of
the acid OH band, decreasing in relative intensity on dilution.

Stretching, O-H (3125-3700 cm⁻¹) observed for phenols.

Strong broad band due to hydrogen bonding, but a sharp peak (near 3645 cm⁻¹)
if large substituents (ortho to the OH) inhibit hydrogen bonding.

Stretching, O-H intra-molecular (3450-3600 cm⁻¹), of symmetric (gaussian) shape.

Near 3250-3450 cm⁻¹, there are TWO N-H stretching bands for the -NH₂ of primary amines,
and ONE N-H stretching band for the >NH of secondary amines.

These bands are good indicators for the presence of N-H.

Similar bands are found for amides, but towards 3150-3450 cm⁻¹.

Organic Spectroscopy Workbook, First Edition. Tom Forrest, Jean-Pierre Rabine, and Michel Rouillard.
© 2011 John Wiley & Sons, Ltd. Published 2011 by John Wiley & Sons, Ltd.

3400 - 3200 cm⁻¹

Stretching, ≡ C-H (3250-3300 cm⁻¹) characteristic of monosubstituted alkynes,
of course, no ≡C- H band is seen for disubstituted alkynes.
This peak is very distinctive,and indicative of monosubstituted alkynes.

Stretching O-H inter-molecular (3200-3400 cm⁻¹).
A band characterized by its symmetric (gaussian) shape, quite different from that
of the acid OH band, decreasing in relative intensity on dilution.

Stretching, O-H (3125-3700 cm⁻¹) observed for phenols.
Strong broad band due to hydrogen bonding, but a sharp peak (near 3645 cm⁻¹)
if large substituents (ortho to the OH) inhibit hydrogen bonding.

Near 3250-3450 cm⁻¹, there are TWO N-H stretching bands for the -NH₂ of primary amines,
and ONE N-H stretching band for the >NH of secondary amines.
These bands are good indicators for the presence of N-H.

Similar bands are found for amides, but towards 3150-3450 cm⁻¹.

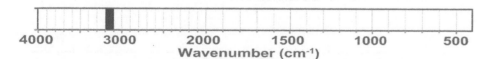

3200 - 3100 cm⁻¹

Stretching, ≡C-H (3250-3300 cm⁻¹) characteristic of monosubstituted alkynes.
As expected, no ≡C-H band is seen for disubstituted alkynes.
This peak is very distinctive, and indicative of monosubstituted alkynes.

Stretching O-H inter-molecular (3200-3400 cm⁻¹).
A band characterized by its symmetric (gaussian) shape, quite different from that
of the acid OH band, decreasing in relative intensity on dilution.

Stretching, O-H (3125-3700 cm⁻¹) observed for phenols.
Strong broad band due to hydrogen bonding, but a sharp peak (near 3645 cm⁻¹)
if large substituents (ortho to the OH) inhibit hydrogen bonding.

Stretching, alkene and aromatic C-H 3000-3100 cm⁻¹,
sometimes submerged in aliphatic C-H stretching bands.

Stretching, O-H of carboxylic acids, extremely broad band due to
extensive hydrogen bonding, between 2500 and 3340 cm⁻¹.
One could not confuse this with the OH stretch of an alcohol.

Near 3100 cm⁻¹, harmonic of the bending band of N-H (1530-1550 cm⁻¹) characteristic
of secondary amides of which the >C=O is in a trans position with respect to an N-H,
peak not to be confused with >N-H stretching!

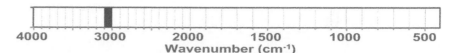

3100 - 3000 cm⁻¹

Stretching, alkene and aromatic C-H, 3000-3100 cm⁻¹,
sometimes submerged in aliphatic C-H stretching bands.

Stretching, O-H of carboxylic acids, (2500 - 3340 cm⁻¹) extremely broad band
due to extensive hydrogen bonding.
This could not be confused with the OH stretch of an alcohol.

Near 3100 cm⁻¹, harmonic of the bending band of N-H (1530-1550 cm⁻¹) characteristic
of secondary amides in which the >C=O is in a trans position with respect to an N-H,
not to be confused with >N-H stretching!

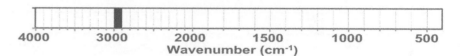

3000 - 2900 cm⁻¹

Stretching symmetric and asymmetric of aliphatic C-H 2850-2980 cm-1
most often indiscernible within the envelope of peaks.

Stretching, =C-H (2800-2900 cm-1) of aldehydes,
frequently hidden in the C-H stretching band, but sometimes distinct,
as in the case of aromatic aldehydes with little alkyl substitution.

Stretching, O-H of carboxylic acids, very broad band due to hydrogen bonding,
between 2500 and 3340 cm-1.
One could not confuse this with the OH stretch of an alcohol.

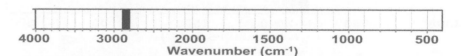

2900 - 2800 cm⁻¹

Stretching symmetric and asymmetric of aliphatic C-H 2850-2980 cm⁻¹
most often indiscernible within a single envelope.

Stretching, =C-H (2800-2900 cm⁻¹) of aldehydes,
frequently hidden in the C-H stretching band, but sometimes distinct,
as in the case of aromatic aldehydes with little alkyl substitution.

Stretching, =C-H (2700-2800 cm⁻¹) of aldehydes, always visible.
This peak constitutes a marker of the aldehyde functional group.
A peak due to the CH₃-O- group (near 2820-2850 cm⁻¹),
is characteristic of aromatic CH₃-O- groups.

Stretching, O-H of carboxylic acids, extremely broad band due to
extensive hydrogen bonding, between 2500 and 3340 cm⁻¹.
One could not confuse this with the OH stretch of an alcohol.

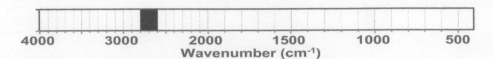

2800 - 2600 cm⁻¹

Stretching, =C-H (2700-2800 cm⁻¹) of aldehydes, always visible.
    This peak constitutes a marker of the aldehyde function.
Stretching, O-H of carboxylic acids, extremely broad band due to
    extensive hydrogen bonding, between 2500 and 3340 cm⁻¹.
    One could not confuse this with the OH stretch of an alcohol.

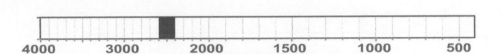

2600 - 2400 cm⁻¹

Stretching, O-H of carboxylic acids, extremely broad band due to
    extensive hydrogen bonding, between 2500 and 3340 cm⁻¹.
    One could not confuse this with the OH stretch of an alcohol.

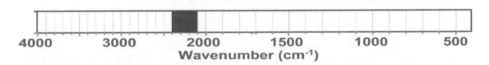

2400 - 2100 cm⁻¹

Stretching, -C≡C- (2050-2250 cm⁻¹), absent in the case of symmetrically disubstituted alkynes.
    The peak is very sensitive to the dipole moment, being very weak in disubstituted
    alkynes. Take care not to confuse it with –C≡N stretching, which is more intense
    than a disubstituted alkyne, and does not have the ≡C-H band (3300 cm⁻¹)
    of the monosubstiyuted alkyne.
Stretching, -C≡N of nitriles (2200-2300 cm⁻¹) with an intensity stronger than the
    -C≡C- stretching, which appears in the same region.

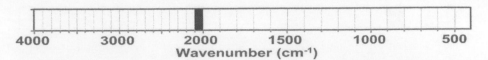

2100 - 2000 cm⁻¹

Stretching, -C≡C- (2050-2250 cm⁻¹), absent in the case of symmetrically disubstituted alkynes.
    The peak is very sensitive to the dipole moment, being very weak in disubstituted
    alkynes. Take care not to confuse it with –C≡N stretching, which is more intense
    than a disubstituted alkyne, and does not have the ≡C-H band (3300 cm⁻¹)
    of the monosubstiyuted alkyne.

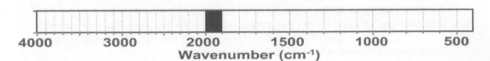

2000 - 1900 cm⁻¹

Possible presence of weak overtone combination bands,
    (fingerprints) for aromatic compounds

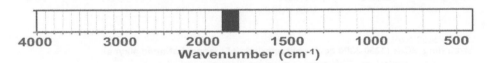

1900 - 1800 cm⁻¹

Stretching, >C=O between 1725 and 1850 cm⁻¹.
    If there are TWO intense bands in this region, they may be due to an anhydride.
    In cyclic anhydrides the higher frequency band is less intense than the other,
    whereas the opposite is true for non-cyclic anhydrides.
Stretching, >C=O (4-member ring lactones) near 1830 cm⁻¹.
Stretching, >C=O (acid halides) near 1760-1830 cm⁻¹.
Possible presence of weak overtone combination bands,
    (fingerprint) for aromatic compounds

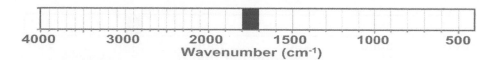

1800 - 1700 cm⁻¹

Stretching, >C=O between 1725 and 1850 cm^{-1}.

    If there are TWO intense bands in this region, they may be due to an anhydride. In cyclic anhydrides the higher frequency band is less intense than the other, whereas the opposite is true for non-cyclic anhydrides.

Stretching, >C=O of ketones (near 1710 cm^{-1}).

    This frequency decreases on conjugation, however the frequency increases for cyclic ketones of a ring size less than six members.

Stretching, >C=O (acid halides) near 1760-1830 cm^{-1}.

Stretching, >C=O (carboxylic acids) 1700-1760 cm^{-1}, very intense.

Stretching, >C=O (esters) near 1740 cm^{-1}, lower frequency if conjugation.

Stretching, >C=O (aldehydes) near 1725 cm^{-1}, lower if conjugated.

Possible presence of weak overtone combination bands,
    (fingerprint) for aromatic compounds

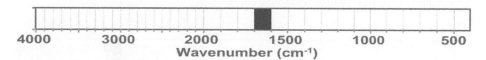

1700 - 1600 cm⁻¹

Stretching, >C=O (carboxylic acids) 1700-1760 cm^{-1}, very intense.

Stretching, >C=O (amides) near 1640-1690 cm^{-1}, intense band.

    The carbonyl frequency is lowered by the resonance donor effect of nitrogen.

Stretching, >C=C< aromatic (1600-1640 cm^{-1}), frequency lowered by conjugation.

Stretching, >C=C< 1580-1660 cm^{-1}, sometimes absent or very weak if the alkene is symmetric or is a trans, (E) alkene.

Deformation in the -NH₂ group, "scissoring band," around 1580-1650 cm^{-1}.

1600 - 1500 cm⁻¹

Stretching, >C=C< 1580-1660 cm^{-1}, sometimes absent or very weak if the alkene is symmetric or is a trans, (E) alkene.

Stretching, >C=C< aromatic (1600-1640 cm^{-1}), frequency lowered by conjugation.

Stretching, >C=C< aromatic (1520-1480 cm^{-1}), of variable intensity.

Deformation in the -NH₂ group, "scissoring band", around 1580-1650 cm^{-1}.

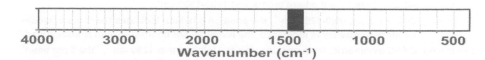

1500 - 1400 cm⁻¹

Stretching, >C=C< aromatic (1520-1480 cm^{-1}), of variable intensity.

Deformation N-H (1530-1550 cm^{-1}) characteristic of secondary amides of which the >C=O is in a trans position with respect to the N-H.

    Its harmonic is observed around 3100 cm^{-1}.

    Take care to not confuse this harmonic with >N-H stretching!

Deformation (scissoring) around 1450 cm^{-1} of -CH₂-, not a particulary useful band as the carbon skeleton bands are found in this region.

Stretching, C-N< (amines) 1000-1400 cm^{-1}, frequency dependant on the number of substituents on the alpha carbon; the more the carbon is substituted, the more the frequency is increased.

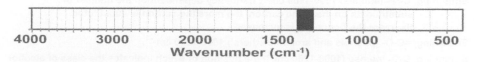

1400 - 1300 cm⁻¹

Stretching, C-N< (amines) 1000-1400 cm^{-1}, frequency dependant on the number of substituents on the alpha carbon; the more the carbon is substituted, the more the frequency is increased.

Deformation corresponding to -CH₃ group around 1380 cm^{-1}.

    It is important to examine this region of the spectrum carefully :

      - a doublet of equal intensity indicates a gem-dimethyl, -CH(CH₃)₂ or >C(CH₃)₂

      - a doublet of different intensities, indicates a tert-butyl -C(CH₃)₃.

Deformation O-H (phenols) 1315-1390 cm^{-1}, of little use for interpretation.

Stretching, =C-O- (phenols and ethers aromatics) 1000-1300 cm^{-1}.

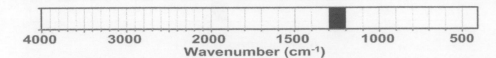

1300 - 1200 cm⁻¹

Stretching, C-N< (amines) 1000-1400 cm^{-1}, frequency dependant on the number of
    substituents on the alpha carbon; the more the carbon is substituted,
    the more the frequency is increased.

Stretching, =C-O of carboxylic acids around 1250 cm^{-1}.

Stretching, =C-O characteristic of acetates around 1250 cm^{-1}.

Stretching, =C-O- (phenols and ethers aromatics) 1000-1300 cm^{-1}.

Stretching, C-O- intense (1000-1260 cm^{-1}) the position of which indicates the class of alcohol.
    The more frequency is increased, the more the OH bearing carbon is substituted.

Stretching, C-O-C symmetric, from epoxide ring stretching, near 1250 cm^{-1}, "the 8 µm band".

Stretching, =C-O-C= and =C-O-C : intense band (1000-1250 cm^{-1}).

Stretching, C-F of alkyl fluorides (1100-1250 cm^{-1}).

C-O- (1100-1210 cm^{-1}) typical tertiary alcohols R$_1$R$_2$R$_3$COH.

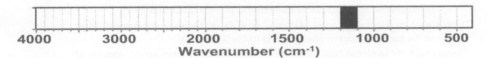

1200 - 1100 cm⁻¹

Stretching, C-N< (amines) 1000-1400 cm^{-1}, frequency dependant on the number of
    substituents on the alpha carbon; the more the carbon is substituted,
    the more the frequency is increased.

Stretching, =C-O- (phenols and aromatic ethers) 1000-1300 cm-$^{-1}$.

Stretching, C-O- intense (1000-1260 cm^{-1}) the position of which indicates the class of alcohol.
    The more frequency is increased, the more the OH bearing carbon is substituted.

Stretching, =C-O-C= and =C-O-C : intense band (1000-1250 cm^{-1}).

Stretching, C-F of alkyl fluorides (1100-1250 cm^{-1}).

C-O- (1100-1210 cm^{-1}) typical of tertiary alcohols R$_1$R$_2$R$_3$COH.

C-O- (1090-1130 cm^{-1}) typical of secondary alcohols >CH-OH.

Stretching, C-O of ethers C-O-C (1000-1150 cm^{-1}).

Stretching, =C-O characteristic of formates around 1180 cm^{-1}.

Stretching, C-N< (amides) between 1050-1175 cm^{-1}.

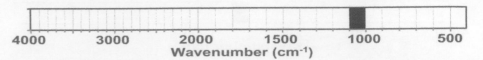

1100 - 1000 cm⁻¹

Stretching, C-N< (amines) 1000-1400 cm^{-1}, frequency dependant on the number of
    substituents on the alpha carbon; the more the carbon is substituted,
    the more the frequency is increased.

Stretching, =C-O- (phenols and aromatic ethers) 1000-1300 cm^{-1}.

Stretching, C-O- intense (1000-1260 cm^{-1}) the position of which indicates the class of alcohol.
    The more frequency is increased, the more the OH bearing carbon is substituted.

Stretching, =C-O-C= and =C-O-C : intense band (1000-1250 cm^{-1}).

Stretching, C-F of alkyl luorides (1100-1250 cm^{-1}).

C-O- (1100-1210 cm^{-1}) typical of tertiary alcohols R$_1$R$_2$R$_3$COH.

C-O- (1090-1130 cm^{-1}) typical of secondary alcohols >CH-OH.

C-O- (1000-1075 cm^{-1}) typical of primary alcohols -CH$_2$OH.

Stretching, C-O of ethers, C-O-C (1000-1150 cm^{-1}).

Stretching, C-N< (amides) between 1050-1175 cm^{-1}.

Stretching, C-Cl of acid chlorides (860-1050 cm^{-1}).

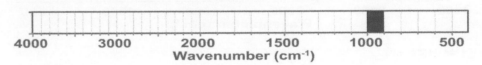

1000 - 900 cm⁻¹

Stretching, C-N< (amines) 1000-1400 cm^{-1}, frequency dependant on the number of
    substituents on the alpha carbon; the more the carbon is substituted,
    the more the frequency is increased.

Stretching, =C-O- (phenols and ethers aromatics) 1000-1300 cm^{-1}.

Stretching, C-O- intense (1000-1260 cm^{-1}) the position of which indicates the class of alcohol.
    The more frequency is increased, the more the OH bearing carbon is substituted.

C-O- (1000-1075 cm-1) typical of primary alcohols -CH$_2$OH.

Stretching, =C-O-C= and =C-O-C : intense band (1000-1250 cm^{-1}).

Stretching, C-O of ethers C-O-C (1000-1150 cm^{-1}).

Stretching, C-Cl of acid chlorides (860-1050 cm^{-1}).

Deformation (2 bands) around 910 and 990 cm^{-1} for monosubstituted alkenes R-CH=CH$_2$.

Deformation (1 band) around 970 cm^{-1} for trans disubstituted alkenes: R-CH=CH-R' (E).

Stretching, asymmetric C-O-C of epoxides near 810-950 cm^{-1}, "the 11 µm band".

Deformation >N-H, (760-920 cm^{-1}) very broad, characteristic of amines I and II.

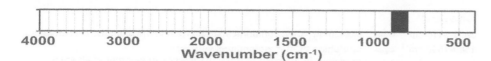

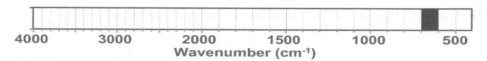

900 - 800 cm⁻¹

Stretching, C-Cl of acid chlorides (860-1050 cm⁻¹).

Deformation OH (carboxylic acids) around 850-960 cm⁻¹, not very useful for identification. The very distinctive broad OH stretching band is more useful.

Stretching, asymmetric C-O-C of epoxides around 810-950 cm⁻¹, "the 11 μm band".

Deformation >N-H, (760-920 cm⁻¹) very broad, characteristic of amines I and II, at a frequency distinctly lower frequency for amides (frequency below 700 cm⁻¹).

Deformation (1 band) about 890 cm⁻¹ for alkenes RR'C=CH₂.

Deformation (1 band) about 790-840 cm⁻¹ for alkenes RR'C=CH-R".

Deformation (800-855 cm⁻¹ characteristic of aromatic rings having 2 adjacent hydrogens on the ring.

Deformation (765-800 cm⁻¹) characteristic of aromatic rings having 3 adjacent hydrogens on the ring.

Stretching, asymmetric C-O-C of epoxides near 750-840 cm⁻¹, "the 12 μm band".

Stretching, C-Cl of alkyl chlorides (600-800 cm⁻¹).

700 - 600 cm⁻¹

Stretching, C-Cl of alkyl chlorides (600-800 cm⁻¹).

Deformation (700-690 cm⁻¹ for 5 H) characteristic of aromatics rings with 5 hydrogens (monosubstituted aromatic rings).

Bending band of monosubstituted alkynes, ≡C-H.

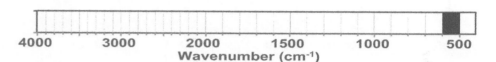

600 - 500 cm⁻¹

Stretching, C-Cl of alkyl chlorides (600-800 cm⁻¹).

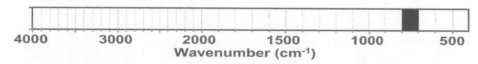

800 - 700 cm⁻¹

Deformation (1 band) around 730 cm⁻¹ for cis disubstituted alkenes R-CH=CH-R' (Z).

Deformation (1 band) between 790-840 cm⁻¹ for trisubstituted alkenes RR'C=CH-R".

Deformation >N-H, (760-920 cm⁻¹) very broad, characteristic of amines I and II, at a frequency distinctly lower for amides (frequency below 700 cm⁻¹).

Deformation (800-855 cm⁻¹ characteristic of aromatic rings having 2 adjacent hydrogens on the ring.

Deformation (765-800 cm⁻¹) characteristic of aromatic rings having 3 adjacent hydrogens on the ring.

Deformation (740-770 for 4 H and 700-690 cm⁻¹ for 5 H) characteristic of aromatic rings with 4 or 5 adjacent hydrogens (aromatics mono-substitued or ortho disubstituted).

Stretching, asymmetric C-O-C of epoxides near 750-840 cm⁻¹, "the 12 μm band".

Stretching, C-Cl of alkyl chlorides (600-800 cm⁻¹).

Rocking, about 720 cm⁻¹ for -CH₂-, a medium intensity band when there are more than 4 methylenes -CH₂- together.

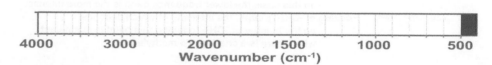

500 - 400 cm⁻¹

No significant bands are observed in this region...

IR Groups: Display of Infrared Database by Functional Group

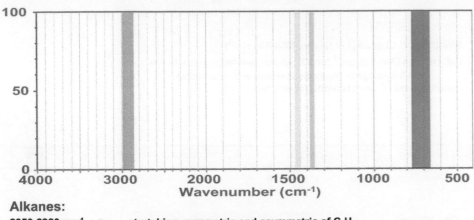

Alkanes:

2850-2980 cm⁻¹	➡	stretching, symmetric and asymmetric of C-H.
near 1450 cm⁻¹	➡	deformation (shearing), band of marginal use !
near 1380 cm⁻¹	➡	deformation corresponding to a -CH₃ group;

doublet of the same intensity, gem-dimethyl >C(CH₃)₂
doublet of different intensities, *tert*-butyl -C(CH₃)₃,
in this case, the lower frequency band is the more intense.

near 720 cm⁻¹ ➡ Rocking of -CH₂- groups; the intensity of which depends
on the number of -CH₂- together, this band is significant
when there is a chain of 4-5 methylenes .

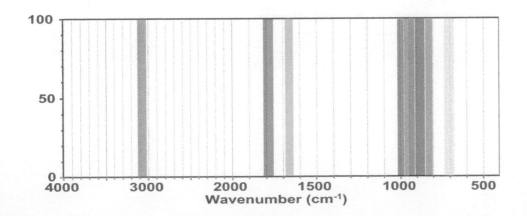

Alkenes:

3000-3110 cm⁻¹	➡	stretching =C-H, sometimes submerged in the aliphatic C-H envelope.
1580-1660 cm⁻¹	➡	stretching >C=C<, sometimes very weak or absent in the case of trans (E) alkenes or symmetric alkenes
near 1800 cm⁻¹	➡	harmonic band at 910 cm⁻¹ (monosubstituted alkenes)
near 1780 cm⁻¹	➡	harmonic band at 890 cm⁻¹ (gem disubstituted alkenes)

The number and position of the deformation bands often allow you to obtain
information about the substitution of the double bond...

R-CH=CH₂	➡	2 deformation bands around 910 and 990 cm⁻¹.
RR'C=CH₂	➡	1 deformation band around 890 cm⁻¹.
R-CH=CH-R' (Z)	➡	1 deformation band around 680-720 cm⁻¹.
R-CH=CH-R' (E)	➡	1 deformation band around 970 cm⁻¹.
RR'C=CH-R"	➡	1 deformation band between 840-790 cm⁻¹.

Of course the spectra will contain peaks attributable to other groups that may be present.
These ranges are fairly representative, but inductive effects and/or conjugation may shift the bands.

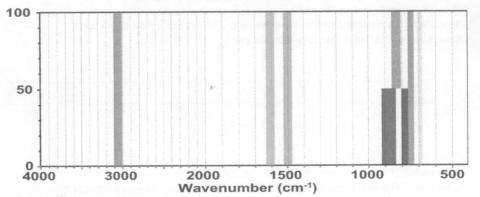

Aromatics:

3000-3100 cm⁻¹	➡	stretching =C-H sometimes buried in the aliphatic C-H stretching.
1620-1580 cm⁻¹	➡	stretching >C=C<, frequency lowered by conjugation.
1520-1480 cm⁻¹	➡	stretching >C=C<, of variable itensity.

The deformation bands often allow one to determine
how the aromatic ring is substituted :

700- 690 cm⁻¹	➡	deformation corresponding to 5 adjacent hydrogens on the ring.
740- 770 cm⁻¹	➡	deformation corresponding to 4 or 5 adjacent hydrogens on the ring.
765- 800 cm⁻¹	➡	deformation corresponding to 3 adjacent hydrogens on the ring.
800- 855 cm⁻¹	➡	deformation corresponding to 2 adjacent hydrogens on the ring.
835- 910 cm⁻¹	➡	deformation corresponding to 1 H between 2 substituents.

Of course the spectra will contain peaks attributable to other groups that may be present.
These ranges are fairly representative, but inductive effects and/or conjugation may shift the bands.

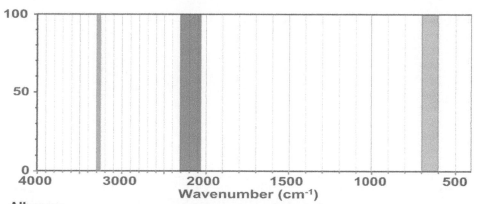

Alkynes:

3250-3300 cm⁻¹ ➡ stretching ≡C- H characteristic of monosubstituted alkynes, absent for disubstitution of alkynes.

2050-2300 cm⁻¹ ➡ stretching -C≡C- , absent for symmetrically disubstituted alkynes. The peak is very sensitive to the dipole moment, it may have a very weak intensity .

Take note:
Because of the possible disubstitution of the -C≡C- bond, and of the effects of substituents, one may not observe any band...

700 - 600 cm⁻¹ ➡ deformation band of ≡C-H.

Of course the spectra will contain peaks attributable to other groups that may be present. These ranges are fairly representative, but inductive effects and/or conjugation may shift the bands.

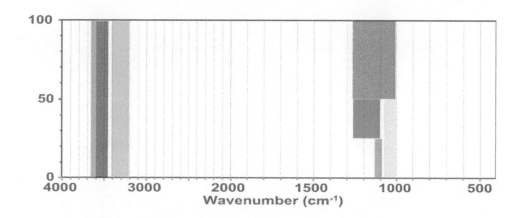

Alcohols:

3600-3645 cm⁻¹ ➡ O-H free (for dilute solutions
3450-3600 cm⁻¹ ➡ of the alcohol).
3200-3400 cm⁻¹ ➡ O-H intra-molecular

Impossible to confuse the OH band of an alcohol with the OH band of an acid.

O-H inter-molecular. Characteristic band (symmetric shape), different shape to that of the OH of carboxylic acids and diminishes on dilution.

1000-1260 cm⁻¹ ➡ stretching C-O- intense band, the position often lets you determine the type of alcohol

➡ C-OH 1100-1260 cm⁻¹,
➡ >CH-OH 1090-1130 cm⁻¹,
➡ -CH₂-OH 1000-1075 cm⁻¹.

The effect is really only exploitable when one compares the spectra of isomers

Of course the spectra will contain peaks attributable to other groups that may be present. These ranges are fairly representative, but inductive effects and/or conjugation may shift the bands.

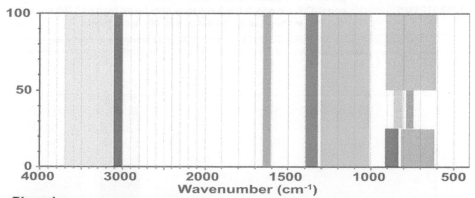

Phenols:

3125-3700 cm⁻¹ ➡ O-H, broad intense band due to the hydrogen bonding sharp (3700 cm⁻¹) if sterically hindered by neighbouring groups.
3000-3100 cm⁻¹ ➡ =C-H sometimes hidden by aliphatic C-H stretching.
1600-1640 cm⁻¹ ➡ stretching >C=C<, frequency lowered by conjugation.
1315-1390 cm⁻¹ ➡ deformation O-H, the deformation of =C-O is below 700 cm⁻¹.
1000-1300 cm⁻¹ ➡ stretching =C-O-.

700- 900 cm⁻¹ ➡ =C-H out-of-plane deformation of aromatics. The number of peaks and their position provide information on the ring substitution.
740- 770 cm⁻¹ ➡ deformation corresponding to 4 or 5 adjacent hydrogens on the ring.
835- 910 cm⁻¹ ➡ deformation corresponding to 1 H between two substituents.
800- 855 cm⁻¹ ➡ deformation corresponding to 2 adjacent hydrogens on the ring.
765- 800 cm⁻¹ ➡ deformation corresponding to 3 adjacent hydrogens on the ring.

Of course the spectra will contain peaks attributable to other groups that may be present. These ranges are fairly representative, but inductive effects and/or conjugation may shift the bands.

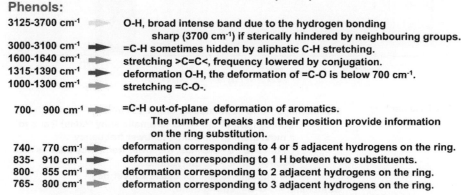

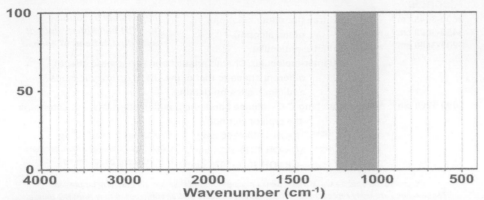

Ethers:

2820-2850 cm⁻¹	group CH₃-O-, aromatics methyl ethers.
1000-1150 cm⁻¹	and C-O-C for which the frequency may be as high as 1250 cm⁻¹ in the case of aromatics and ethylenic ethers.

Of course the spectra will contain peaks attributable to other groups that may be present.
These ranges are fairly representative, but inductive effects and/or conjugation may shift the bands.

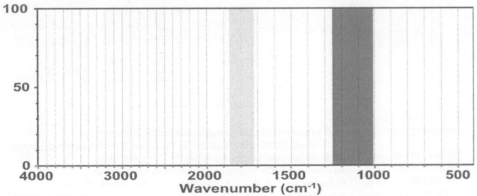

Anhydrides:

1725-1850 cm⁻¹	2 stretchings >C=O separated by about 65 cm⁻¹. Conjugation of the >C=O (aromatic anhydrides) leads to a movement of the bands to lower frequency. Of the two >C=O bands, the lower frequency band is less intense for acyclic anhydrides. The opposite is true for cyclic anhydrides.
1000-1250 cm⁻¹	Stretching, =C-O-C=, strong band.

Of course the spectra will contain peaks attributable to other groups that may be present.
These ranges are fairly representative, but inductive effects and/or conjugation may shift the bands.

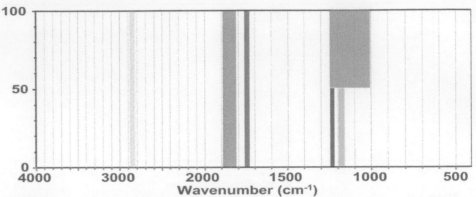

Esters:

near 2850 cm⁻¹	Sharp band characteristic of a methoxy group, CH₃-O.
near 1740 cm⁻¹	stretching >C=O, lower if conjugated.
1000-1250 cm⁻¹	stretching bands =C-O and O-C.
near 1180 cm⁻¹	band characteristic of formates, H-COO-R,
near 1250 cm⁻¹	band characteristic of acetates, CH₃-COO-R.

Lactones

1740-1830 cm⁻¹ ➡ stretching >C=O,

The frequency depends on the ring size.
For strained rings the frequency increases (4-membered, 1830 cm⁻¹; 5-membered, 1770 cm⁻¹) while 6-membered and higher have normal ester frequencies (1740 cm⁻¹).

Of course the spectra will contain peaks attributable to other groups that may be present.
These ranges are fairly representative, but inductive effects and/or conjugation may shift the bands.

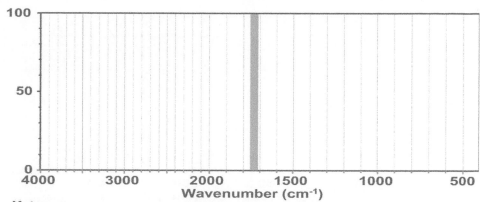

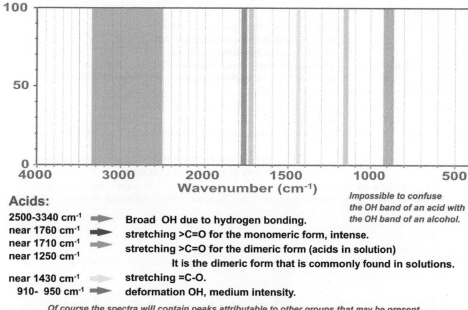

Acids:

2500-3340 cm⁻¹	➡	Broad OH due to hydrogen bonding.
near 1760 cm⁻¹	➡	stretching >C=O for the monomeric form, intense.
near 1710 cm⁻¹	➡	stretching >C=O for the dimeric form (acids in solution)
near 1250 cm⁻¹		It is the dimeric form that is commonly found in solutions.
near 1430 cm⁻¹	➡	stretching =C-O.
910- 950 cm⁻¹	➡	deformation OH, medium intensity.

Impossible to confuse the OH band of an acid with the OH band of an alcohol.

Of course the spectra will contain peaks attributable to other groups that may be present.
These ranges are fairly representative, but inductive effects and/or conjugation may shift the bands.

Ketones:

near 1710 cm⁻¹	➡	stretching >C=O.

The frequency is lowered by conjugation, but it is increased for ketones in rings of a size less than 6 members.
Of all carbonyl compounds, the ketones are least easily identified.
Only after eliminating the other carbonyl functional groups such as aldehydes, acids, esters, anhydrides, etc, can the assignment of the carbonyl band to a ketone be made.

Of course the spectra will contain peaks attributable to other groups that may be present.
These ranges are fairly representative, but inductive effects and/or conjugation may shift the bands.

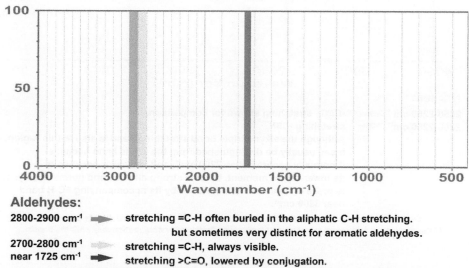

Aldehydes:

2800-2900 cm⁻¹	➡	stretching =C-H often buried in the aliphatic C-H stretching. but sometimes very distinct for aromatic aldehydes.
2700-2800 cm⁻¹	➡	stretching =C-H, always visible.
near 1725 cm⁻¹	➡	stretching >C=O, lowered by conjugation.

Of course the spectra will contain peaks attributable to other groups that may be present.
These ranges are fairly representative, but inductive effects and/or conjugation may shift the bands.

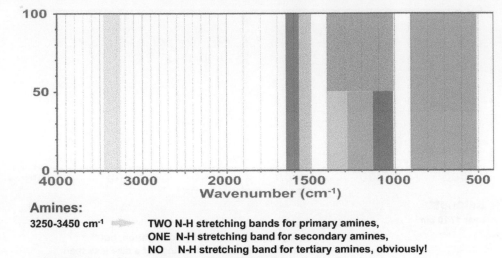

Amines:

3250-3450 cm⁻¹ ➭ TWO N-H stretching bands for primary amines,
ONE N-H stretching band for secondary amines,
NO N-H stretching band for tertiary amines, obviously!

1580-1650 cm⁻¹ ➭ R-NH₂, scissors deformation,
1490-1580 cm⁻¹ ➭ RR'NH , same, but more difficult to detect.

1000-1400 cm⁻¹ ➭ stretching C-N<, frequency increased by carbon substitution
in the alpha position:

➭ RCH₂-NH₂,   This effect is really only
➭ RR'CH-NH₂,   useful when comparing
➭ RR'R"C-NH₂   spectra of isomers.

near 900 cm⁻¹ ➭ deformation >N-H, broad band characteristic of I and II amines.
The frequency is distinctly lower for amides..

Of course the spectra will contain peaks attributable to other groups that may be present.
These ranges are fairly representative, but inductive effects and/or conjugation may shift the bands.

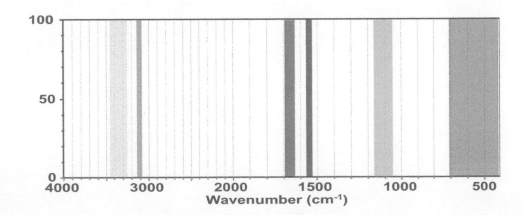

Amides:

3250-3450 cm⁻¹ ➭ TWO N-H stretching bands for primary amides,
ONE N-H band for secondary amides,
NO N-H stretch for tertiary amides.

near 3100 cm⁻¹ ➭ harmonic of the N-H deformation of amides II, trans,
different from >N-H stretching (not the same intensity).

1640-1690 cm⁻¹ ➭ stretching >C=O.

1530-1550 cm⁻¹ ➭ deformation N-H characteristic of amides II, trans.
1055-1175 cm⁻¹ ➭ stretching C-N<.
< 700 cm⁻¹ ➭ deformation >N-H, broad band typical of amides I and II.

**The amides are the most difficult carbonyl compound
to identify in an infrared spectrum!**

Of course the spectra will contain peaks attributable to other groups that may be present.
These ranges are fairly representative, but inductive effects and/or conjugation may shift the bands.

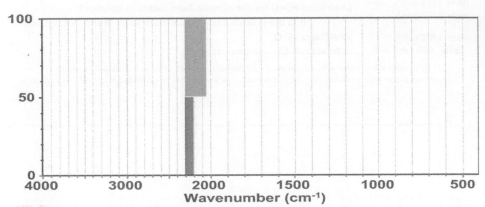

Nitriles:

2050-2260 cm⁻¹ ➭ -C≡C- stretching shown for comparison.
2200-2260 cm⁻¹ ➭ stretching -C≡N.

Although the alkyne triple bond stretching band is also in this region,
it can usually be distinguished from the nitrile band because
it normally occurs below 2200 cm⁻¹, and is weaker because of
its lower dipole moment. When a strong alkyne band (terminal alkyne)
is present, it can be distinguished by Its accompanying ≡C-H band
near 3300 cm⁻¹.

Of course the spectra will contain peaks attributable to other groups that may be present.
These ranges are fairly representative, but inductive effects and/or conjugation may shift the bands.

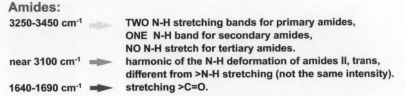

^{1}H NMR Data

Tables of Chemical Shift

Reference

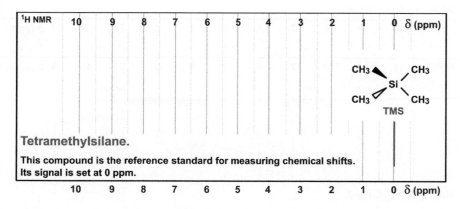

Tetramethylsilane.

This compound is the reference standard for measuring chemical shifts.
Its signal is set at 0 ppm.

General Tables of Functional Groups

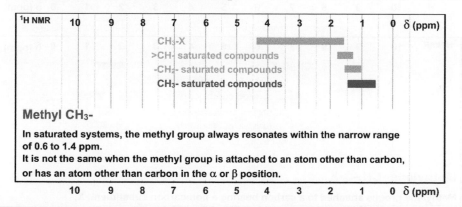

Methyl CH$_3$-

In saturated systems, the methyl group always resonates within the narrow range of 0.6 to 1.4 ppm.
It is not the same when the methyl group is attached to an atom other than carbon, or has an atom other than carbon in the α or β position.

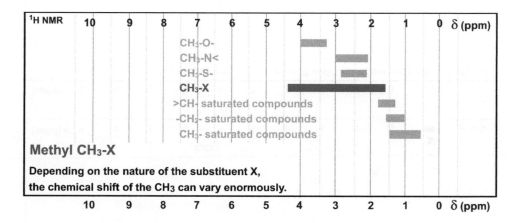

Methyl CH$_3$-X

Depending on the nature of the substituent X,
the chemical shift of the CH$_3$ can vary enormously.

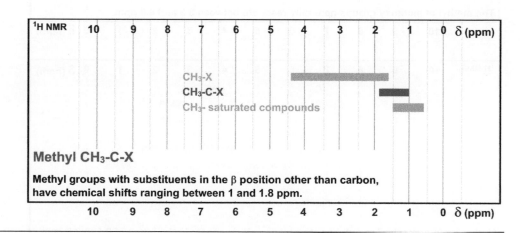

Methyl CH$_3$-C-X

Methyl groups with substituents in the β position other than carbon, have chemical shifts ranging between 1 and 1.8 ppm.

Organic Spectroscopy Workbook, First Edition. Tom Forrest, Jean-Pierre Rabine, and Michel Rouillard.
© 2011 John Wiley & Sons, Ltd. Published 2011 by John Wiley & Sons, Ltd.

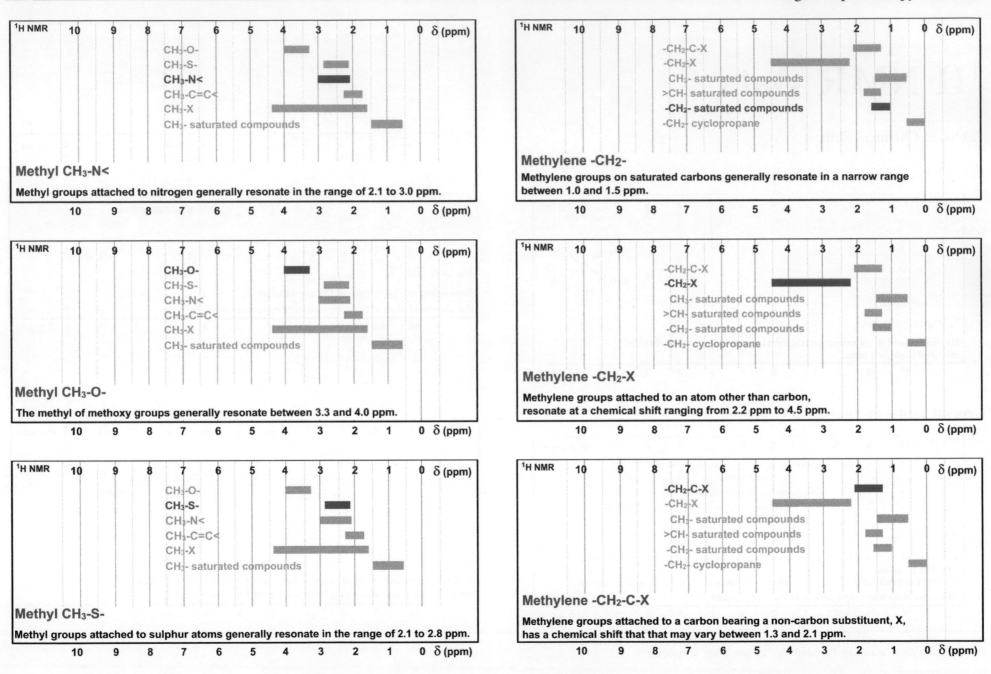

Methyl CH₃-N<

Methyl groups attached to nitrogen generally resonate in the range of 2.1 to 3.0 ppm.

Methyl CH₃-O-

The methyl of methoxy groups generally resonate between 3.3 and 4.0 ppm.

Methyl CH₃-S-

Methyl groups attached to sulphur atoms generally resonate in the range of 2.1 to 2.8 ppm.

Methylene -CH₂-

Methylene groups on saturated carbons generally resonate in a narrow range between 1.0 and 1.5 ppm.

Methylene -CH₂-X

Methylene groups attached to an atom other than carbon, resonate at a chemical shift ranging from 2.2 ppm to 4.5 ppm.

Methylene -CH₂-C-X

Methylene groups attached to a carbon bearing a non-carbon substituent, X, has a chemical shift that that may vary between 1.3 and 2.1 ppm.

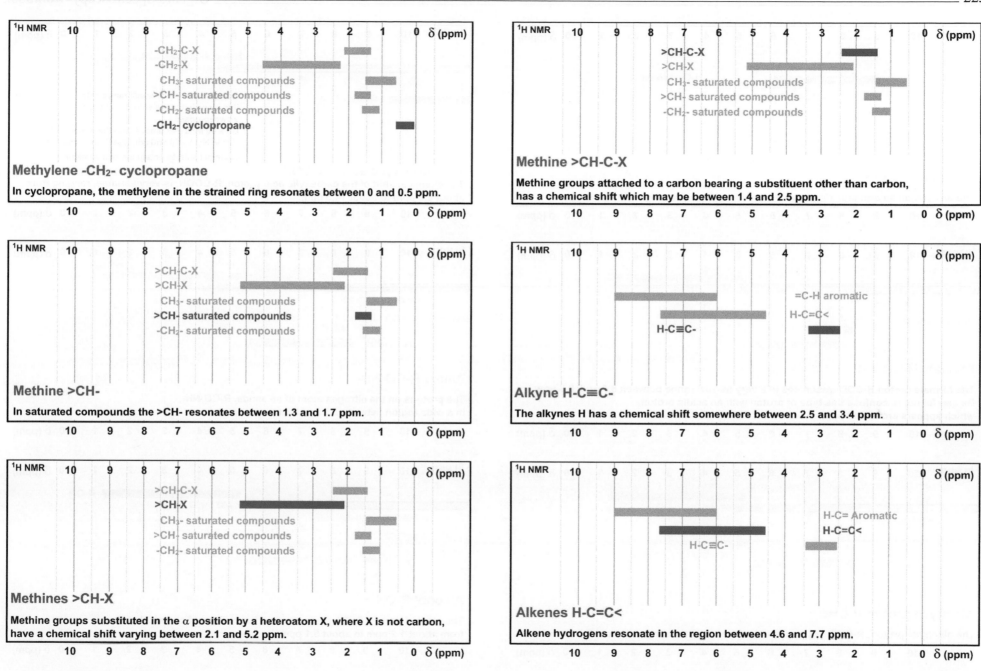

Methylene -CH₂- cyclopropane

In cyclopropane, the methylene in the strained ring resonates between 0 and 0.5 ppm.

Methine >CH-C-X

Methine groups attached to a carbon bearing a substituent other than carbon, has a chemical shift which may be between 1.4 and 2.5 ppm.

Methine >CH-

In saturated compounds the >CH- resonates between 1.3 and 1.7 ppm.

Alkyne H-C≡C-

The alkynes H has a chemical shift somewhere between 2.5 and 3.4 ppm.

Methines >CH-X

Methine groups substituted in the α position by a heteroatom X, where X is not carbon, have a chemical shift varying between 2.1 and 5.2 ppm.

Alkenes H-C=C<

Alkene hydrogens resonate in the region between 4.6 and 7.7 ppm.

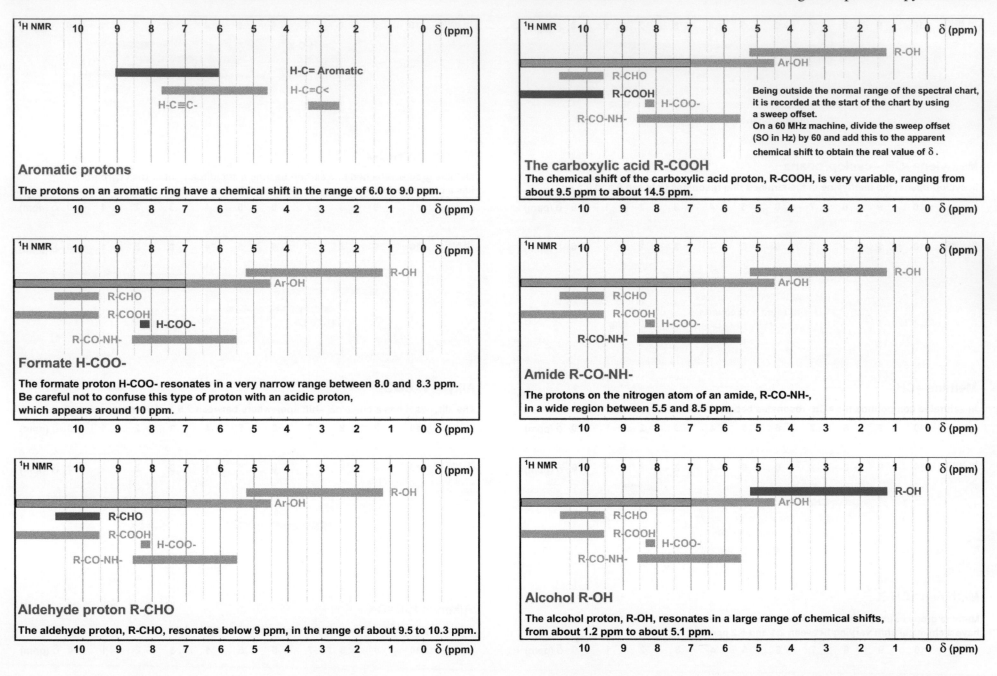

Aromatic protons

The protons on an aromatic ring have a chemical shift in the range of 6.0 to 9.0 ppm.

The carboxylic acid R-COOH

The chemical shift of the carboxylic acid proton, R-COOH, is very variable, ranging from about 9.5 ppm to about 14.5 ppm.

Being outside the normal range of the spectral chart, it is recorded at the start of the chart by using a sweep offset.
On a 60 MHz machine, divide the sweep offset (SO in Hz) by 60 and add this to the apparent chemical shift to obtain the real value of δ.

Formate H-COO-

The formate proton H-COO- resonates in a very narrow range between 8.0 and 8.3 ppm. Be careful not to confuse this type of proton with an acidic proton, which appears around 10 ppm.

Amide R-CO-NH-

The protons on the nitrogen atom of an amide, R-CO-NH-, in a wide region between 5.5 and 8.5 ppm.

Aldehyde proton R-CHO

The aldehyde proton, R-CHO, resonates below 9 ppm, in the range of about 9.5 to 10.3 ppm.

Alcohol R-OH

The alcohol proton, R-OH, resonates in a large range of chemical shifts, from about 1.2 ppm to about 5.1 ppm.

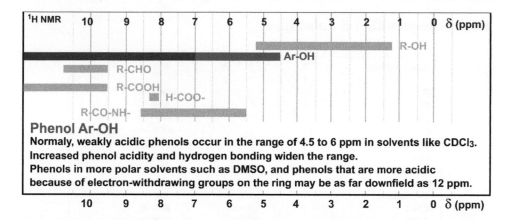

Phenol Ar-OH

Normaly, weakly acidic phenols occur in the range of 4.5 to 6 ppm in solvents like CDCl₃.
Increased phenol acidity and hydrogen bonding widen the range.
Phenols in more polar solvents such as DMSO, and phenols that are more acidic
because of electron-withdrawing groups on the ring may be as far downfield as 12 ppm.

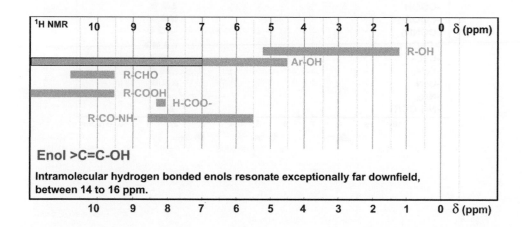

Enol >C=C-OH

Intramolecular hydrogen bonded enols resonate exceptionally far downfield,
between 14 to 16 ppm.

Note

When a chemical shift δ is being outside the normal range of the spectral chart, it is recorded
at the start of the chart by using a sweep offset (SO).

To obtain the chemical shift, convert the sweep offset to ppm (on a 100-MHz machine,
divide the sweep offset in Hz by 100) and add this to the apparent chemical shift to obtain
the real value of δ.

Comparison of δ of Alkyl Groups Bearing a Substituent in α or β Position

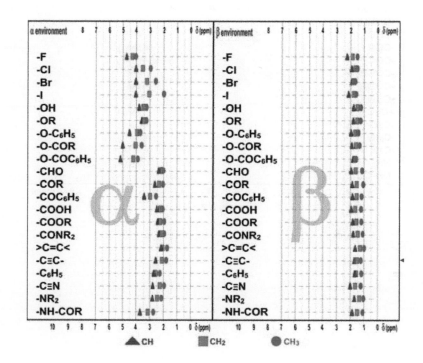

Alkyl Groups with an α Substituent

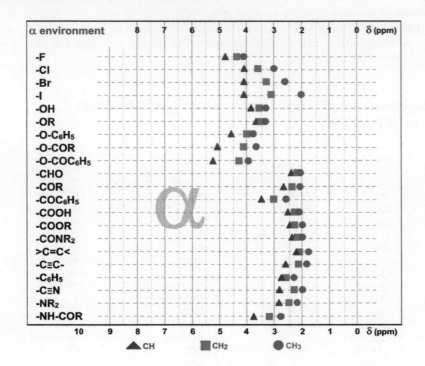

Alkyl Groups with a β Substituent

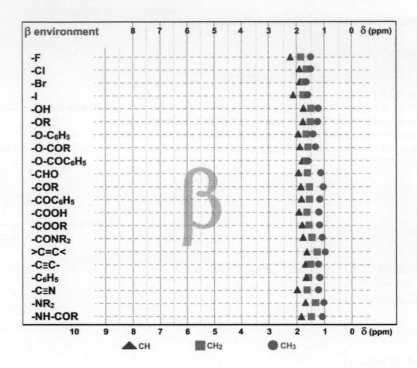

Tables of δ for –CH$_2$– Doubly Substituted in α Position

Values calculated according to the Shoolery rules using the substituent constants σ_1 and σ_2

$$\delta = 0.23 + \sigma_1 + \sigma_2$$

		NHCOR	NO$_2$	NRAr	NR$_2$	NH$_2$	CAr	C=C	CN	C≡C	CONR$_2$	COOR	COOH	COAr	COR	CHO	OCOCF$_3$	OCOAr	OCOR	OAr	OR	OH	I	Br	Cl	F
		1.32	*1.32*	*1.64*	*1.57*	*1.57*	*1.85*	*1.70*	*1.7*	*1.44*	*1.59*	*1.55*	*1.55*	*1.9*	*1.7*	*1.7*	*3.2*	*3.8*	*3.13*	*3.23*	*2.36*	*2.56*	*1.82*	*2.33*	*2.53*	*3.2*
F	*3.2*	4.75	4.75	5.07	5.00	5.00	5.28	5.13	5.13	4.87	5.02	4.98	4.98	5.33	5.13	5.13	6.63	7.23	6.56	6.66	5.79	5.99	5.25	5.76	5.96	6.63
Cl	*2.53*	4.08	4.08	4.40	4.33	4.33	4.61	4.46	4.46	4.20	4.35	4.31	4.31	4.66	4.46	4.46	5.96	6.56	5.89	5.99	5.12	5.32	4.58	5.09	5.29	
Br	*2.33*	3.88	3.88	4.20	4.13	4.13	4.41	4.26	4.26	4.00	4.15	4.11	4.11	4.46	4.26	4.26	5.76	6.36	5.69	5.79	4.92	5.12	4.38	4.89		
I	*1.82*	3.37	3.37	3.69	3.62	3.62	3.90	3.75	3.75	3.49	3.64	3.60	3.60	3.95	3.75	3.75	5.25	5.85	5.18	5.28	4.41	4.61	3.87			
OH	*2.56*	4.11	4.11	4.43	4.36	4.36	4.64	4.49	4.49	4.23	4.38	4.34	4.34	4.69	4.49	4.49	5.99	6.59	5.92	6.02	5.15	5.35				
OR	*2.36*	3.91	3.91	4.23	4.16	4.16	4.44	4.29	4.29	4.03	4.18	4.14	4.14	4.49	4.29	4.29	5.79	6.39	5.72	5.82	4.95					
OAr	*3.23*	4.78	4.78	5.10	5.03	5.03	5.31	5.16	5.16	4.90	5.05	5.01	5.01	5.36	5.16	5.16	6.66	7.26	6.59	6.69						
OCOR	*3.13*	4.68	4.68	5.00	4.93	4.93	5.21	5.06	5.06	4.80	4.95	4.91	4.91	5.26	5.06	5.06	6.56	7.16	6.49							
OCOAr	*3.8*	5.35	5.35	5.67	5.60	5.60	5.88	5.73	5.73	5.47	5.62	5.58	5.58	5.93	5.73	5.73	7.23	7.83								
OCOCF$_3$	*3.2*	4.75	4.75	5.07	5.00	5.00	5.28	5.13	5.13	4.87	5.02	4.98	4.98	5.33	5.13	5.13	6.63									
CHO	*1.7*	3.25	3.25	3.57	3.50	3.50	3.78	3.63	3.63	3.37	3.52	3.48	3.48	3.83	3.63	3.63										
COR	*1.7*	3.25	3.25	3.57	3.50	3.50	3.78	3.63	3.63	3.37	3.52	3.48	3.48	3.83	3.63											
COAr	*1.9*	3.45	3.45	3.77	3.70	3.70	3.98	3.83	3.83	3.57	3.72	3.68	3.68	4.03												
COOH	*1.55*	3.10	3.10	3.42	3.35	3.35	3.63	3.48	3.48	3.22	3.37	3.33	3.33													
COOR	*1.55*	3.10	3.10	3.42	3.35	3.35	3.63	3.48	3.48	3.22	3.37	3.33														
CONR$_2$	*1.59*	3.14	3.14	3.46	3.39	3.39	3.67	3.52	3.52	3.26	3.41															
C≡C	*1.44*	2.99	2.99	3.31	3.24	3.24	3.52	3.37	3.37	3.11																
CN	*1.7*	3.25	3.25	3.57	3.50	3.50	3.78	3.63	3.63																	
C=C	*1.32*	2.87	2.87	3.19	3.12	3.12	3.40	3.25																		
CAr	*1.85*	3.40	3.40	3.72	3.65	3.65	3.93																			
NH$_2$	*1.57*	3.12	3.12	3.44	3.37	3.37																				
NR$_2$	*1.57*	3.12	3.12	3.44	3.37																					
NRAr	*1.64*	3.19	3.19	3.51																						
NO$_2$	*1.32*	2.87	2.87																							
NHCOR	*1.32*	2.87																								

σ_1 and σ_2 values of the substituents.

Values calculated according to an approximation using familiar methylene shifts: $\delta = \delta CH_2 - X + \delta\,CH_2 - Y - 1.4$

δCH_2–X. δCH_2–Y being the chemical shifts of CH_2 in α position to X and to Y.

1.4 ppm being the mean chemical shift of CH_2 group in alkanes.

$\delta = \delta CH_2 - X + \delta\,CH_2 - Y - 1.4$

		NHCOR	NO₂	NRAr	NR₂	NH₂	CAr	C=C	CN	C≡C	CONR₂	COOR	COOH	COAr	COR	CHO	OCOCF₃	OCOAr	OCOR	OAr	OR	OH	I	Br	Cl	F
		3.25	*4.30*	*3.10*	*2.50*	*2.60*	*2.70*	*2.05*	*2.40*	*2.20*	*2.15*	*2.30*	*2.35*	*3.00*	*2.40*	*2.30*	*4.45*	*4.35*	*4.20*	*4.00*	*3.40*	*3.55*	*3.10*	*3.30*	*3.60*	*4.35*
F	*4.35*	6.20	7.25	6.05	5.45	5.55	5.65	5.00	5.35	5.15	5.10	5.25	5.30	5.95	5.35	5.25	7.40	7.30	7.15	6.95	6.35	6.50	6.05	6.25	6.55	7.30
Cl	*3.6*	5.45	6.50	5.30	4.70	4.80	4.90	4.25	4.60	4.40	4.35	4.50	4.55	5.20	4.60	4.50	6.65	6.55	6.40	6.20	5.60	5.75	5.30	5.50	5.80	
Br	*3.3*	5.15	6.20	5.00	4.40	4.50	4.60	3.95	4.30	4.10	4.05	4.20	4.25	4.90	4.30	4.20	6.35	6.25	6.10	5.90	5.30	5.45	5.00	5.20		
I	*3.1*	4.95	6.00	4.80	4.20	4.30	4.40	3.75	4.10	3.90	3.85	4.00	4.05	4.70	4.10	4.00	6.15	6.05	5.90	5.70	5.10	5.25	4.80			
OH	*3.55*	5.40	6.45	5.25	4.65	4.75	4.85	4.20	4.55	4.35	4.30	4.45	4.50	5.15	4.55	4.45	6.60	6.50	6.35	6.15	5.55	5.70				
OR	*3.4*	5.25	6.30	5.10	4.50	4.60	4.70	4.05	4.40	4.20	4.15	4.30	4.35	5.00	4.40	4.30	6.45	6.35	6.20	6.00	5.40					
OAr	*4.00*	5.85	6.90	5.70	5.10	5.20	5.30	4.65	5.00	4.80	4.75	4.90	4.95	5.60	5.00	4.90	7.05	6.95	6.80	6.60						
OCOR	*4.2*	6.05	7.10	5.90	5.30	5.40	5.50	4.85	5.20	5.00	4.95	5.10	5.15	5.80	5.20	5.10	7.25	7.15	7.00							
OCOAr	*4.35*	6.20	7.25	6.05	5.45	5.55	5.65	5.00	5.35	5.15	5.10	5.25	5.30	5.95	5.35	5.25	7.40	7.30								
OCOCF₃	*4.45*	6.30	7.35	6.15	5.55	5.65	5.75	5.10	5.45	5.25	5.20	5.35	5.40	6.05	5.45	5.35	7.50									
CHO	*2.3*	4.15	5.20	4.00	3.40	3.50	3.60	2.95	3.30	3.10	3.05	3.20	3.25	3.90	3.30	3.20										
COR	*2.4*	4.25	5.30	4.10	3.50	3.60	3.70	3.05	3.40	3.20	3.15	3.30	3.35	4.00	3.40											
COAr	*3.00*	4.85	5.90	4.70	4.10	4.20	4.30	3.65	4.00	3.80	3.75	3.90	3.95	4.60												
COOH	*2.35*	4.20	5.25	4.05	3.45	3.55	3.65	3.00	3.35	3.15	3.10	3.25	3.30													
COOR	*2.3*	4.15	5.20	4.00	3.40	3.50	3.60	2.95	3.30	3.10	3.05	3.20														
CONR₂	*2.15*	4.00	5.05	3.85	3.25	3.35	3.45	2.80	3.15	2.95	2.90															
C≡C	*2.2*	4.05	5.10	3.90	3.30	3.40	3.50	2.85	3.20	3.00																
CN	*2.4*	4.25	5.30	4.10	3.50	3.60	3.70	3.05	3.40																	
C=C	*2.05*	3.90	4.95	3.75	3.15	3.25	3.35	2.70																		
CAr	*2.7*	4.55	5.60	4.40	3.80	3.90	4.00																			
NH₂	*2.60*	4.45	5.50	4.30	3.70	3.80																				
NR₂	*2.5*	4.35	5.40	4.20	3.60																					
NRAr	*3.10*	4.95	6.00	4.80																						
NO₂	*4.3*	6.15	7.20																							
NHCOR	*3.25*	5.10																								

$\delta\ CH_2$–X. $\delta\ CH_2$–Y generally observed for only one substituent in α position.

Tables of δ for Aromatics

A substituent on an aromatic ring makes different contributions to the chemical shift of a ring proton, depending on its position on the ring relative to the proton. Each substituent contribution must be added or subtracted, depending on the sign, to the value of the chemical shift.

For the first substituent in the table, H, the chemical shift of the proton is that of benzene, 7.27 ppm, the base value to be used for all other calculations.

Contributions

Substituent	ortho	meta	para
H	0.00	0.00	0.00
F	−0.30	−0.02	−0.22
Cl	0.03	−0.06	−0.04
Br	0.22	−0.13	−0.03
I	0.40	−0.26	−0.03
C_6H_5	0.18	0.00	0.08
CH_3	−0.17	−0.09	−0.18
CH_2-CH_3	−0.15	−0.06	−0.18
$CH(CH_3)_2$	−0.14	−0.09	−0.18
$C(CH_3)_3$	0.01	−0.10	−0.24
CH_2OH	−0.10	−0.10	−0.10
CH_2NH_2	0.00	0.00	0.00
CH_2Cl	0.00	0.01	0.00
$CHCl_2$	−0.10	−0.06	−0.10

Substituent	ortho	meta	para
CCl_3	0.80	0.20	0.20
CHO	0.58	0.21	0.27
$COCH_3$	0.64	0.09	0.30
COCl	0.83	0.16	0.30
COOH	0.80	0.14	0.20
CO_2CH_3	0.74	0.07	0.20
OCH_3	−0.43	−0.09	−0.37
$OCOCH_3$	−0.21	−0.02	−0.13
OH	−0.50	−0.14	−0.40
SCH_3	−0.03	0.00	0.00
CN	0.27	−0.11	−0.30
NO_2	0.95	0.17	0.33
NH_2	−0.75	−0.24	−0.63
$N(CH_3)_2$	−0.60	−0.10	−0.62

Effect of One Substituent

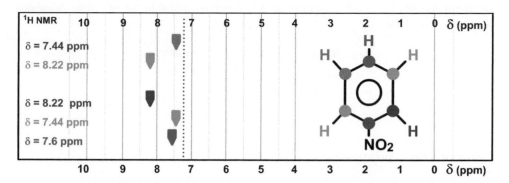

The NO_2 group has a deshielding effect of 0.95 in the ortho position, +0.17 in the meta position and +0.33 in the para position. To estimate the chemical shifts of a proton, one adds the appropriate value of the effect to 7.27, (the chemical shift of benzene protons).

$$\delta_{\text{H blue}} = 7.27 + 0.95 = 8.22\,\text{ppm}$$
$$\delta_{\text{H green}} = 7.27 + 0.17 = 7.44\,\text{ppm}$$
$$\delta_{\text{H red}} = 7.27 + 0.33 = 7.60\,\text{ppm}$$

Effect of Two Substituents

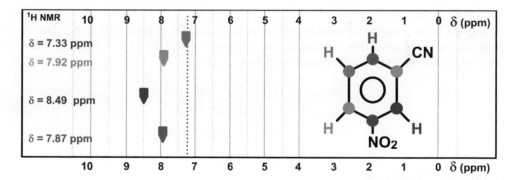

If a cyano group is added to the ring in the meta position to the NO_2 group, the additional contributions (+0.27, −0.11, −0.30) of the CN group are added to the preceding calculated chemical shifts. One gets:

$$\delta_{H\ blue} = 8.22 + 0.27 = 8.49\,ppm$$

$$\delta_{H\ red} = 7.60 + 0.27 = 7.87\,ppm$$

$$\delta_{H\ grey} = 7.44 - 0.11 = 7.33\,ppm$$

$$\delta_{H\ cyan} = 8.22 - 0.30 = 7.92\,ppm$$

Effect of Three Substituents

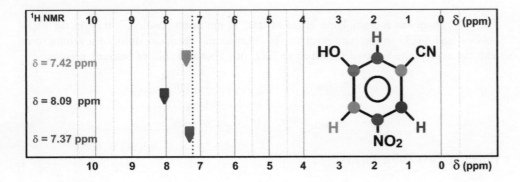

Using the same principle, if another substituent is added, for example OH in the position meta to the NO_2 group, the additional contributions (−0.50, −0.14, −0.40) of the new substituent are added to the appropriate positions.

$$\delta_{H\ blue} = 8.49 - 0.40 = 8.09\,ppm$$

$$\delta_{H\ red} = 7.87 - 0.50 = 7.37\,ppm$$

$$\delta_{H\ cyan} = 7.92 - 0.50 = 7.42\,ppm$$

Coupling Constants

The coupling constant J is the separation between two adjacent lines of the signal from a proton, (or more than one isochronous proton) caused by the splitting of the signal which would otherwise be expected to be a single line.

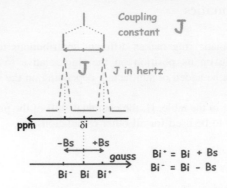

The splitting results from the perturbation of the signal by the influence of the spin of a neighbouring nucleus. This is referred to as a spin–spin interaction. The coupling constant could be expressed in magnetic field units (gauss), or in frequency units (Hertz). The latter is used in NMR spectroscopy; however, gauss (milligauss) is used in electron paramagnetic resonance (EPR).

^1J coupling

In ^1J coupling, a single bond separates the two nuclei; it is principally encountered in the case of heteronuclear coupling, for example, in ^{13}C–^1H coupling. The covalent C–H bond is formed by the overlap of a 1s hydrogen orbital with a hybridized orbital of a carbon atom. The s character of the hybrid orbital is indicated by the s/p ratio in the hybridized orbital.

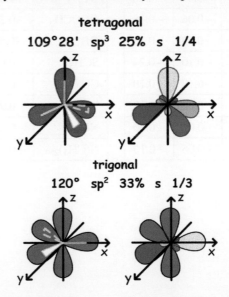

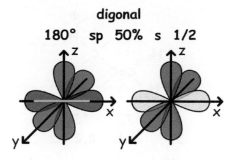

digonal

180°   sp   50%   s   1/2

The size of the coupling constant increases with increasing s character of the hybridized orbital of the carbon atom, i.e., the percentage of s orbital participating in the sigma bond.

hybridation C		Character s	^1J ^{13}C-H
sp³ (tetragonal)	→C-H	0.50	125 Hz
sp² (trigonal)	=C-H	0.57	150 Hz
sp (digonal)	≡C-H	0.70	225 Hz

Thus, the coupling is dependent on the electronic transmission by the participating s orbitals, the only ones that have a nonzero density at the nucleus.

Geminal Coupling ^2J

Geminal coupling is defined as a spin–spin interaction between two non isochronous nuclei separated by two bonds, as for example, the two protons of a CH₂ group.

When it involves two protons, geminal coupling is normally only observed in the case where the two protons are chemically nonequivalent: in the case of the axial and equatorial protons of a cyclohexane ring the coupling is observed with a value of $^2J = 8$–14 Hz.

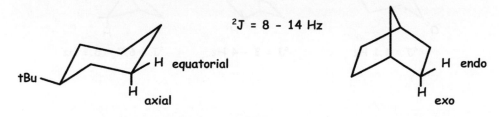

$^2J = 8 - 14$ Hz

tBu — H equatorial / H axial

H endo / H exo

- in the case of the protons of a terminal methylene group. >C=CH₂ ($^2J \approx 2$ Hz).
- in the case of protons of a sp³ methylene group in an asymmetric environment ($^2J = 10$ to 20 Hz).

Vicinal Coupling ^3J

Vicinal coupling is designated as a spin–spin interaction between two non–isochronous nuclei separated by three bonds: H–C–C–H.

This coupling is most frequently observed in a two-carbon chain bearing nonisochronous hydrogen atoms.

The value of the coupling constant depends on many parameters: molecular geometry, electronegativity of neighbouring atoms, hybridization of the atoms involved and of their neighbours.

In the case of vicinal coupling, one geometric factor is particularly significant; that is the dihedral angle ϕ formed by the planes containing the C–C bond and the C–H bonds involved in the coupling H–C–C–H.

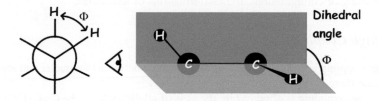

Dihedral angle

For example.

the angle is	0°	for the H–C = C–H $^3J_{cis}$ coupling.
and	180°	for the H–C = C–H $^3J_{trans}$ coupling.

In these two cases, the Karplus curves of the dihedral angles show that one could expect to see strong coupling.

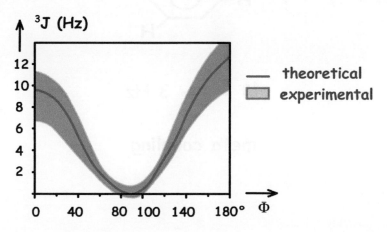

^3J (Hz)

— theoretical
▢ experimental

Φ

The general form of the Karplus equations are:

$$^3J = A . \cos^2 f + B \text{ for } 0° < \phi < 90° \text{ and}$$
$$^3J = A' . \cos^2 f + B' \text{ for } 90° < \phi < 180°$$

with $A < A'$, these values are also dependent on other effects (electronegativity. etc.). When there is free rotation about the C–C bond, the mean value of the vicinal coupling is about 6 to 8 Hz.

For more rigid structures, the coupling constant can vary from 0 to 15 Hz depending on the value of the dihedral angle ϕ(H–C–C–H) (the value of J is at a maximum for 0° and 180°, and zero for 90°, see Karplus curves).

The dependence of the coupling constant on the dihedral angles provides a very important tool in the study of configurations and conformations of molecules.

Long-Range Coupling

Long-range coupling is defined as spin–spin interaction between two nuclei that are separated by more than three bonds. It is observed infrequently.

^4J long-range coupling is observed in: the case of allylic systems, H–C=C–CH<, terminal acetylenes, H–C≡C–CH<, aromatics with hydrogens meta to one another.

The absolute value varies between 0 and 3 Hz. Although quite weak, the presence of long-range coupling can offer a good indication of the presence of particular molecular fragments.

Φ	$^4J_{ab}$	$^4J_{ac}$
90°	1.8	2.2 Hz

Allylic coupling

$^4J_{ab}$ cisoïd = 3 Hz
$^4J_{ac}$ transoïd = 3 Hz

Homoallylic coupling

Long-range coupling is also observed in cases where a rigid geometry provides a coplanar 'W' or 'M' arrangement of the coupling pathway, that is, for a zig–zag H–C–C–C–H arrangement:

$^4J = 1.1$ Hz $^4J = 3 - 4$ Hz $^4J = 6.7 - 8.1$ Hz

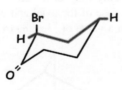

$^4J = 1 - 3$ Hz

meta coupling

^{13}C NMR Data

Chemical Shift Scales... Note the large range of chemical shifts of ^{13}C compared to those of ^{1}H

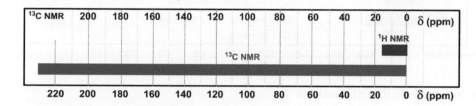

The spectrum is separated into four broad ranges:

- from 20 to 70 ppm, sp^3 carbons (bearing no more than one electron–withdrawing group)
- from 70 to 100 ppm, sp carbons (carbon–carbon triple bond)
- from 100 to 220 ppm, sp^2 carbons, separated into two regions
  - 100 to 155 ppm, carbon–carbon double bonds
  - 155 to 220 ppm, carbon double bonded to oxygen or another heteroatom

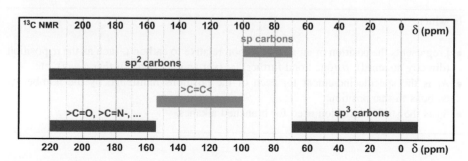

General Ranges for CH$_3$–, –CH$_2$–, >CH–...

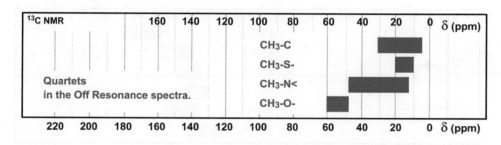

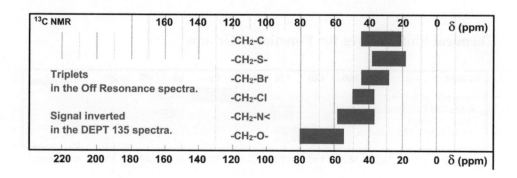

Organic Spectroscopy Workbook, First Edition Tom Forrest, Jean-Pierre Rabine, and Michel Rouillard.
© 2011 John Wiley & Sons, Ltd. Published 2011 by John Wiley & Sons, Ltd.

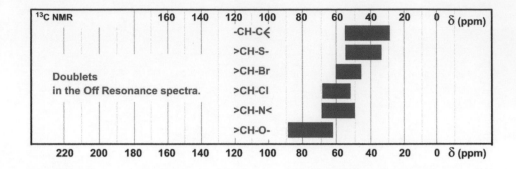

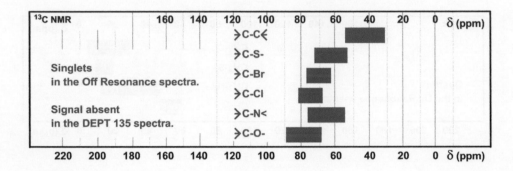

Chemical Shift Ranges for Functional Groups

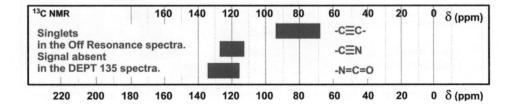

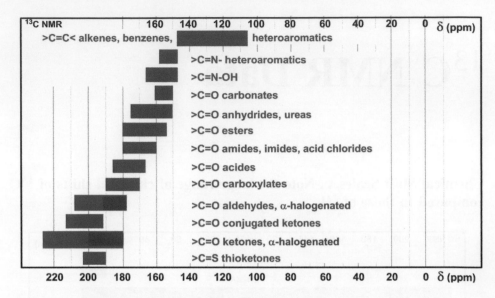

Chemical Shift Estimations

Alkanes

The chemical shift δ_i of a carbon i in a linear alkane can be estimated using the following equation

$$\delta_i = -2.3 + \Sigma A_j.n_j + S_{ij}$$

In which

- j represents the position of an alkane carbon relative to carbon i, such as the α position (directly attached), β (one bond further), γ (yet another bond further away), etc.
- A_j is the specific increment for each of the positions multiplied by the number of carbons in that position.
- S_{ij} is the steric correction used for branched alkanes.

The A_j increments are simple to use; you just add the following contributions per carbon of each type present:

α: +9.1 ppm	β: +9.4 ppm	γ: −2.5 ppm	δ: +0.3 ppm	ε: +0.1 ppm

The steric corrections are for branched alkanes, the S_{ij} terms represent corrections accounting for the steric compression of the carbon being evaluated and depend on the nature of the carbon being calculated (I, II, III or IV, i.e. primary, secondary, tertiary or quaternary, i.e. CH$_3$, CH$_2$, CH or C), and of the nature (primary, secondary, tertiary or quaternary) of the carbons which are directly attached.

	CH$_3$ or I	CH$_2$ or II	CH or III	C or IV
CH$_3$ or I			−1.1	−3.4
CH$_2$ or II			−2.5	−7.5
CH or III		−3.7	−9.5	−15.0
C or IV	−1.5	−8.4	−15.0	−25.0

Example:

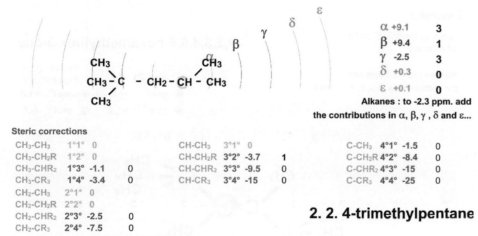

α +9.1   **3**
β +9.4   **1**
γ -2.5   **3**
δ +0.3   **0**
ε +0.1   **0**

Alkanes : to -2.3 ppm. add the contributions in α, β, γ , δ and ε...

Steric corrections

CH$_3$-CH$_3$	1°1° 0		CH-CH$_3$	3°1° 0	C-CH$_3$	4°1° -1.5 0
CH$_3$-CH$_2$R	1°2° 0		CH-CH$_2$R	3°2° -3.7 1	C-CH$_2$R	4°2° -8.4 0
CH$_3$-CHR$_2$	1°3° -1.1 0		CH-CHR$_2$	3°3° -9.5 0	C-CHR$_2$	4°3° -15 0
CH$_3$-CR$_3$	1°4° -3.4 0		CH-CR$_3$	3°4° -15 0	C-CR$_3$	4°4° -25 0
CH$_2$-CH$_3$	2°1° 0					
CH$_2$-CH$_2$R	2°2° 0					
CH$_2$-CHR$_2$	2°3° -2.5 0					
CH$_2$-CR$_3$	2°4° -7.5 0					

Note, there is one steric contribution!

3 carbons on the α layer... 1 carbon on the β layer... 3 carbons on the γ layer... a steric contribution CH-CH$_2$R **3°2° of -3.7...**

δ = -2.3 +3*9.1 +1*9.4 -5*2.5 = 23.2 ppm

δ Experimental value = 24.7 ppm

2. 2. 4-trimethylpentane

Alcohols

Increments to be added to the chemical shifts of the corresponding alkane (due to the insertion of O in a C–H bond)

	α	β	γ
1°-alcohol	48.3	10.2	−5.8
2°-alcohol	44.5	9.7	−3.3
3°-alcohol	39.7	7.3	−1.8

The variations observed for the different classes of alcohols arise from changes in conformation due to the increased steric effect. Thus the γ effect is weaker for molecules already sterically hindered. We see a strongly deshielding α effect (range of δ from 50 to 90 ppm), a smaller deshielding β effect of about 10 ppm and a slightly shielding γ effect.

Other Substitutions

Depending on the position of the substituent, terminal or internal, the increment may be different. We see again that the substitution is deshielding in the α or β position, but shielding in the γ position.

A set of different increments have been determined for the calculation of the chemical shifts of alkenes with functional groups.

Groups	α terminal	β terminal	α internal	β internal	γ
−CH$_3$	9	10	6	8	−2
−CH = CH$_2$	20	6	0	0	−0.5
−C≡C–H	4.5	5.5	0	0	−3.5
−C$_6$H$_5$	23	9	17	7	−2
−CN	4	3	1	3	−3
−COOH	21	3	16	2	−2
−COO$^-$	25	5	20	3	−2
−COOR	20	3	17	2	−2
−COCl	33	0	28	2	0

(continued overleaf)

(continued)

Groups	α terminal	β terminal	α internal	β internal	γ
–CONH$_2$	22	2.5	0	0	−0.5
–COR	30	1	24	1	−2
–CHO	31	0	0	0	−2
–OH	48	10	41	8	−5
–OR	58	8	51	5	−4
–OCOR	51	6	45	5	−3
–SH	11	12	11	11	−4
–SR	20	7	0	0	−3
–NH$_2$	29	11	24	10	−5
–NH$_3^+$	26	8	24	6	−5
–NHR	37	8	31	6	−4
–NR$_3{}^+$	31	5	0	0	−7
–NO$_2$	63	4	57	4	0
–F	68	9	63	6	−4
–Cl	31	11	32	10	−4
–Br	20	11	25	10	−3
–I	−6	11	4	12	−1

Alkenes

For alkenes, a system of incremental adjustments analogous to those for the alkane is available.

The algorithm provided here starts with a base value of 122.1 ppm and adds increments for different substituents. The value of the substituent constant depends on the position of the substituent relative to the double bond (α. β, γ or δ), and to the carbon in question. It may be directly bonded (on the σ carbon), or on the other carbon of the double bond (i.e., the π carbon). The two positions of the substituent chain, directly bonded and on opposite sides of the double bond, have different factors for the calculation of chemical shifts. The increments are positive in α, β and δ; negative in γ for the substituent directly attached the carbon being calculated (red colour on the diagram below); negative in β, γ; and positive in δ and ε if the substituent chain is on the other carbon of the double bond.

There are also steric correction terms that are added, notably the presence of two substituents in cis, geminal (on the same sp^2 carbon), and whether they are σ or π relative to the calculated carbon.

The values of the different increments are shown on the structural diagram.

$\delta_{C(k)} = 122.1 + \sum A_i n_{ki} + S$			
I	A_i	Substitution	S
α^σ	+11.0	cis (Z)	−1.2
β^π	−7.1	gem$^\alpha$	−4.9
β^σ	+6.0	gem$^\beta$	+1.2
γ^π	−1.9	mult$^\alpha$	+1.3
γ^σ	−1.0	mult$^\pi$	−0.7
δ^π	+1.1		
δ^σ	+0.7		
ε	+0.2		

Example:

Table of chemical shifts of alkenes

Alkenes : to 122.1 ppm add the contributions in α, β, γ , δ and ε...

2,2,3,4,5,5-hexamethylhex-3-ene

Steric Corrections

○ cis (Z) -1.2
◉ gem$^\alpha$ -4.9   ◉ mul$^\sigma$ +1.3
◉ gem$^\beta$ +1.2   ◉ mul$^\pi$ -0.7

ε +0.2 δ^π+1.1 γ^π−1.9 β^π−7.1 α^σ+11.0 β^σ+6.0 γ^σ−1.0 δ^σ+0.7

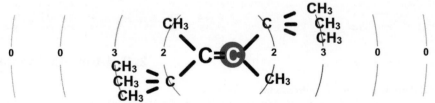

There is 1 carbon on α^σ, 3 carbon on β^π, 1 carbon on β^π, 3 carbons on γ^π.

Don't forget the steric contribution : gem$^\alpha$, gem$^\beta$, the mul$^\alpha$ and mul$^\beta$ branching.

$\delta = 122.1 + 2*11.0 + 3*6.0 - 2*7.1 -3*1.9 - 4.9 + 1.2 +1.3 - 0.7 = 139.1$ ppm

δ Experimental value = 137.3 ppm

Monosubstituted Alkenes

For the substituted alkenes, the substituent effects are found to be very different, depending on whether the substituent is in the α or β position. For the molecule $X–C_\alpha H=C_\beta H_2$ the following substituent effects are observed for the vinyl carbons as compared to ethylene:

$$\delta_i = 123.5 + \Delta\delta_i$$

Substituent X	$\Delta\delta_\alpha$	$\Delta\delta_\beta$
$–CH_3$	10.6	−7.9
$–CH_2–CH_3$	15.5	−9.7
$–CH(CH_3)_2$	20.4	−11.5
$–C(CH_3)_3$	25.3	−13.3
$–CH_2Cl$	10.2	−6.0
$–CH_2Br$	10.9	−4.5
$–CH_2I$	14.2	−4.0
$–CH_2OH$	14.2	−8.4
$–C_6H_5$	12.5	−11.0
$–OCH_3$	29.4	−38.9
$–OCOCH_3$	18.4.	−26.7
F	24.9	−34.3
Cl	2.6	−6.1
Br	−7.9	−1.4
I	−38.1	7.0
$–CN$	−15.1	14.2
$–NO_2$	22.3	−0.9
$–CHO$	13.1	12.7
$–COCH_3$	15.0	5.8
$–COOH$	4.2	8.9
$–COOCH_2CH_3$	6.3	7.0

Alkynes

For the alkynes, there is a similar incremental system.

$\delta_{C(k)} = 71.9 + \Sigma A_i n_{ki}$	
I	A_i
α	+6.93
β	+4.75
γ	−0.13
δ	+0.51
α'_π	−5.69
β'_π	+2.32
γ'_π	−1.32
δ'_π	+0.56

Example:

Table of chemical shifts of alkynes

2,2,5,5-tetramethylhex-3-yne

Alkynes : to 71.9 ppm add the contributions in $\alpha, \beta, \gamma, \delta$ and ϵ...

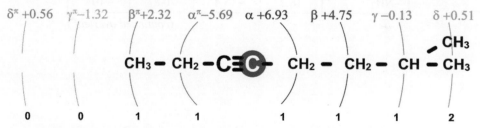

There is 1 carbon on the α layer, 1 on the β layer, 1 on the γ layer, 2 on the δ layer. 1 on the α^π layer and 1 on the β^π layer.

$\delta = 71.9 + 1*6.93 + 1*4.75 - 1*5.69 + 1*2.32 - 1*1.32 + 2*0.56 = 80.01$ ppm

δ Experimental value = 79.56 ppm

Substituted Benzenes

The calculation of the ^{13}C chemical shifts of the aromatic carbons is based on the series of increments obtained from the differences between the observed chemical shifts for a series of mono-substituted benzenes and that of benzene (128.5 ppm).

The formula is : $\delta_i = 128.5 + \Delta\delta_i$

Substituent	$\Delta\delta\ C_{ipso}$	$\Delta\delta\ C_{ortho}$	$\Delta\delta\ C_{meta}$	$\Delta\delta\ C_{para}$
$-CH_3$	9.2	0.7	−0.1	−3.1
$-CH_2-CH_3$	15.6	−0.5	0	−2.7
$-CH(CH_3)_2$	20.1	−2.0	0	−2.5
$-C(CH_3)_3$	22.1	−3.4	−0.4	−3.1
$-CH_2-CN$	1.7	0.5	−0.8	−0.7
$-CH_2COCH_3$	6.0	1.0	0.2	−1.6
$-CH_2NH_2$	14.9	1.4	−0.1	−1.9
$-CH_2OH$	12.4	−1.2	0.2	−1.1
$-CH_2OCH_3$	11.0	0.5	−0.4	−0.5
$-OH$	26.9	−12.8	1.4	−7.4
$-OCH_3$	31.9	−14.4	1.0	−7.7
$-OCOCH_3$	22.4	−7.1	0.4	−3.2
$-NH_2$	18.2	−13.4	0.8	−10

(continued overleaf)

(continued)

Substituent	$\Delta\delta\ C_{ipso}$	$\Delta\delta\ C_{ortho}$	$\Delta\delta\ C_{meta}$	$\Delta\delta\ C_{para}$
$-N(CH_3)_2$	22.5	−15.4	0.9	−11.5
$-NHCOCH_3$	9.7	−8.1	0.2	−4.4
F	34.8	−13.0	1.6	−4.4
Cl	6.3	0.4	1.4	−1.9
Br	5.8	3.2	1.6	−1.6
I	−34.1	8.9	1.6	−1.1
$-CF_3$	2.5	−3.2	0.3	3.3
$-CN$	−15.7	3.6	0.7	4.3
$-NO_2$	19.9	−4.9	0.9	6.1
$-CHO$	8.4	1.2	0.5	5.7
$-COCH_3$	8.9	0.0	−0.1	4.4
$-COC_6H_5$	9.3	1.6	−0.3	3.7
$-CONH_2$	5.0	−1.2	0.1	3.4
$-COOH$	2.1	1.6	−0.1	5.2
$-COOCH_3$	3.0	1.2	−0.1	4.3
$-COOCH_2CH_3$	2.1	1.0	−0.5	3.9

Polysubstituted Benzenes

An estimation of the chemical shift of the different carbons can be made by adding the increments observed for each substituent in the monosubstituted cases.

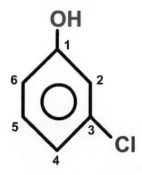

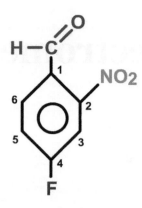

	1	2	3	4	5	6
	128.5	128.5	128.5	128.5	128.5	128.5
OH	26.9$_{ipso}$	−12.8$_o$	1.4$_m$	−7.4$_p$	1.4$_m$	−12.8$_o$
Cl	1.4$_m$	0.4$_o$	6.3$_{ipso}$	0.4$_o$	1.4$_m$	−1.9$_p$
Calc.	156.8	116.1	136.2	121.5	131.3	113.8
Exp.	156.3	116.1	135.1	121.5	130.5	113.9

	1	2	3	4	5	6
	128.5	128.5	128.5	128.5	128.5	128.5
CHO	8.4$_{ipso}$	1.2$_o$	0.5$_m$	5.7$_p$	0.5$_m$	1.2$_o$
NO$_2$	−4.9$_o$	19.9$_{ipso}$	−4.9$_o$	0.9$_m$	6.1$_p$	0.9$_m$
F	−4.4$_p$	1.6$_m$	−13.0$_o$	34.8$_{ipso}$	−13.0$_o$	1.6$_m$
Calc.	127.6	151.2	111.1	169.9	122.1	132.2
Exp.	126.4	150.0	112.0	163.8	120.5	132.4

Mass Spectrometry Data

Main Fragments Observed

Alkanes

Fragments observed for alkanes.

The masses of fragments of alkanes correspond to the mass of a methyl group CH_3, that is 15, increased by 14 for each CH_2 group.
One has fragments of mass $15 + 14n = 29, 43, 57, \ldots$

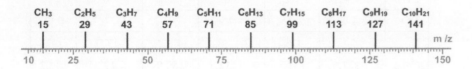

Alkenes

Fragments observed for alkenes.
Fragments observed for alkanes.

The masses of fragments of alkenes are lower by 2u
compared to the masses of fragments of the corresponding alkanes (2H less).

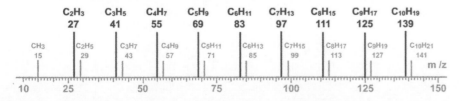

Alkynes

Fragments observed for alkynes.
Fragments observed for alkenes.
Fragments observed for alkanes.

The masses of fragments of alkynes are
4u less than the masses corresponding fragments of alkanes (4H less).

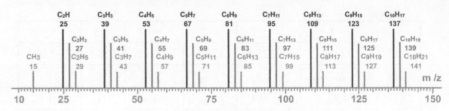

Aromatics

Fragments observed for aromatics.
Fragments observed for alkenes.
Fragments observed for alkanes.

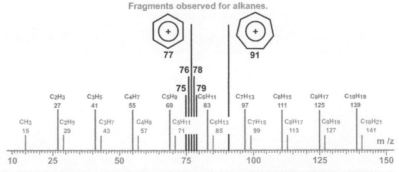

Organic Spectroscopy Workbook, First Edition. Tom Forrest, Jean-Pierre Rabine, and Michel Rouillard.
© 2011 John Wiley & Sons, Ltd. Published 2011 by John Wiley & Sons, Ltd.

Alcohols

Fragments observed for alcohols, ethers.
Fragments observed for alkenes.
Fragments observed for alkanes.

The masses of ions containing an oxygen (31, 45, 59, ...) are greater by 2u
than the masses of fragments of alkanes having an extra carbon
(exchange of a CH_2 of mass 14 for an oxygen of mass 16).

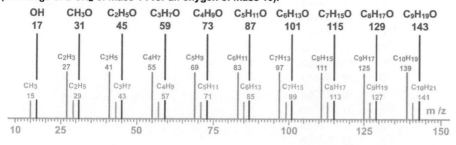

Ketones

Fragments observed for ketones.
Fragments observed for alkanes.

The masses of fragments of ketones correspond to the mass of a methyl CH_3, 15,
plus that of a carbonyl, CO, 28, increased by 14 for each CH_2, present.
One has fragments of mass 43 + 14n = 43, 57, 71...

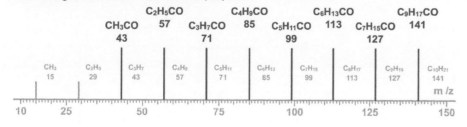

Aldehydes

Fragments observed for aldehydes.
Fragments observed for ketones.
Fragments observed for alkanes.

The masses of fragments of aldehydes corresponding to the mass of HCO,
that is 29, increased by 14 for each CH_2 group present.
One has fragments of mass 29 + 14n = 29, 43, 57, 71...

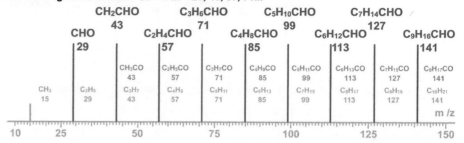

Acids

Fragments observed for acids.
Fragments observed for alcohols, ethers
Fragments observed for alkanes.

The masses of fragments of acids correspond to COOH (12 + 32 + 1 = 45)
and to homologous fragments containing CH_2 groups: 45 + 14n = 45, 59, 73...

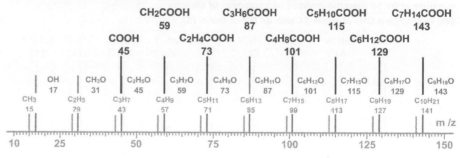

Esters

Fragments observed for esters.
Fragments observed for alkanes, alcohols, ethers, and acids.

The masses of fragments of esters correspond to $C_2O_2H_3$ (24 + 32 + 3 = 59)
and to derivatives having one or more CH_2 units : 59 + 14n = 73, 87, 101...

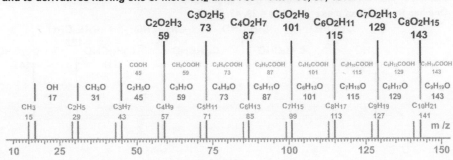

Amines

Fragments observed for amines.
Fragments observed for alcohols, ethers.
Fragments observed for alkanes.

The masses of ions containing nitrogen (30, 44, 58, ...)
are greater by 1u than the masses of fragments of alkanes
(we exchange a CH_2 of mass 14 with an NH of mass 15).

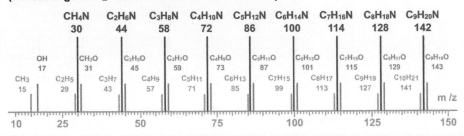

Atomic Masses

^{1}H:	1.007825
^{2}H:	2.0140
^{12}C:	12.000000
^{13}C:	13.003355
^{14}N:	14.003074
^{15}N:	15.000108
^{16}O:	15.994915
^{17}O:	16.999131
^{18}O:	17.999160
^{19}F:	18.998403
^{31}P:	30.973762
^{32}S:	31.972070
^{33}S:	32.971456
^{34}S:	33.967866
^{36}S:	35.967080
^{35}Cl:	34.968852
^{37}Cl:	36.965903
^{79}Br:	78.918336
^{81}Br:	80.916289
^{127}I:	126.90447

Isotopic Abundances in %

12	M+1	M+2	Mass
C	1.11	0.00	12.0000

13	M + 1	M + 2	Mass
CH	1.13	0.00	13.0078

14	M + 1	M + 2	Mass
N	0.37	0.00	14.0031
CH_2O	1.14	0.00	14.0157

15	M + 1	M + 2	Mass
HN	0.39	0.00	15.0109
CH_3	1.16	0.00	15.0235

16	M + 1	M + 2	Mass
H_2N	0.40	0.00	16.0187
CH_4	1.17	0.00	16.0313

17	M + 1	M + 2	Mass
HO	0.06	0.20	17.0027
H_3N	0.42	0.00	17.0266

18			
H_2O	0.07	0.20	18.0106

19, 20, 21, 22, 23
No fragment containing C, H, O or N
corresponds to these masses...

24	M + 1	M + 2	Mass
C_2	2.22	0.01	24.0000

25	M + 1	M + 2	Mass
C_2H	2.24	0.01	25.0078

26	M + 1	M + 2	Mass
CN	1.48	0.00	26.0031
C_2H_2	2.25	0.01	26.0157

27	M + 1	M + 2	Mass
HCN	1.50	0.00	27.0109
C_2H_3	2.27	0.01	27.0235

28	M + 1	M + 2	Mass
N_2	0.74	0.00	28.0062
CO	1.15	0.20	27.9949
CH_2N	1.51	0.00	28.0187
C_2H_4	2.28	0.01	28.0313

29	M + 1	M + 2	Mass
HN_2	0.76	0.00	29.0140
CHO	1.17	0.20	29.0027
CH_3N	1.53	0.00	29.0266
C_2H_5	2.30	0.01	29.0391

30	M + 1	M + 2	Mass
NO	0.41	0.20	29.9980
H_2N_2	0.77	0.00	30.0218
CH_2O	1.18	0.20	30.0106
CH_4N	1.54	0.01	30.0344
C_2H_6	2.31	0.01	30.0470

31	M + 1	M + 2	Mass
HNO	0.43	0.20	31.0058
H_3N_2	0.79	0.00	31.0297
CH_3O	1.20	0.20	31.0184
CH_5N	1.56	0.01	31.0422

32	M + 1	M + 2	Mass
O_2	0.08	0.40	31.9898
H_2NO	0.44	0.20	32.0136
H_4N_2	0.80	0.00	32.0375
CH_4O	1.21	0.20	32.0262

33	M + 1	M + 2	Mass
HO_2	0.10	0.40	32.9976
H_3NO	0.46	0.20	33.0215

34			
H_2O_2	0.11	0.40	34.0054

35

No fragment containing C, H, O or N
corresponds to this mass...

36	M + 1	M + 2	Mass
C_3	3.33	0.04	36.0000

37	M + 1	M + 2	Mass
C_3H	3.35	0.04	37.0078

38	M + 1	M + 2	Mass
C_2N	2.59	0.02	38.0031
C_3H_2	3.36	0.04	38.0157

39	M + 1	M + 2	Mass
C_2HN	2.61	0.02	39.0109
C_3H_3	3.38	0.04	39.0235

40	M + 1	M + 2	Mass
CN_2	1.85	0.01	40.0062
C_2O	2.26	0.21	39.9949
C_2H_2N	2.62	0.02	40.0187
C_3H_4	3.39	0.04	40.0313

41	M + 1	M + 2	Mass
CHN_2	1.87	0.01	41.0140
C_2HO	2.28	0.21	41.0027
C_2H_3N	2.64	0.02	41.0266
C_3H_5	3.41	0.04	41.0391

42	M + 1	M + 2	Mass
N_3	1.11	0.00	42.0093
CNO	1.52	0.21	41.9980
CH_2N_2	1.88	0.01	42.0218
C_2H_2O	2.29	0.21	42.0106
C_2H_4N	2.65	0.02	42.0344
C_3H_6	3.42	0.04	42.0470

43	M + 1	M + 2	Mass
HN_3	1.13	0.00	43.0171
$CHNO$	1.54	0.21	43.0058
CH_3N_2	1.90	0.01	43.0297
C_2H_3O	2.31	0.21	43.0184
C_2H_5N	2.67	0.02	43.0422
C_3H_7	3.44	0.04	43.0548

44	M + 1	M + 2	Mass
N_2O	0.78	0.20	44.0011
H_2N_3	1.14	0.00	44.0249
CO_2	1.19	0.40	43.9898
CH_2NO	1.55	0.21	44.0136
CH_4N_2	1.91	0.01	44.0375
C_2H_4O	2.32	0.21	44.0262
C_2H_6N	2.68	0.02	44.0501
C_3H_8	3.45	0.04	44.0626

45	M + 1	M + 2	Mass
HN_2O	0.80	0.20	45.0089
H_3N_3	1.16	0.00	45.0328
CHO_2	1.21	0.40	44.9976
CH_3NO	1.57	0.21	45.0215
CH_5N_2	1.93	0.01	45.0453
C_2H_5O	2.34	0.21	45.0340
C_2H_7N	2.70	0.02	45.0579

46	M + 1	M + 2	Mass
NO_2	0.45	0.40	45.9929
H_2N_2O	0.81	0.20	46.0167
H_4N_3	1.17	0.01	46.0406
CH_2O_2	1.22	0.40	46.0054
CH_4NO	1.58	0.21	46.0293
CH_6N_2	1.94	0.01	46.0532
C_2H_6O	2.35	0.22	46.0419

47	M + 1	M + 2	Mass
HNO_2	0.47	0.40	47.0007
H_3N_2O	0.83	0.20	47.0248
H_5N_3	1.19	0.01	47.0484
CH_3O_2	1.24	0.40	47.0133
CH_5NO	1.60	0.21	47.0371

48	M + 1	M + 2	Mass
O_3	0.12	0.60	47.9847
H_2NO_2	0.48	0.40	48.0085
H_4N_2O	0.84	0.20	48.0324
CH_4O_2	1.25	0.40	48.0211
C_4	4.44	0.07	48.0000

49	M + 1	M + 2	Mass
HO_3	0.14	0.60	48.9925
H_3NO_2	0.50	0.40	49.0164
C_4H	4.46	0.07	49.0078

50	M + 1	M + 2	Mass
H_2O_3	0.15	0.60	50.0003
C_3N	3.70	0.05	50.0031
C_4H_2	4.47	0.07	50.0157

51	M + 1	M + 2	Mass
C_3HN	3.72	0.05	51.0109
C_4H_3	4.49	0.08	51.0235

52	M + 1	M + 2	Mass
C_2N_2	2.96	0.03	52.0062
C_3O	3.37	0.24	51.9949
C_3H_2N	3.73	0.05	52.0187
C_4H_4	4.50	0.08	52.0313

53	M + 1	M + 2	Mass
C_2HN_2	2.98	0.03	53.0140
C_3HO	3.39	0.24	53.0027
C_3H_3N	3.75	0.05	53.0266
C_4H_5	4.52	0.08	53.0391

54	M + 1	M + 2	Mass
CN_3	2.22	0.02	54.0093
C_2NO	2.63	0.22	53.9980
$C_2H_2N_2$	2.98	0.03	54.0218
C_3H_2O	3.40	0.24	54.0106
C_3H_4N	3.76	0.05	54.0344
C_4H_6	4.53	0.08	54.0470

55	M + 1	M + 2	Mass
CHN_3	2.24	0.02	55.0171
C_2HNO	2.65	0.22	55.0058
$C_2H_3N_2$	3.01	0.03	55.0297
C_3H_3O	3.42	0.24	55.0184
C_3H_5N	3.78	0.05	55.0422
C_4H_7	4.55	0.08	55.0548

56	M + 1	M + 2	Mass
N_4	1.48	0.01	56.0124
CN_2O	1.89	0.21	56.0011
CH_2N_3	2.25	0.02	56.0249
C_2O_2	2.30	0.41	55.9898
C_2H_2NO	2.66	0.22	56.0136
$C_2H_4N_2$	3.02	0.03	56.0375
C_3H_4O	3.43	0.24	56.0262
C_3H_6N	3.79	0.05	56.0501
C_4H_8	4.56	0.08	56.0626

57	M + 1	M + 2	Mass
HN_4	1.50	0.01	57.0202
CHN_2O	1.91	0.21	57.0089
CH_3N_3	2.27	0.02	57.0328
C_2HO_2	2.32	0.41	56.9976
C_2H_3NO	2.68	0.22	57.0215
$C_2H_5N_2$	3.04	0.03	57.0453
C_3H_5O	3.45	0.24	57.0340
C_3H_7N	3.81	0.05	57.0579
C_4H_9	4.58	0.08	57.0705

58	M + 1	M + 2	Mass
N_3O	1.15	0.20	58.0042
H_2N_4	1.51	0.01	58.0280
CNO_2	1.56	0.41	57.9929
CH_2N_2O	1.92	0.21	58.0167
CH_4N_3	2.28	0.02	58.0406
$C_2H_2O_2$	2.33	0.42	58.0054
C_2H_4NO	2.69	0.22	58.0293
$C_2H_6N_2$	3.05	0.03	58.0532
C_3H_6O	3.46	0.24	58.0419
C_3H_8N	3.82	0.05	58.0657
C_4H_{10}	4.59	0.08	58.0783

59	M + 1	M + 2	Mass
HN_3O	1.17	0.20	59.0120
H_3N_4	1.53	0.01	59.0359
$CHNO_2$	1.58	0.41	59.0007
CH_3N_2O	1.94	0.21	59.0246
CH_5N_3	2.30	0.02	59.0484
$C_2H_3O_2$	2.35	0.42	59.0133
C_2H_5NO	2.71	0.22	59.0371
$C_2H_7N_2$	3.07	0.03	59.0610
C_3H_7O	3.48	0.24	59.0497
C_3H_9N	3.84	0.05	59.0736

60	M + 1	M + 2	Mass
N_2O_2	0.82	0.40	59.9960
H_2N_3O	1.18	0.20	60.0198
H_4N_4	1.54	0.01	60.0437
CO_3	1.23	0.60	59.9847
CH_2NO_2	1.59	0.41	60.0085
CH_4N_2O	1.95	0.21	60.0324
CH_6N_3	2.31	0.02	60.0563
$C_2H_4O_2$	2.36	0.42	60.0211
C_2H_6NO	2.72	0.22	60.0449
$C_2H_9N_2$	3.08	0.03	60.0688
C_3H_9O	3.49	0.24	60.0575
C_5	5.55	0.12	60.0000

61	M + 1	M + 2	Mass
HN_2O_2	0.84	0.40	61.0038
H_3N_3O	1.20	0.21	61.0277
H_5N_4	1.56	0.01	61.0515
CHO_3	1.25	0.60	60.9925
CH_3NO_2	1.61	0.41	61.0164
CH_5N_2O	1.97	0.21	61.0402
CH_7N_3	2.33	0.02	61.0641
$C_2H_5O_2$	2.38	0.42	61.0289
C_2H_7NO	2.74	0.22	61.0528
C_5H	5.57	0.12	61.0078

62	M + 1	M + 2	Mass
NO_3	0.49	0.60	61.9878
$H_2N_2O_2$	0.85	0.40	62.0116
H_4N_3O	1.21	0.42	62.0368
H_6N_4	1.57	0.01	62.0594
CH_2O_3	1.26	0.60	62.0003
CH_4NO_2	1.62	0.41	62.0242
CH_6N_2O	1.98	0.21	62.0480
$C_2H_6O_2$	2.39	0.42	62.0368
C_4N	4.81	0.09	62.0031
C_5H_2	5.58	0.12	62.0157

63			
HNO_3	0.51	0.60	62.9956
$H_3N_2O_2$	0.87	0.40	63.0195
H_5N_3O	1.23	0.21	63.0433
CH_3O_3	1.28	0.60	63.0082
CH_5NO_2	1.64	0.41	63.0320
C_4HN	4.83	0.09	63.0109
C_5H_3	5.60	0.12	63.0235

64	M + 1	M + 2	Mass
O_4	0.16	0.80	63.9796
H_2NO_3	0.52	0.60	64.0034
$H_4N_2O_2$	0.88	0.40	64.0273
CH_4O_3	1.29	0.60	64.0160
C_3N_2	4.07	0.06	64.0062
C_4O	4.48	0.27	63.9949
C_4H_2N	4.84	0.09	64.0187
C_5H_4	5.61	0.12	64.0313

65	M + 1	M + 2	Mass
HO_4	0.18	0.80	64.9874
H_3NO_3	0.54	0.60	65.0113
C_3HN_2	4.09	0.06	65.0140
C_4HO	4.50	0.27	65.0027
C_4H_3N	4.86	0.09	65.0266
C_5H_5	5.63	0.12	65.0391

66	M + 1	M + 2	Mass
H_2O_4	0.19	0.80	65.9953
C_2N_3	3.33	0.04	66.0093
C_3NO	3.74	0.25	65.9980
$C_3H_2N_2$	4.10	0.06	66.0218
C_4H_2O	4.51	0.27	66.0106
C_4H_4N	4.87	0.09	66.0344
C_5H_6	5.64	0.12	66.0470

67	M + 1	M + 2	Mass
C_2HN_3	3.35	0.04	67.0171
C_3HNO	3.76	0.25	67.0058
$C_3H_3N_2$	4.12	0.06	67.0297
C_4H_3O	4.53	0.27	67.0184
C_4H_5N	4.89	0.09	67.0422
C_5H_7	5.66	0.12	67.0548

68	M + 1	M + 2	Mass
CN_4	2.59	0.02	68.0124
C_2N_2O	3.00	0.23	68.0011
$C_2H_2N_3$	3.36	0.04	68.0249
C_3O_2	3.41	0.44	67.9898
C_3H_2NO	3.77	0.25	68.0136
$C_3H_4N_2$	4.13	0.06	68.0375
C_4H_4O	4.54	0.28	68.0262
C_4H_6N	4.90	0.09	68.0501
C_5H_8	5.67	0.13	68.0626

69	M + 1	M + 2	Mass
CHN_4	2.61	0.03	69.0202
C_2HN_2O	3.02	0.23	69.0089
$C_2H_3N_3$	3.38	0.04	69.0328
C_3HO_2	3.43	0.44	68.9976
C_3H_3NO	3.79	0.25	69.0215
$C_3H_5N_2$	4.15	0.06	69.0453
C_4H_5O	4.56	0.28	69.0340
C_4H_7N	4.92	0.09	69.0579
C_5H_9	5.69	0.13	69.0705

70	M + 1	M + 2	Mass
CN_3O	2.26	0.22	70.0042
CH_2N_4	2.62	0.03	70.0280
C_2NO_2	2.67	0.42	69.9929
$C_2H_2N_2O$	3.03	0.23	70.0167
$C_2H_4N_3$	3.39	0.04	70.0406
$C_3H_2O_2$	3.44	0.44	70.0054
C_3H_4NO	3.80	0.25	70.0293
$C_3H_6N_2$	4.16	0.07	70.0532
C_4H_6O	4.57	0.28	70.0419
C_4H_8N	4.93	0.09	70.0657
$C_5H_{10}O$	5.70	0.13	70.0783

71	M + 1	M + 2	Mass
CHN_3O	2.28	0.22	71.0120
CH_3N_4	2.64	0.03	71.0359
C_2HNO_2	2.69	0.42	71.0007
$C_2H_3N_2O$	3.05	0.23	71.0246
$C_2H_5N_3$	3.41	0.04	71.0484
$C_3H_3O_2$	3.46	0.44	71.0133
C_3H_5NO	3.82	0.25	71.0371
$C_3H_7N_2$	4.18	0.07	71.0610
C_4H_7O	4.59	0.28	71.0497
C_4H_9N	4.95	0.10	71.0736
C_5H_{11}	5.72	0.13	71.0861

72	M + 1	M + 2	Mass
N_4O	1.52	0.21	72.0073
CN_2O_2	1.93	0.41	71.9960
CH_2N_3O	2.29	0.22	72.0198
CH_4N_4	2.65	0.03	72.0437
C_2O_3	2.34	0.62	71.9847
$C_2H_2NO_2$	2.70	0.42	72.0085
$C_2H_4N_2O$	3.06	0.23	72.0324
$C_2H_6N_3$	3.42	0.04	72.0563
$C_3H_4O_2$	3.47	0.44	72.0211
C_3H_6NO	3.83	0.25	72.0449
$C_3H_8N_2$	4.19	0.07	72.0688
C_4H_8O	4.60	0.28	72.0575
$C_4H_{10}N$	4.96	0.09	72.0814
C_5H_{12}	5.73	0.13	72.0939
C_6	6.66	0.18	72.0000

73	M + 1	M + 2	Mass
HN_4O	1.54	0.21	73.0151
CHN_2O_2	1.95	0.41	73.0038
CH_3N_3O	2.31	0.22	73.0277
CH_5N_4	2.67	0.03	73.0515
C_2HO_3	2.36	0.62	72.9925
$C_2H_3NO_2$	2.72	0.42	73.0164
$C_2H_5N_2O$	3.08	0.23	73.0402
$C_2H_7N_3$	3.44	0.04	73.0641
$C_3H_5O_2$	3.49	0.44	73.0289
C_3H_7NO	3.85	0.25	73.0528
$C_3H_9N_2$	4.21	0.07	73.0767
C_4H_9O	4.62	0.28	73.0653
$C_4H_{11}N$	4.98	0.09	73.0892
C_6H	6.68	0.18	73.0078

74	M + 1	M + 2	Mass
N_3O_2	1.19	0.41	73.9991
H_2N_4O	1.55	0.21	74.0229
CNO_3	1.60	0.61	73.9878
$CH_2N_2O_2$	1.96	0.41	74.0116
CH_4N_3O	2.32	0.22	74.0355
CH_6N_4	2.68	0.03	74.0594
$C_2H_2O_3$	2.37	0.62	74.0003
$C_2H_4NO_2$	2.73	0.42	74.0242
$C_2H_6N_2O$	3.09	0.23	74.0480
$C_2H_8N_3$	3.45	0.05	74.0719
$C_3H_6O_2$	3.50	0.44	74.0368
C_3H_8NO	3.86	0.25	74.0606
$C_3H_{10}N_2$	4.22	0.07	74.0845
$C_4H_{10}O$	4.63	0.28	74.0732
C_5N	5.92	0.14	74.0031
C_6H_2	6.69	0.18	74.0157

75	M + 1	M + 2	Mass
HN_3O_2	1.21	0.41	75.0069
H_3N_4O	1.57	0.21	75.0308
$CHNO_3$	1.62	0.61	74.9956
$CH_3N_2O_2$	0.98	0.41	75.0195
CH_5N_3O	12.34	0.22	75.0433
CH_7N_4	2.70	0.03	75.0672

$C_2H_3O_3$	2.39	0.62	75.0082
$C_2H_5NO_2$	2.75	0.43	75.0320
$C_2H_7N_2O$	3.11	0.23	75.0559
$C_2H_9N_3$	3.47	0.05	75.0798
$C_3H_7O_2$	3.52	0.44	75.0446
C_3H_9NO	3.88	0.25	75.0684
C_5HN	5.94	0.14	75.0109
C_6H_3	6.71	0.18	75.0235

76	M + 1	M + 2	Mass
N_2O_3	0.86	0.60	75.9909
$H_2N_3O_2$	1.22	0.41	76.0147
H_4N_4O	1.58	0.21	76.0386
CO_4	1.27	0.80	75.9796
CH_2NO_3	1.63	0.61	76.0034
$CH_4N_2O_2$	1.99	0.41	76.0273
CH_6N_3O	2.35	0.22	76.0511
CH_6N_4	2.71	0.03	76.0750
$C_2H_4O_3$	2.40	0.62	76.0160
$C_2H_6NO_2$	2.76	0.43	76.0399
$C_2H_8N_2O$	3.12	0.24	76.0637
$C_3H_8O_2$	3.53	0.44	76.0524
C_4N_2	5.18	0.10	76.0062
C_5O	5.59	0.32	75.9949
C_5H_2N	5.95	0.14	76.0187
C_6H_4	6.72	0.19	76.0313

77	M + 1	M + 2	Mass
HN_2O_3	0.88	0.60	76.9987
$H_3N_3O_2$	1.24	0.41	77.0226
H_5N_4O	1.60	0.21	77.0464
CHO_4	1.29	0.80	76.9874
CH_3NO_3	1.65	0.61	77.0113
$CH_5N_2O_2$	2.01	0.41	77.0351
CH_7N_3O	2.37	0.22	77.0590
$C_2H_5O_3$	2.42	0.62	77.0238
$C_2H_7NO_2$	2.78	0.43	77.0477
C_4HN_2	5.20	0.11	77.0140
C_5HO	5.61	0.32	77.0027
C_5H_3N	5.97	0.15	77.0266
C_6H_5	6.74	0.19	77.0391

78	M + 1	M + 2	Mass
NO_4	0.53	0.80	77.9827
$H_2N_2O_3$	0.89	0.60	78.0065
$H_4N_3O_2$	1.25	0.41	78.0304
H_6N_4O	1.61	0.21	78.0542
CH_2O_4	1.30	0.80	77.9953
CH_4NO_3	1.66	0.61	78.0191
$CH_6N_2O_2$	2.02	0.41	78.0429
$C_2H_6O_3$	2.43	0.62	78.0317
C_3N_3	4.44	0.08	78.0093
C_4NO	4.85	0.29	77.9980
$C_4H_2N_2$	5.21	0.11	78.0218
C_5H_2O	5.62	0.32	78.0106
C_5H_4N	5.98	0.14	78.0344
C_6H_6	6.75	0.19	78.0470

79	M + 1	M + 2	Mass
HNO_4	0.55	0.80	78.9905
$H_3N_2O_3$	0.91	0.60	79.0144
$H_5N_3O_2$	1.27	0.41	79.0382
CH_3O_4	1.32	0.80	79.0031
CH_5NO_3	1.68	0.61	79.0269
C_3HN_3	4.46	0.08	79.0171
C_4HNO	4.87	0.29	79.0058
$C_4H_3N_2$	5.23	0.11	79.0297
C_5H_3O	5.64	0.32	79.0184
C_5H_5N	6.00	0.14	79.0422
C_6H_7	6.77	0.19	79.0548

80	M + 1	M + 2	Mass
H_2NO_4	0.56	0.80	79.9983
$H_4N_2O_3$	0.92	0.60	80.0222
CH_4O_4	1.33	0.80	80.0109
C_2N_4	3.70	0.05	80.0124
C_3N_2O	4.11	0.26	80.0011
$C_3H_2N_3$	4.47	0.08	80.0249
C_4O_2	4.52	0.47	79.9898
C_4H_2NO	4.88	0.29	80.0136
$C_4H_4N_2$	5.24	0.11	80.0375
C_5H_4O	5.65	0.32	80.0262
C_5H_6N	6.0	10.14	80.0501
C_6H_8	6.78	0.19	80.0626

81	M + 1	M + 2	Mass
H_3NO_4	0.58	0.80	81.0062
C_2HN_4	3.72	0.05	81.0202
C_3HN_2O	4.13	0.26	81.0089
$C_3H_3N_3$	4.49	0.08	81.0328
C_4HO_2	4.54	0.48	80.9976
C_4H_3NO	4.90	0.29	81.0215
$C_4H_5N_2$	5.26	0.11	81.0453
C_5H_5O	5.67	0.32	81.0340
C_5H_7N	6.03	0.14	81.0579
C_6H_9	6.80	0.19	81.0705

82	M + 1	M + 2	Mass
C_2N_3O	3.37	0.24	82.0042
$C_2H_2N_4$	3.73	0.05	82.0280
C_3NO_2	3.78	0.45	81.9929
$C_3H_2N_2O$	4.14	0.26	82.0167
$C_3H_4N_3$	4.50	0.08	82.0406
$C_4H_2O_2$	4.55	0.48	82.0054
C_4H_4NO	4.91	0.29	82.0293
$C_4H_6N_2$	5.27	0.11	82.0532
C_5H_6O	5.68	0.32	82.0419
C_5H_8N	6.04	0.14	82.0657
C_6H_{10}	6.81	0.19	82.0783

83	M + 1	M + 2	Mass
C_2HN_3O	3.39	0.24	83.0120
$C_2H_3N_4$	3.75	0.06	83.0359
C_3HNO_2	3.80	0.45	83.0007
$C_3H_3N_2O$	4.16	0.27	83.0246
$C_3H_5N_3$	4.52	0.08	83.0484
$C_4H_3O_2$	4.57	0.48	83.0133
C_4H_5NO	4.93	0.29	83.0371
$C_4H_7N_2$	5.29	0.11	83.0610
C_5H_7O	5.70	0.33	83.0497
C_5H_9N	6.06	0.15	83.0736
C_6H_{11}	6.83	0.19	83.0861

84	M + 1	M + 2	Mass
CN_4O	2.63	0.23	84.0073
$C_2H_2N_3O$	3.40	0.24	84.0198
$C_2N_2O_2$	3.04	0.43	83.9960

$C_2H_4N_4$	3.76	0.06	84.0437
C_3O_3	3.45	0.64	83.9847
$C_3H_2NO_2$	3.81	0.45	84.0085
$C_3H_4N_2O$	4.17	0.27	84.0324
$C_3H_6N_3$	4.53	0.08	84.0563
$C_4H_4O_2$	4.58	0.48	84.0211
C_4H_6NO	4.94	0.29	84.0449
$C_4H_8N_2$	5.30	0.11	84.0688
C_5H_8O	5.71	0.33	84.0575
$C_5H_{10}N$	6.07	0.15	84.0814
C_6H_{12}	6.84	0.19	84.0939
C_7	7.77	0.26	84.0000

85	M + 1	M + 2	Mass
CHN_4O	2.65	0.23	85.0151
$C_2HN_2O_2$	3.06	0.43	85.0038
$C_2H_3N_3O$	3.42	0.24	85.0277
$C_2H_5N_4$	3.78	0.06	85.0515
C_3HO_3	3.47	0.64	84.9925
$C_3H_3NO_2$	3.83	0.45	85.0164
$C_3H_5N_2O$	4.19	0.27	85.0402
$C_3H_7N_3$	4.55	0.08	85.0641
$C_4H_5O_2$	4.60	0.48	85.0289
C_4H_7NO	4.96	0.29	85.0528
$C_4H_9N_2$	5.32	0.11	85.0767
C_5H_9O	5.73	0.33	85.0653
$C_5H_{11}N$	6.09	0.16	85.0892
C_6H_{13}	6.86	0.20	85.1018
C_7H	7.79	0.26	85.0078

86	M + 1	M + 2	Mass
CN_3O_2	2.30	0.41	85.9991
CH_2N_4O	2.66	0.21	86.0229
C_2NO_3	2.71	0.62	85.9878
$C_2H_2N_2O_2$	3.07	0.43	86.0116
$C_2H_4N_3O$	3.43	0.24	86.0355
$C_2H_6N_4$	3.79	0.06	86.0594
$C_3H_2O_3$	3.48	0.64	86.0003
$C_3H_4NO_2$	3.84	0.45	86.0242
$C_3H_6N_2O$	4.20	0.27	86.0480
$C_3H_8N_3$	4.56	0.08	86.0719
$C_4H_6O_2$	4.61	0.48	86.0368

C_4H_8NO	4.97	0.30	86.0606
$C_4H_{10}N_2$	5.33	0.11	86.0845
$C_5H_{10}O$	5.74	0.33	86.0732
$C_5H_{12}N$	6.10	0.16	86.0970
C_6H_{14}	6.87	0.21	86.1096
C_6N	7.03	0.21	86.0031
C_7H_2	7.80	0.26	86.0157

87	M + 1	M + 2	Mass
CHN_3O_2	2.32	0.42	87.0069
CH_3N_4O	2.68	0.23	87.0308
C_2HNO_3	2.73	0.62	86.9956
$C_2H_3N_2O_2$	3.09	0.43	87.0195
$C_2H_5N_3O$	3.45	0.25	87.0433
$C_2H_7N_4$	3.81	0.06	87.0672
$C_3H_3O_3$	3.50	0.64	87.0082
$C_3H_5NO_2$	3.86	0.45	87.0320
$C_3H_7N_2O$	4.22	0.27	87.0559
$C_3H_9N_3$	4.58	0.08	87.0798
$C_4H_7O_2$	4.63	0.48	87.0446
C_4H_9NO	4.99	0.30	87.0684
$C_4H_{11}N_2$	5.35	0.11	87.0923
$C_5H_{11}O$	5.76	0.33	87.0810
$C_5H_{13}N$	6.12	0.15	87.1049
C_6HN	7.05	0.21	87.0109
C_7H_3	7.82	0.26	87.0235

88	M + 1	M + 2	Mass
N_4O_2	1.56	0.41	88.0022
CN_2O_3	1.97	0.61	87.9909
$CH_2N_3O_2$	2.33	0.42	88.0147
CH_4N_4O	2.69	0.23	88.0386
C_2O_4	2.38	0.82	87.9796
$C_2H_2NO_3$	2.74	0.63	88.0034
$C_2H_4N_2O_2$	3.10	0.43	88.0273
$C_2H_6N_3O$	3.46	0.25	88.0511
$C_2H_8N_4$	3.82	0.06	88.0750
$C_3H_4O_3$	33.51	0.64	88.0160
$C_3H_6NO_2$	3.87	0.45	88.0399
$C_3H_8N_2O$	4.23	0.27	88.0637
$C_3H_{10}N_3$	4.59	0.08	88.0876

	M + 1	M + 2	Mass
$C_4H_8O_2$	4.64	0.48	88.0524
$C_4H_{10}NO$	5.00	0.30	88.0763
$C_4H_{11}N_2$	5.36	0.11	88.1001
$C_5H_{12}O$	5.77	0.33	88.0888
C_5N_2	6.29	0.16	88.0062
C_6O	6.70	0.38	87.9949
C_6H_2N	7.06	0.21	88.0187
C_7H_4	7.83	0.26	88.0313

89	M + 1	M + 2	Mass
HN_4O_2	1.58	0.41	89.0100
CHN_2O_3	1.99	0.61	88.9987
$CH_3N_3O_2$	2.35	0.42	89.0226
CH_5N_4O	2.71	0.23	89.0464
C_2HO_4	2.40	0.82	88.9874
$C_2H_3NO_3$	2.76	0.63	89.0113
$C_2H_5N_2O_2$	3.12	0.44	89.0351
$C_2H_7N_3O$	3.48	0.25	89.0590
$C_2H_9N_4$	3.84	0.06	89.0829
$C_3H_5O_3$	3.53	0.64	89.0238
$C_3H_7NO_2$	3.89	0.46	89.0477
$C_3H_9N_2O$	4.25	0.27	89.0715
$C_3H_{11}N_3$	4.61	0.08	89.0954
$C_4H_9O_2$	4.66	0.48	89.0603
$C_4H_{11}NO$	5.02	0.30	89.0841
C_5HN_2	6.31	0.16	89.0140
C_6HO	6.72	0.38	89.0027
C_6H_3N	7.08	0.21	89.0266
C_7H_5	7.85	0.26	89.0391

90	M + 1	M + 2	Mass
N_3O_3	1.23	0.60	89.9940
$H_2N_4O_2$	1.59	0.40	90.0178
CNO_4	1.64	0.80	89.9827
$CH_2N_2O_3$	2.00	0.61	90.0065
$CH_4N_3O_2$	2.36	0.42	90.0304
CH_6N_4O	2.72	0.23	90.0542
$C_2H_2O_4$	2.41	0.82	89.9953
$C_2H_4NO_3$	2.77	0.63	90.0191
$C_2H_6N_2O_2$	3.13	0.44	90.0429
$C_2H_8N_3O$	3.49	0.25	90.0668
$C_2H_{10}N_4$	3.85	0.06	90.0907

	M + 1	M + 2	Mass
$C_3H_6O_3$	3.54	0.64	90.0317
$C_3H_8NO_2$	3.90	0.46	90.0555
$C_3H_8N_2O$	4.26	0.27	90.0794
$C_4H_{10}O_2$	4.67	0.48	90.0681
C_4N_3	5.55	0.13	90.0093
C_5NO	5.96	0.34	89.9980
$C_5H_2N_2$	6.32	0.17	90.0218
C_6H_2O	6.73	0.38	90.0106
C_6H_4N	7.09	0.21	90.0344
C_6H_6	7.86	0.26	90.0470

91	M + 1	M + 2	Mass
HN_3O_3	1.25	0.60	91.0018
$H_3N_4O_2$	1.61	0.41	91.0257
$CHNO_4$	1.66	0.81	90.9905
$CH_3N_2O_3$	2.02	0.61	91.0144
$CH_5N_3O_2$	2.38	0.42	91.0382
CH_7N_4O	2.74	0.23	91.0621
$C_2H_3O_4$	2.43	0.82	91.0031
$C_2H_5NO_3$	2.79	0.63	91.0269
$C_2H_7N_2O_2$	3.15	0.44	91.0508
$C_2H_9N_3O$	3.51	0.25	91.0746
$C_3H_7O_3$	3.56	0.64	91.0395
$C_3H_9NO_2$	3.92	0.46	91.0634
C_4HN_3	5.57	0.13	91.0171
C_5HNO	5.98	0.34	91.0058
$C_5H_3N_2$	6.34	0.17	91.0297
C_6H_3O	6.75	0.38	91.0184
C_6H_5N	7.11	0.21	91.0422
C_7H_7	7.88	0.26	91.0548

92	M + 1	M + 2	Mass
N_2O_4	0.90	0.80	91.9858
$H_2N_3O_3$	1.26	0.60	92.0096
$H_4N_4O_2$	1.62	0.41	92.0335
CH_2NO_4	1.67	0.81	91.9983
$CH_4N_2O_3$	2.03	0.61	92.0222
$CH_6N_3O_2$	2.39	0.42	92.0460
CH_8N_4O	2.75	0.23	92.0699
$C_2H_4O_4$	2.44	0.82	92.0109
$C_2H_6NO_3$	2.80	0.63	92.0348
$C_2H_8N_2O_2$	3.16	0.44	92.0586

C$_3$H$_3$O$_3$	3.57	0.64	92.0473
C$_3$N$_4$	4.81	0.09	92.0124
C$_4$N$_2$O	5.22	0.31	92.0011
C$_4$H$_2$N$_3$	5.58	0.13	92.0249
C$_5$O$_2$	5.63	0.52	91.9898
C$_5$H$_2$NO	5.99	0.34	92.0136
C$_5$H$_4$N$_2$	6.35	0.17	92.0375
C$_6$H$_4$O	5.76	0.38	92.0262
C$_6$H$_6$N	7.12	0.21	92.0501
C$_7$H$_8$	7.89	0.27	92.0626

93	M + 1	M + 2	Mass
HN$_2$O$_4$	0.92	0.80	92.9936
H$_3$N$_3$O$_3$	1.28	0.60	93.0175
H$_5$N$_4$O$_2$	1.64	0.41	93.0413
CH$_3$NO$_4$	1.69	0.81	93.0062
CH$_5$N$_2$O$_3$	2.05	0.61	93.0300
CH$_7$N$_3$O	2.41	0.42	93.0539
C$_2$H$_5$O$_4$	2.46	0.82	93.0187
C$_2$H$_7$NO$_3$	2.82	0.63	93.0426
C$_3$HN$_4$	4.83	0.09	93.0202
C$_4$HN$_2$O	5.24	0.31	93.0089
C$_4$H$_3$N$_3$	5.60	0.13	93.0328
C$_5$HO$_2$	5.65	0.52	92.9976
C$_5$H$_3$NO	6.01	0.35	93.0215
C$_5$H$_5$N$_2$	6.37	0.17	93.0453
C$_6$H$_5$O	6.78	0.38	93.0340
C$_6$H$_7$N	7.14	0.22	93.0579
C$_7$H$_9$	7.91	0.27	93.0705

94	M + 1	M + 2	Mass
H$_2$N$_2$O$_4$	0.93	0.80	94.0014
H$_4$N$_3$O$_3$	1.29	0.61	94.0253
H$_6$N$_4$O$_2$	1.65	0.41	94.0491
CH$_4$NO$_4$	1.70	0.81	94.0140
CH$_6$N$_2$O$_3$	2.06	0.62	94.0379
C$_2$H$_6$O$_4$	2.47	0.82	94.0266
C$_3$N$_3$O	4.48	0.28	94.0042
C$_3$H$_2$N$_4$	4.84	0.09	94.0280
C$_4$NO$_2$	4.89	0.49	93.9929
C$_4$H$_2$N$_2$O	5.25	0.31	94.0167
C$_4$H$_4$N$_3$	5.61	0.13	94.0406

C$_5$H$_2$O$_2$	5.66	0.52	94.0054
C$_5$H$_4$NO	6.02	0.35	94.0293
C$_5$H$_6$N$_2$	6.38	0.17	94.0532
C$_6$H$_6$O	6.79	0.38	94.0419
C$_6$H$_8$N	7.15	0.22	94.0657
C$_7$H$_{10}$	7.92	0.27	94.0783

95	M + 1	M + 2	Mass
H$_3$N$_2$O$_4$	0.95	0.80	95.0093
H$_5$N$_3$O$_3$	1.31	0.60	95.0331
C$_3$HN$_3$O	1.72	0.81	95.0218
C$_3$H$_3$N$_4$	4.50	0.28	95.0120
C$_4$HNO$_2$	4.86	0.10	95.0359
C$_4$H$_3$N$_2$O	4.91	0.49	95.0007
C$_4$H$_5$N$_3$	5.27	0.31	95.0246
C$_5$H$_3$O$_2$	5.63	0.13	95.0484
C$_5$H$_5$NO	5.68	0.52	95.0133
C$_5$H$_7$N$_2$	6.04	0.35	95.0371
C$_6$H$_7$O	6.40	0.17	95.0610
C$_6$H$_9$N	6.81	0.39	95.0497
C$_7$H$_{11}$	7.17	0.22	95.0736

96	M + 1	M + 2	Mass
H$_3$N$_2$O$_4$	0.96	0.80	96.0171
C$_2$N$_4$O	3.74	0.26	96.0073
C$_3$N$_2$O$_2$	4.15	0.47	95.9960
C$_3$H$_2$N$_3$O	4.51	0.28	96.0198
C$_3$H$_4$N$_4$	4.87	0.10	96.0437
C$_4$O$_3$	4.56	0.67	95.9847
C$_4$H$_2$NO$_2$	4.92	0.49	96.0085
C$_4$H$_4$N$_2$O	5.28	0.31	96.0324
C$_4$H$_6$N$_3$	5.64	0.13	96.0563
C$_5$H$_4$O$_2$	5.69	0.53	96.0211
C$_5$H$_6$NO	6.05	0.35	96.0449
C$_5$H$_8$N$_2$	6.41	0.17	96.0688
C$_6$H$_8$O	6.82	0.39	96.0575
C$_6$H$_{10}$N	7.18	0.22	96.0814
C$_7$H$_{12}$	7.95	0.27	96.0939
C$_8$	8.88	0.34	96.0000

97	M + 1	M + 2	Mass
C_2HN_4O	3.76	0.26	97.0151
$C_3HN_2O_2$	4.17	0.47	97.0038
$C_3H_3N_3O$	4.53	0.28	97.0277
$C_3H_5N_4$	4.89	0.10	97.0515
C_4HO_3	4.58	0.68	96.9925
$C_4H_3NO_2$	4.94	0.49	97.0164
$C_4H_5N_2O$	5.30	0.31	97.0402
$C_4H_7N_3$	5.66	0.13	97.0641
$C_5H_5O_2$	5.71	0.53	97.0289
C_5H_7NO	6.07	0.35	97.0528
$C_5H_9N_2$	6.43	0.17	97.0767
C_6H_9O	6.84	0.39	97.0653
$C_6H_{11}N$	7.20	0.22	97.0892
C_7H_{13}	7.97	0.27	97.1018
C_8H	8.90	0.34	97.0078

98	M + 1	M + 2	Mass
$C_2N_3O_2$	3.41	0.44	97.9991
$C_2H_2N_4O$	3.77	0.26	98.0229
C_3NO_3	3.82	0.65	97.9878
$C_3H_2N_2O_2$	4.18	0.47	98.0116
$C_3H_4N_3O$	4.54	0.28	98.0355
$C_3H_6N_4$	4.90	0.10	98.0594
$C_4H_2O_3$	4.59	0.68	98.0003
$C_4H_4NO_2$	4.95	0.49	98.0242
$C_4H_6N_2O$	5.31	0.31	98.0480
$C_4H_8N_3$	5.67	0.13	98.0719

$C_5H_6O_2$	5.72	0.53	98.0368
C_5H_8NO	6.08	0.35	98.0606
$C_5H_{10}N_2$	6.44	0.17	98.0645
$C_6H_{10}O$	6.85	0.39	98.0732
$C_6H_{12}N$	7.21	0.21	98.0970
C_7H_{14}	7.98	0.26	98.1096
C_7N	8.14	0.27	98.0031
C_8H_2	8.91	0.33	98.0157

99	M + 1	M + 2	Mass
$C_2HN_3O_2$	3.43	0.44	99.0069
$C_2H_3N_4O$	3.79	0.25	99.0308
C_3HNO_3	3.84	0.65	98.9956
$C_3H_3N_2O_2$	4.20	0.47	99.0195
$C_3H_5N_3O$	4.56	0.28	99.0433
$C_3H_7N_4$	4.92	0.10	99.0672
$C_4H_3O_3$	4.61	0.68	99.0082
$C_4H_5NO_2$	4.97	0.49	99.0320
$C_4H_7N_2O$	5.33	0.31	99.0559
$C_4H_9N_3$	5.69	0.13	99.0798
$C_5H_7O_2$	5.74	0.53	99.0446
C_5H_9NO	6.11	0.35	99.0684
$C_5H_{11}N_2$	6.46	0.17	99.0923
$C_6H_{11}O$	6.86	0.39	99.0810
$C_6H_{13}N$	7.23	0.22	99.1049
C_7H_{15}	8.00	0.27	99.1174
C_7HN	8.16	0.29	99.0109
C_8H_3	8.93	0.35	99.0235

Internet Resources

Multispectroscopy

Internet access to the exercises in this workbook may be obtained through the
   Multispectroscopy on-line exercises at:
http://spectros.unice.fr

Spectral Databases

Spectral Database for Organic Compounds, MS, NMR, IR
http://riodb01.ibase.aist.go.jp/sdbs/cgi-bin/cre_index.cgi

Sigma-Aldrich, NMR, IR
http://www.sigmaaldrich.com

NIST Chemistry Webbook, MS, IR
http://webbook.nist.gov/chemistry/

Acros Organics, IR
http://www.acros.com/portal/alias__Rainbow/lang__en-US/tabID__28/DesktopDefault.aspx

Biological Magnetic Resonance Databank
http://www.bmrb.wisc.edu/metabolomics/

MassBank, high resolution mass spectral database
http://www.massbank.jp/index.html?lang=en

Other Resources

A comprehensive set of chemical shift and coupling data
http://www.chem.wisc.edu/areas/reich/handouts/nmr-h/hdata.htm

NMR acronyms
http://www.chem.ox.ac.uk/spectroscopy/nmr/acropage.htm

Organic Spectroscopy Workbook, First Edition. Tom Forrest, Jean-Pierre Rabine, and Michel Rouillard.
© 2011 John Wiley & Sons, Ltd. Published 2011 by John Wiley & Sons, Ltd.

Glossary of Terms Used in the Exercises

Mass Spectrometry

α-cleavage: cleavage of a bond attached to an atom which is adjacent to the one assumed to carry the charge. **β-cleavage:** cleavage of a bond one further removed from the charge site. An example of α-cleavage is found in the fragmentation of the molecular ion of ethyl ether in which a methyl radical is lost.

HRMS: high-resolution mass spectrometry.

M or M$^+$: molecular ion – It is the ion formed by the removal, without fragmentation, of one or more electrons from the molecule to form a positive ion, or the addition of same to form a negative ion. The molecular ion is considered to be the ion containing atoms of the most abundant naturally occurring isotopes.

McLafferty rearrangement: a β-cleavage involving a six-membered transition state in which a γ-hydrogen atom is transferred to the atom of a double bond. An example is found in the fragmentation by which butyraldehyde loses a molecule of ethylene.

P or P$^+$: parent or precursor ion – P is sometimes used to refer to the molecular ion, but it may also refer to any charged fragment ion which may dissociate to form fragments.

retro Diels–Alder: a fragmentation by a rearrangement reaction which involves the cleavage of two sigma bonds to give products equivalent to the starting components of a 4+2 pericyclic cycloaddition reaction. An example of a retro Diels–Alder rearrangement is the loss of ethylene from the molecular ion of cyclohexene.

Organic Spectroscopy Workbook, First Edition. Tom Forrest, Jean-Pierre Rabine, and Michel Rouillard.
© 2011 John Wiley & Sons, Ltd. Published 2011 by John Wiley & Sons, Ltd.

NMR

J: coupling constant: a scalar through-bond coupling independent of the applied magnetic field, expressed in Hertz.

δ: chemical shift: the difference in resonance frequency (expressed in ppm) of a nucleus compared to a standard, normally TMS, (tetramethylsilane).

Isochronous, chemical shift equivalent: refers to nuclei that have identical chemical shifts. Nuclei that are exchangeable by a symmetry operation, or by a rapid process such as bond rotation, are chemical shift equivalent.

Magnetic equivalence: refers to the relationship between nuclei that have identical chemical shifts, and also have identical spin–spin coupling with the nuclei of any neighbouring groups. A system of two pairs of magnetically equivalent nuclei could be labelled as A_2B_2, whereas a pair with chemical shift equivalence only would be labelled $AA'BB'$ because J_{AB} is not identical to $J_{AB'}$.

BB: broad band decoupled spectra: cabon-13 spectra in which all $^{13}C-^1H$ coupling is removed by applying a broad band of irradiation covering the complete proton range.

DEPT: distortionless enhanced polarization transfer. All DEPT spectra presented in these exercises are DEPT-135 spectra, in which the CH_3 and CH signals are positive, CH_2 signals are negative, and quaternary carbons have no signals.

Downfield: towards higher δ values (to the left of the charts used in these exercises). It refers to the need to apply a lower field to cause resonance when some change decreases the shielding of the nucleus. The term originates from the techniques used in continuous wave spectrometry.

OR: off resonance: refers to carbon-13 spectra in which the $^{13}C-^1H$ coupling is reduced to the extent that only $^1J_{CH}$ coupling remains observable.

Quaternary carbon: in ^{13}C NMR spectroscopy, refers to a carbon having no directly attached hydrogens. Tertiary, secondary, and primary in this sense refer to the number of bonds to the carbon (including π bonds) other than C–H bonds.

Shield, deshield: to affect the degree to which the nucleus is shielded from the applied magnetic field. This is a term that is used to describe a change in any of the factors that affect the chemical shift. For example, a change which increases the electron density at the nucleus, is 'shielding', and results in an upfield shift. A decrease in the electron density at the nucleus would be 'deshielding', causing a downfield shift.

TMS: tetramethylsilane, a compound sometimes added to the sample as a reference standard. TMS is not necessarily present, as solvent peaks may be used as a secondary reference.

Upfield: towards lower δ values (to the right of the charts used in these exercises). It refers to the need to apply a higher field to cause resonance when some change increases the shielding of the nucleus. The term originates from the techniques used in continuous wave spectrometry.

NMR abbreviations: accepted for reporting spectral data δ, chemical shift in ppm downfield from the standard, normally tetramethylsilane.

d, doublet; t, triplet: q, quartet; m, multiplet; br, broadened, Ar, aryl.

Infrared

IR abbreviations accepted for reporting spectral data

b, broad; m, medium; s, strong; w, weak.

Answers

Organic Spectroscopy Workbook, First Edition. Tom Forrest, Jean-Pierre Rabine, and Michel Rouillard.
© 2011 John Wiley & Sons, Ltd. Published 2011 by John Wiley & Sons, Ltd.

(001)	(002)	(003)	(004)	(005)	(006)	(007)	(008)	(009)	(010)
CH₃-CH₂ O CH₃-CH₂	H-C≡C-(CH₂)₂-OH	CH₃-C=O O CH₃-CH₂	CH₃-C(=O)-N(CH₃)CH₃	CH₃ CH₃-C-C=O CH₃ H	CH₃ CH-CH₂-C=O CH₃ H	CH₃C=O O-(CH₂)₂-CH₃	CH₃-C(=O)-N(CH₃)-CH₂-CH₃	CH₃ CH₃ CH-O-CH CH₃ CH₃	CH₃-CH₂-CH₂ O CH₃-CH₂-CH₂

(011)	(012)	(013)	(014)	(015)	(016)	(017)	(018)	(019)	(020)
CH₃-CH₂ CH₃-CH₂—N CH₃-CH₂	⬡-CH₃	CH₃ ⬡ CH₃ CH₃	C=O O-CH₂-CH₃ O-CH₂-CH₃	CH₃-CH₂-O C=O CH₃-CH₂-O	CH₃-CH₂-CH₂-C≡C-H	CH₃-CH₂-C≡C-CH₃	⬡ (cyclohexene)	CH₂ ⬡ C=O CH₂	CH₂-CH=CH₂ H-N CH₂-CH=CH₂

(021)	(022)	(023)	(024)	(025)	(026)	(027)	(028)	(029)	(030)
CH=CH₂ (CH₂)₃ CH₃	CH₃ CH-CH₂-CH₂ CH₃ O C=O H	CH₃ (CH₂)₄ O=C NH₂	⬡-C(=O)H	⬡-O-CH₃	(CH₂)₂-CH₃ O=C O O=C (CH₂)₂-CH₃	⬡-CH₂-C≡N	CH₃O C=O ⬡ O=C OCH₃	CH₃O C(=O)H ⬡ CH₃O OCH₃	CH₃O C(=O)H CH₃O ⬡ OCH₃

(031)	(032)	(033)	(034)	(035)	(036)	(037)	(038)	(039)	(040)
CH₃ ⬡ CH₃ CH-CH₃ CH₃	⬡ CH₃ CH-CH₃ CH₃	CH₂-(CH₂)₂-C≡C-H CH₂-(CH₂)₂-C≡C-H	CH₃ CH₃ ⬡ CH₃ CH₃	CH₂-CH₂-C≡C-CH₃ CH₂-CH₂-C≡C-CH₃	(CH₃)₃C C=O O C=O (CH₃)₃C	O₂N-⬡-O-CH₂ C=O OH	⬡⬡-CH₃	CH₃-(CH₂)₉-C≡N	CO-CH₃ ⬡ CH₃-OC CO-CH₃

(041)	(042)	(043)	(044)	(045)	(046)	(047)	(048)	(049)	(050)
O C=O F-CH₂ O CH₂-CH₃	O H₂C C O-CH₂-CH₃ H₂C O-CH₂-CH₃ O	O O C C-CH₃ O H₂C CH₂	Br ⬡-⬡ Br	CH₃ CH₂-CO-OH CH₃-C CH₃ CH₂-CO-OH	CH₃ C=O ⬡ Cl	O C=O O-CH₃ ⬡	NH₂ ⬡ O=C O-CH₂CH₃	H₂N-C=O ⬡ OCH₂CH₃	CH₃ CH-O-C CH₃ O O C-CH₃ O

(051)	(052)	(053)	(054)	(055)	(056)	(057)	(058)	(059)	(060)
Cl, Cl, Cl, Cl-CH₂, CH₂-Cl, Cl	CH₂Br / CH₂Br	CH₂Cl / ClCH₂	CCl₃ / Cl₃C	CH₃-O-CH₂, O=C, CH₂, O=C, O-CH₃	CH₃-O-C=O, CH₂, CH-CH₃, O=C, O-H	H-O-C=O, (H₂C)₄, C=O, H-O	H-O-C=O, CH₂, CH₃-HC, H-O-C=O	CH₂-N-H, CH, CH₃	CH₃-(CH₂)₄, N
(061)	(062)	(063)	(064)	(065)	(066)	(067)	(068)	(069)	(070)
CH₃, CH, CH₃, NH₂, CH₃	CH₃, CH-N-CH₃, CH₃	CH₂CH₃, N, CH₂CH₃	COO-CH₂CH₃, COO-CH₂CH₃	COO-CH₃, COO-CH₃	CH₃-(CH₂)₅-NH₂	CH₃, CH-CH-CH, CH₃, CH₃, CH₃	O, CH₃-N-N-CH₃, CH₃, CH₃	O, H-N-N-H, CH₃-CH₂, CH₂-CH₃	HO, CH-CH₂-CH₃, CH₃
(071)	(072)	(073)	(074)	(075)	(076)	(077)	(078)	(079)	(080)
H₃C, O-CH₃, O	O, O, O, O	HO, O, OH	F, H, H, H, H, NO₂	(CH₂)₂COOCH₃, H₂C, (CH₂)₂COOCH₃	OOC(CH₂)₂CH₃, H-C-CH₃, COOCH₂CH₃	O-CH₃, O, HO-C-C-(CH₂)₂-C-OH, CH₃	H, H, H, Cl, CH₂, H, H	O, O, H₂C, C-O-CH₃, H₂C, CH, H	(fused rings)
(081)	(082)	(083)	(084)	(085)	(086)	(087)	(088)	(089)	(090)
(fused rings)	Cl, O-H	HS, OH, HO, SH	H, I, CH₂, H-C, H	N, CH₃, S	CH₃-(CH₂)₁₇-I	SH, NH₂	H, N, S	I, NH₂	N-C=S, NH₂
(091)	(092)	(093)	(094)	(095)	(096)	(097)	(098)	(099)	(100)
Br	N(CH₃)₂, S=C, N(CH₃)₂	S, CH₃	CH₃CH₂, CH₂CH₃, CH-CH, CH₃CH₂, CH₂CH₃	O, O	(fused rings)	C=C	OH	H, C=C, H	CH=CH₂, Br

Index

General Index

Compound Index

Organic Spectroscopy Workbook, First Edition. Tom Forrest, Jean Pierre Rabine, and Michel Rouillard.
© 2011 John Wiley & Sons, Ltd. Published 2011 by John Wiley & Sons, Ltd.